易海博 著

多变量公钥密码芯片
技术原理和实践

人民邮电出版社

北京

图书在版编目（ＣＩＰ）数据

多变量公钥密码芯片技术原理和实践 / 易海博著
． －－ 北京 ：人民邮电出版社，2018.7（2018.11重印）
ISBN 978-7-115-47895-5

Ⅰ．①多… Ⅱ．①易… Ⅲ．①公钥密码系统 Ⅳ．
①TN918.4

中国版本图书馆CIP数据核字(2018)第027065号

内 容 提 要

鉴于量子计算机被证明能够在多项式时间内解决大整数因子分解和离散对数问题，因此传统的公钥密码体制（比如RSA、ECC等加密算法）变得不再安全。作为少数能抵御量子计算机攻击的公钥密码，多变量公钥密码的重要性日益显现，研究多变量公钥密码芯片也成为量子计算机时代的重要课题之一。

本书的目的是介绍能够抵御量子计算机攻击的公钥密码芯片技术，使之应用于多个关键领域，保护使用者的个人信息安全。本书共分为 7 章，内容涵盖了有限域和有限域计算的相关知识，密码学中重要的概念和主要的密码算法，芯片设计相关的知识（包括涉及的工具、技术、编程语言、编程环境），多变量公钥密码的发展过程、密码体制、算法等，基于有限域并行运算结构的多变量公钥密码算法的快速实现技术，基于优化多项式的多变量公钥密码算法的高效实现技术，以及基于精简指令集和模运算单元的多变量密码处理器技术。

本书适合集成电路领域和信息安全领域的从业人员阅读，也可作为高校、科研机构的教学用书或参考书。

◆ 著　　易海博
　　责任编辑　傅道坤
　　责任印制　焦志炜

◆ 人民邮电出版社出版发行　北京市丰台区成寿寺路 11 号
　　邮编　100164　电子邮件　315@ptpress.com.cn
　　网址　http://www.ptpress.com.cn
　　固安县铭成印刷有限公司印刷

◆ 开本：800×1000　1/16
　　印张：23.75
　　字数：492 千字　　　　　　　2018 年 7 月第 1 版
　　印数：1 201－1 500 册　　　　2018 年 11 月河北第 2 次印刷

定价：99.00 元
读者服务热线：(010)81055410　印装质量热线：(010)81055316
反盗版热线：(010)81055315
广告经营许可证：京东工商广登字 20170147 号

作 者 简 介

易海博，博士研究生学历，湖南湘潭人，毕业于华南理工大学信息安全专业，2012~2014年担任美国辛辛那提大学（University of Cincinnati）访问学者，现任职于深圳职业技术学院，承担"C语言程序设计"、"面向对象程序设计"、"云计算技术概论"、"大数据技术与应用"、"Web Design"、"网络操作系统（Linux）"等专业课程的授课工作，主要研究云计算、大数据、信息安全、微电子、计算机与互联网等方向。主持国家、省、市级项目8项，在国内外重要学术期刊和会议上发表36篇论文（第一作者SCI论文9篇，EI论文16篇），获得34项中国、美国、欧洲等国发明专利和实用新型专利（已授权17项），45项著作权，担任SCI期刊和国际会议审稿人。获得工业和信息化部举办的第五届中国电子信息博览会"CITE2017创新产品与应用奖"，广东省教育厅举办的广东省计算机教育软件评审活动一等奖（第一名）等多项奖项。指导学生获得2017年"挑战杯—彩虹人生"广东省职业学校创新创效创业大赛特等奖等多项奖项。

前　　言

近年来，美国麻省理工学院学者 Peter Shor 证明了量子计算机能够在多项式时间内解决大整数因子分解和离散对数问题，由于 RSA、ECC 等大部分的公钥密码的安全性基于这类问题，这些密码算法将不再安全。RSA 和 ECC 等公钥密码在公交卡、银行 IC 卡、智能车锁、监控摄像头、无人机等芯片产品中应用广泛，随着量子计算机的不断发展，这些产品的安全性面临着量子计算机的巨大威胁。考虑到"不把所有鸡蛋放到一个篮子里"，芯片产品需要基于其他数学问题的公钥密码来设计。

求解多元多项式方程组问题是一类 NP-Hard 问题，目前量子计算机对于这类问题并没有较好的求解方法。基于这类问题，多变量公钥密码（Multivariate Public Key Cryptography）的安全性优势在于，相比 RSA 和 ECC 等公钥密码，它可以抵御量子计算攻击；相比量子密码，它可以在电子计算机上使用。作为少数能够抵御量子计算机攻击的公钥密码，多变量公钥密码的重要性日益显现，研究多变量公钥密码芯片成为量子计算机时代的重要课题之一。撰写本书的目的是介绍能够抵御量子计算机攻击的公钥密码芯片技术，即设计多变量公钥密码芯片，使之能运用于多个关键领域，保护使用者的个人信息安全。

作者针对密码芯片设计的 3 个方向（时间优化、面积优化和性能优化），设计了 3 种多变量公钥密码芯片技术：快速芯片显著地缩短了密码运算所需的时间，比 RSA 和 ECC 等公钥密码的运算时间短，可以运用于实时性要求高的环境；小面积芯片采用处理器技术，基于精简指令集和模运算逻辑单元，将公钥密码芯片的面积减小了一半以上，可以运用于面积受限的环境；高效芯片专注于时间和面积两方面的优化，既缩短了运算时间，也减小了它需要的面积，使之可运用于资源受限的环境。

本书组织结构

本书分为 7 章，包括数学基础、密码学基础、芯片设计基础、多变量公钥密码技术、多变量公钥密码快速芯片技术、多变量公钥密码高效芯片技术和多变量公钥密码处理器技术。

第 1 章 "数学基础"，介绍有限域及有限域计算的相关概念和知识。
第 2 章 "密码学基础"，介绍密码学中的重要概念和主要的密码算法。
第 3 章 "芯片设计基础"，介绍芯片设计的工具、技术、编程语言、编程环境等知识。
第 4 章 "多变量公钥密码技术"，介绍多变量公钥密码的发展过程、密码体制、算法等。
第 5 章 "多变量公钥密码快速芯片技术"，介绍基于有限域并行运算结构的多变量公钥密码算法的快速实现技术。

第 6 章 "多变量公钥密码高效芯片技术"，介绍基于优化多项式的多变量公钥密码算法的高效实现技术。

第 7 章 "多变量公钥密码处理器技术"，介绍基于精简指令集和模运算单元的多变量公钥密码处理器技术。

每章的参考文献可以帮助读者更好地了解本书的内容。另外，本书给出了作者编写的代码，并补充了来自网络的部分代码。这些代码包括 DES、AES、Rainbow、HFE 等密码算法的 C、VHDL、Verilog 语言的软硬件实现代码，供读者参考。

本书前 4 章的内容是多变量公钥密码芯片技术的相关理论和实践基础，后 3 章的内容是作者多年来研究多变量公钥密码芯片技术的主要成果。在写作本书的过程中，参考了大量国内外论文、专著和资料，有选择地把一些重要知识纳入本书。由于作者能力有限，本书难免存在不足之处，望广大读者不吝赐教，可将意见或建议通过邮件发送至本人的邮箱 haiboyi@126.com。

本书读者对象

本书面向信息安全和集成电路领域，可作为高校、科研机构的老师和学生的教学用书、参考书，也可作为企业的研究人员、工程师的工具书。

资源与支持

本书由异步社区出品，社区（https://www.epubit.com/）为您提供相关资源和后续服务。

提交勘误

作者和编辑尽最大努力来确保书中内容的准确性，但难免会存在疏漏。欢迎您将发现的问题反馈给我们，帮助我们提升图书的质量。

当您发现错误时，请登录异步社区，按书名搜索，进入本书页面，点击"提交勘误"，输入勘误信息，点击"提交"按钮即可。本书的作者和编辑会对您提交的勘误进行审核，确认并接受后，您将获赠异步社区的 100 积分。积分可用于在异步社区兑换优惠券、样书或奖品。

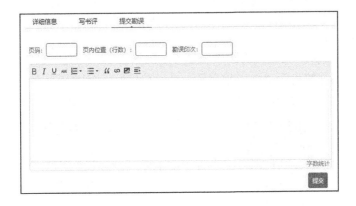

扫码关注本书

扫描下方二维码，您将会在异步社区微信服务号中看到本书信息及相关的服务提示。

与我们联系

我们的联系邮箱是 contact@epubit.com.cn。

如果您对本书有任何疑问或建议，请您发邮件给我们，并请在邮件标题中注明本书书名，以便我们更高效地做出反馈。

如果您有兴趣出版图书、录制教学视频，或者参与图书翻译、技术审校等工作，可以发邮件给我们；有意出版图书的作者也可以到异步社区在线提交投稿（直接访问www.epubit.com/selfpublish/submission 即可）。

如果您是学校、培训机构或企业，想批量购买本书或异步社区出版的其他图书，也可以发邮件给我们。

如果您在网上发现有针对异步社区出品图书的各种形式的盗版行为，包括对图书全部或部分内容的非授权传播，请您将怀疑有侵权行为的链接发邮件给我们。您的这一举动是对作者权益的保护，也是我们持续为您提供有价值的内容的动力之源。

关于异步社区和异步图书

"**异步社区**"是人民邮电出版社旗下 IT 专业图书社区，致力于出版精品 IT 技术图书和相关学习产品，为作译者提供优质出版服务。异步社区创办于 2015 年 8 月，提供大量精品 IT 技术图书和电子书，以及高品质技术文章和视频课程。更多详情请访问异步社区官网 https://www.epubit.com。

"**异步图书**"是由异步社区编辑团队策划出版的精品 IT 专业图书的品牌，依托于人民邮电出版社近 30 年的计算机图书出版积累和专业编辑团队，相关图书在封面上印有异步图书的 LOGO。异步图书的出版领域包括软件开发、大数据、AI、测试、前端、网络技术等。

异步社区

微信服务号

目　　录

第 1 章　数学基础 ·· 1
1.1　代数基础 ·· 1
　　1.1.1　集合 ·· 1
　　1.1.2　群 ·· 1
　　1.1.3　环 ·· 2
　　1.1.4　域 ·· 2
1.2　有限域概念 ·· 3
　　1.2.1　有限域定义 ·· 3
　　1.2.2　常用有限域形式 ·· 3
　　1.2.3　不可约多项式 ·· 4
1.3　有限域元素 ·· 4
　　1.3.1　多项式基 ·· 4
　　1.3.2　正规基 ·· 4
　　1.3.3　对偶基 ·· 5
1.4　有限域基础运算 ·· 5
　　1.4.1　有限域加法 ·· 5
　　1.4.2　有限域乘法 ·· 6
　　1.4.3　有限域平方 ·· 9
　　1.4.4　有限域乘方 ··· 10
　　1.4.5　有限域求逆 ··· 10
　　1.4.6　有限域除法 ··· 13
　　1.4.7　求解线性方程组 ······································· 14
1.5　其他数学问题 ·· 22
　　1.5.1　MQ 问题 ·· 22
　　1.5.2　线性仿射变换 ·· 23
　　1.5.3　随机数发生器 ·· 24
1.6　本章小结 ·· 25
1.7　本章参考文献 ·· 26

第 2 章　密码学基础 ·· 32
2.1　密码和密码学 ·· 32
　　2.1.1　密码 ·· 32
　　2.1.2　密码学 ·· 32
　　2.1.3　密码系统 ·· 33
2.2　密码体制 ·· 34
　　2.2.1　对称密码 ·· 34
　　2.2.2　公钥密码 ·· 35
　　2.2.3　数字签名 ·· 36
2.3　常用的密码算法 ·· 39
　　2.3.1　DES ·· 39
　　2.3.2　AES ··· 41
　　2.3.3　RSA ··· 42
　　2.3.4　ECC ··· 42
2.4　互联网与信息安全 ··· 43
　　2.4.1　信息安全 ·· 43
　　2.4.2　信息安全产业 ·· 43
　　2.4.3　云计算安全 ··· 44
　　2.4.4　公钥基础设施 ·· 45
　　2.4.5　身份与访问管理 ······································· 46
　　2.4.6　后量子密码 ··· 47
　　2.4.7　散列 ·· 48
2.5　本章参考代码 ·· 49
　　2.5.1　DES ·· 49
　　2.5.2　AES ··· 59
　　2.5.3　RSA ··· 104
　　2.5.4　ECC ··· 120
2.6　本章小结 ·· 144
2.7　本章参考文献 ·· 144

第 3 章　芯片设计基础 146

3.1　数字电路基础 146
3.1.1　现场可编程逻辑门电路 146
3.1.2　专用集成电路 147
3.1.3　硬件编程语言 148
3.1.4　有限状态机技术 149

3.2　硬件编程语言 VHDL 151
3.2.1　VHDL 概述 151
3.2.2　标识符 151
3.2.3　数据类型 152
3.2.4　数据对象 153
3.2.5　运算符 154
3.2.6　VHDL 的结构 155

3.3　Altera FPGA 开发环境 Quartus II 156
3.3.1　Quartus II 介绍 156
3.3.2　Quartus II 使用例解 157

3.4　集成电路产业 178
3.4.1　集成电路 178
3.4.2　产业发展现状 179
3.4.3　产业发展前景 180

3.5　本章参考代码 183
3.5.1　VHDL 参考例子 183
3.5.2　Verilog 参考例子 187

3.6　本章小结 188
3.7　本章参考文献 188

第 4 章　多变量公钥密码技术 192

4.1　多变量公钥密码概述 192
4.1.1　多变量加密 192
4.1.2　多变量签名 193
4.1.3　多变量公钥密码芯片 193

4.2　多变量公钥密码系统 194
4.3　多变量公钥密码典型算法 .. 195
4.3.1　MI 密码算法 196
4.3.2　PMI+密码算法 196
4.3.3　HFE 密码算法 198
4.3.4　ℓ-IC 密码算法 199
4.3.5　TTM 密码算法 200
4.3.6　TTS 签名算法 201
4.3.7　en-TTS 签名算法 202
4.3.8　油醋签名算法 204
4.3.9　UOV 签名算法 205
4.3.10　Rainbow 签名算法 207

4.4　多变量公钥密码分析方法 .. 209
4.4.1　暴力攻击 209
4.4.2　直接攻击 209
4.4.3　线性化方程攻击 210
4.4.4　秩攻击 210
4.4.5　差分攻击 211

4.5　本章参考代码 211
4.5.1　Rainbow 211
4.5.2　HFE 282

4.6　本章小结 286
4.7　本章参考文献 286

第 5 章　多变量公钥密码快速芯片技术 315

5.1　本章概述 315
5.2　快速多变量签名方案 316
5.3　不可约多项式 319
5.4　加速二元和三元乘法运算 .. 319
5.5　加速求逆运算 320
5.6　加速求解线性方程组运算 .. 320
5.7　加速可逆仿射变换运算 324
5.8　加速多元二次多项式求值运算 325
5.9　技术实现 325
5.10　实现对比 326
5.11　本章小结 327
5.12　本章参考文献 327

第 6 章　多变量公钥密码高效芯片技术 ································ 329

6.1　本章概述 ··· 329
6.2　高效多变量签名方案 ································ 329
6.3　选择特定有限域的不可约多项式 ·················· 333
6.4　优化特定有限域的乘法 ····························· 333
6.5　优化特定有限域的求逆 ····························· 335
6.6　优化特定有限域的求解线性方程组 ··············· 336
6.7　技术实现 ·· 337
6.8　实现对比 ·· 337
6.9　本章小结 ·· 338
6.10　本章参考文献 ·· 339

第 7 章　多变量公钥密码处理器技术 ································ 340

7.1　本章概述 ·· 340
7.2　架构设计 ·· 341
7.3　多变量数字签名方案和参数的选择 ··············· 342
7.4　模运算逻辑单元 ······································· 349
7.5　RAM 和寄存器 ······································· 355
7.6　微控制器和指令集 ···································· 356
7.7　多变量公钥密码的基本密码运算 ·················· 358
7.8　技术实现 ·· 361
7.9　实现对比 ·· 365
7.10　本章小结 ··· 366
7.11　本章参考文献 ··· 367

第 1 章 数学基础

1.1 代数基础

1.1.1 集合

集合是数学中一个基本概念,由一个或多个确定的元素构成的整体叫作集合。

定义 1.1 集合:假定 G 非空,它的元素满足以下 3 个特征,则称 G 是一个集合。

1)确定性:集合中的元素必须是确定的。
2)互异性:集合中的元素互不相同。
3)无序性:集合中的元素没有先后之分。

集合可以分成多种不同的类型,例如群、环、域。其中,有限域是域的一种特殊形式,是多变量公钥密码以及其他密码体制使用的一种数域。有限域理论作为现代代数的重要分支,在密码学、编码理论、组合理论、大规模集成电路设计等诸多领域都发挥着重要作用。它的应用极大地推动了这些学科的发展,其中,许多相关领域的研究热点都可以归结为有限域理论中的关键问题,这使得有限域理论日益得到重视、充实和推动。

在介绍有限域的概念之前,我们先介绍群和环的概念,这是因为有限域是在两者的基础上发展而来的。

1.1.2 群

定义 1.2 群:假定一个集合 G 非空,我们在集合内定义了一种代数运算,这个集合可以被定义为群的条件如下。

1)封闭性:对于集合中任意的两个元素,它们的运算结果也必须在集合中;若 $a,b \in G$,则 $a \cdot b \in G$ 成立。

2)结合律:对于集合中任意的 3 个元素,三者的运算必须满足结合律;若 $a,b,c \in G$,则 $a \cdot (b \cdot c) = (a \cdot b) \cdot c$ 成立。

3)单位元:对于集合中任意的一个元素,集合中必须存在这么一个元素,与之运算的结果还是本身;若 $a \in G$,一定存在 $e \in G$,使得 $a \cdot e = e \cdot a = a$ 成立。

4）逆元：对于集合中任意的一个元素，存在它的逆元，即它们的运算结果恒等于集合中的某个元素；若 $a \in G$，一定存在 $a' \in G$，使得 $a \cdot a' = a' \cdot a = e$ 成立。

满足上述所有条件的集合可以被定义为群。

若群中任意两个元素的运算结果与两者交换前后顺序的运算结果相同，那么它可以被称为交换群，它的定义描述如下。

定义 1.3 交换群：假定一个群 G 还满足以下条件。

交换律：对于任意的 $a,b \in G$，则 $a \cdot b = b \cdot a$。

满足上述所有条件的集合可以被定义为交换群。

1.1.3 环

环是在交换群的基础上发展而来的，下面我们来看它的定义。

定义 1.4 环：定义 + 表示二元加法，• 表示二元乘法，R 表示非空集合，R 对于+是一个交换群，即满足定义 1.2 和定义 1.3 的所有原则，还满足以下条件。

1）乘法的封闭性：对于任意的 $a,b \in R$，$a \cdot b \in R$ 成立。

2）乘法的结合律：对于任意的 $a,b,c \in R$，$a \cdot (b \cdot c) = (a \cdot b) \cdot c$ 成立。

3）分配律：对于任意的 $a,b,c \in R$，$a \cdot (b + c) = a \cdot b + a \cdot c$ 和 $(a + b) \cdot c = a \cdot c + b \cdot c$ 成立。

满足上述所有条件的集合可以被定义为环。

若环 R 的乘法满足交换律，它可以被定义为交换环，它的定义如下。

定义 1.5 交换环：假定一个环的乘法还满足以下条件。

交换律：对于任意的 $a,b \in G$，有 $a \cdot b = b \cdot a$。

满足上述所有条件的集合可以被定义为交换环。

交换环满足一定条件可以定义成整环，它的定义如下。

定义 1.6 整环：交换环 G 中存在非零的乘法单位元，即存在 G 中的一个元素，记作 e，还满足以下条件。

1）e 不等于 0，且对任意 $a \in G$，有 $e \cdot a = a \cdot e = a$。

2）对于 a，$b \in G$，若 $a \cdot b = 0$ 可以推出 $a = 0$ 或 $b = 0$。

满足上述所有条件的集合可以被定义为整环。

1.1.4 域

域是在整环的基础上发展而来的，下面我们来看它的定义。

定义 1.7 域：定义 + 表示二元加法，• 表示二元乘法，F 表示非空集合，如果满足以下条件。

1）F 是一个整环，即满足定义 1.6 的所有原则。

2）$a \cdot a^{-1} = (a^{-1}) \cdot a = 1$ 成立。

满足上述所有条件的集合可以被定义为域。

1.2 有限域概念

1.2.1 有限域定义

有限域是仅含有限多个元素的域，它首先由伽罗瓦所发现，因而又被称为伽罗瓦域。下面我们来看有限域的定义。

定义 1.8 有限域：假定一个域非空，我们定义了它的加法和乘法，+表示二元加法，•表示二元乘法，它满足以下条件。

它只含有有限个元素，即对于域$\{F,+,\cdot\}$，若F中的元素个数为P，且$P < \infty$，则域$\{F,+,\cdot\}$为有限域。

满足上述条件的域可以被定义为有限域，它常用$GF(p)$表示，G是迦罗瓦（Galois）的英文首字母，F是域（Field）的英文首字母，P是有限域元素的个数。

复合有限域是有限域的一种特殊形式，又被称作复合域，它的定义如下。

定义 1.9 复合有限域：若一个有限域非空，它可以被定义为复合有限域的条件是它可以被表示成$GF(((a^b)^{\cdots})^c)$的形式，其中a、b和c均是正整数。

满足上述条件的有限域可以被定义为复合有限域。

1.2.2 常用有限域形式

常用的有限域形式一般有$GF(p)$和$GF(2^n)$，其中n是大于 1 的正整数，p一般是素数。

1）$GF(p)$的范围是$0,1,2,\ldots,p-1$，它的元素可以直接用十进制表示。

2）$GF(2^n)$的范围是$0,1,2,\ldots,2^n-1$，它的元素可以用多项式的形式表示，也可以用二进制的形式表示。以$GF(2^n)$的元素a为例，它的多项式形式是$a_{n-1}x^{n-1} + a_{n-2}x^{n-2} + \ldots + a_0$，它的二进制形式是$(a_{n-1},a_{n-2},\ldots,a_0)_2$，这里$a_i$是 0 或 1，$i = n-1, n-2,\ldots,0$。

常见的复合有限域的形式一般有$GF((2^n)^2)$，我们把$GF(2^n)$称为它的子域。以$GF((2^n)^2)$的元素a为例，它的元素可以被表示为多项式的形式$a_1x + a_0$，或者系数的形式$(a_1,a_0)_n$，其中，a_1, a_0均是子域$GF(2^n)$的元素。

1.2.3 不可约多项式

有限域的最重要的多项式是不可约多项式,它参与了所有的有限域运算。

定义 1.10 不可约多项式:一个多项式可以被定义为不可约多项式的条件如下。

1)它的次数大于零。

2)除了常数和常数与本身的乘积之外,它不可能被域的其他多项式除尽。

满足上述所有条件的多项式可以被定义为不可约多项式。对于每一个有限域,一般不少于一个多项式可以被选择为不可约多项式。

1.3 有限域元素

有限域 $GF(2^n)$ 是密码算法常用的域形式。$GF(2^n)$ 的基有以下几种:多项式基、正规基、对偶基、移位多项式基和弱对偶基等。

设 $\beta = \{\beta_0, \beta_1, \cdots, \beta_{n-1}\}$ 为有限域 $GF(2^n)$ 上的一组线性无关的元素,则 β 称为有限域 $GF(2^n)$ 上的一组基。有限域 $GF(2^n)$ 中的任意一个元素对于基 β 可以表示成以下形式:

$$\alpha = \alpha_0 \beta_0 + \alpha_1 \beta_1 + \ldots + \alpha_{n-1} \beta_{n-1}。$$

α 也可以写成向量形式:

$$(\alpha_0, \alpha_1, \ldots, \alpha_{n-1})。$$

1.3.1 多项式基

多项式基也被称为标准基。

设 $\beta = \{x^0, x^1, \ldots, x^{n-1}\}$ 是有限域 $GF(2^n)$ 上的一组基,其中 x 属于有限域 $GF(2^n)$,$G(x)$ 为有限域 $GF(2^n)$ 中的不可约多项式,使得有限域 $GF(2^n)$ 中的任意一个元素 α 都可以表示为 $\alpha = \alpha_0 + \alpha_1 x + \ldots + \alpha_{n-1} x^{n-1}$ 的形式,则 β 是有限域 $GF(2^n)$ 上基于不可约多项式 $G(x)$ 的一组多项式基。由于有限域 $GF(2^n)$ 的基域为 2,所以 $\alpha_i \in \{0,1\}$。

1.3.2 正规基

设 $\beta = \{x^{2^0}, x^{2^1}, \ldots, x^{2^{n-1}}\}$ 是有限域 $GF(2^n)$ 上的一组基,其中 x 属于有限域 $GF(2^n)$,

$G(x)$ 为有限域 $GF(2^n)$ 中的不可约多项式,使得有限域 $GF(2^n)$ 中的任意一个元素 α 都可以表示为 $\alpha = \alpha_0 x + \alpha_1 x^2 + \ldots + \alpha_{n-1} x^{2^{n-1}}$ 的形式,则 β 是有限域 $GF(2^n)$ 上基于不可约多项式 $G(x)$ 的一组正规基。由于有限域 $GF(2^n)$ 的基域为 2,所以 $\alpha_i \in \{0,1\}$。

1.3.3 对偶基

设 $\beta = \{\beta_0, \beta_1, \ldots, \beta_{n-1}\}$ 和 $\lambda = \{\lambda_0, \lambda_1, \ldots, \lambda_{n-1}\}$ 是有限域 $f(\alpha \beta_i \lambda_j) = 0$ 上的两组基,线性仿射 f 是从有限域 $GF(2^n)$ 到有限域 $GF(2)$ 的线性函数,如果对于有限域 $GF(2^n)$ 中的元素 α,满足以下条件:

1)当 i 等于 j 时,$f(\alpha \beta_i \lambda_j) = 1$ 成立;
2)当 i 不等于 j 时,$f(\alpha \beta_i \lambda_j) = 0$ 成立。

则 β 是多项式基,而 λ 是其对偶基。

1.4 有限域基础运算

有限域基础运算在信息安全、存储、通信中被广泛运用,常用的运算包括有限域的加法、乘法、平方、乘方、求逆、除法等。

1.4.1 有限域加法

有限域加法又被称作模加,我们下面以 $GF(2^n)$ 的多项式形式为例说明加法的运算。我们假定 $a(x)$ 和 $b(x)$ 是有限域 $GF(2^n)$ 的元素,则 $a(x)$ 和 $b(x)$ 的有限域加法的计算如下:

$$a(x) = a_{n-1} x^{n-1} + a_{n-2} x^{n-2} + \ldots + a_0$$
$$b(x) = b_{n-1} x^{n-1} + b_{n-2} x^{n-2} + b_0$$
$$c(x) = a(x) + b(x)$$
$$c(x) = (a_{n-1} + b_{n-1}) x^{n-1} + (a_{n-2} + b_{n-2}) x^{n-2} + \ldots + (a_0 + b_0)$$

这里 $c(x) = \sum_{i=0}^{n-1} c_i x^i$ 是 $a(x)$ 和 $b(x)$ 的加法结果。

以下是有限域 $GF(2^8)$ 加法的 VHDL 代码示例,其中 a 和 b 是两个运算数,c 是运算结果。

```
LIBRARY IEEE;
USE IEEE.STD_LOGIC_1164.ALL;

ENTITY GF_add IS
      PORT (a,b:  IN STD_LOGIC_VECTOR(7 DOWNTO 0);
               c: OUT STD_LOGIC_VECTOR(7 DOWNTO 0));
END GF_add;

ARCHITECTURE rtl OF GF_add IS
BEGIN
       c<=a XOR b;
END rtl;
```

1.4.2 有限域乘法

有限域乘法又被称作模乘,是有限域最重要的运算,我们下面以 $GF(2^n)$ 的多项式形式为例说明乘法的运算。我们假定 $a(x)$ 和 $b(x)$ 是有限域 $GF(2^n)$ 的元素,则 $a(x)$ 和 $b(x)$ 的有限域乘法的计算如下:

$$c(x) = (a(x) \times b(x)) \bmod f(x)$$

这里 $f(x)$ 是我们选定的 $GF(2^n)$ 的不可约多项式。

1. 多项式基乘法

有限域乘法的算法很多,我们下面介绍一种常见的多项式基的乘法。

假定 $a(x)$ 和 $b(x)$ 是有限域 $GF(2^n)$ 的元素,$f(x)$ 是选定的 $GF(2^n)$ 的不可约多项式。

首先,对于 $i = 0, 1, \ldots, 2(n-1)$ 和 $j = 0, 1, \ldots, n-1$,我们计算 v_{ij} 的过程如下:

$$x^i \bmod f(x) = \sum_{j=0}^{n-1} v_{ij} x^j$$

然后,对于 $i = 0, 1, \ldots, 2(n-1)$,我们计算 S_i 的过程如下:

$$S_i = \sum_{j+k=i} a_j b_k$$

这里,j,k 的取值范围分别是 $j = 0, 1, \ldots, n-1$ 和 $k = 0, 1, \ldots, n-1$。

最后,对于 $i = 0, 1, \ldots, n-1$,我们计算 c_i 的过程如下:

$$c_i = \sum_{j=0}^{2(n-1)} v_{ji} s_j$$

乘法结果 $a(x) \times b(x) \bmod f(x)$ 是 $\sum_{i=0}^{n-1} c_i x^i$。

以下是有限域 $GF(2^8)$ 乘法的 VHDL 代码示例，其中 a 和 b 是两个运算数，c 是运算结果。

```vhdl
LIBRARY IEEE;
USE IEEE.STD_LOGIC_1164.ALL;
use IEEE.STD_LOGIC_ARITH.ALL;
use IEEE.STD_LOGIC_UNSIGNED.ALL;

ENTITY gfmult IS
PORT(a,b:IN BIT_VECTOR(7 DOWNTO 0);
     c:OUT BIT_VECTOR(7 DOWNTO 0));
END gfmult;

ARCHITECTURE rtl OF gfmult IS
SIGNAL ly:BIT_VECTOR(14 DOWNTO 0);
SIGNAL lym:BIT_VECTOR(7 DOWNTO 0);
BEGIN
ly(14)<=a(7) AND b(7);
ly(13)<=(a(7) AND b(6)) XOR (a(6) AND b(7));
ly(12)<=(a(7) AND b(5))XOR(a(6) AND b(6))XOR(a(5) AND b(7));
ly(11)<=(a(7) AND b(4))XOR(a(6) AND b(5))XOR(a(5) AND b(6))XOR(a(4) AND b(7));
ly(10)<=(a(7) AND b(3))XOR(a(6) AND b(4))XOR(a(5) AND b(5))XOR(a(4) AND b(6))XOR(a(3) AND b(7));
ly(9)<=(a(7) AND b(2))XOR(a(6) AND b(3))XOR(a(5) AND b(4))XOR(a(4) AND b(5))XOR(a(3) AND b(6))XOR(a(2) AND b(7));
ly(8)<=(a(7) AND b(1))XOR(a(6) AND b(2))XOR(a(5) AND b(3))XOR(a(4) AND b(4))XOR(a(3) AND b(5))XOR(a(2) AND b(6))XOR(a(1) AND b(7));
ly(7)<=(a(7) AND b(0))XOR(a(6) AND b(1))XOR(a(5) AND b(2))XOR(a(4) AND b(3))XOR(a(3) AND b(4))XOR(a(2) AND b(5))XOR(a(1) AND b(6))XOR(a(0) AND b(7));
ly(6)<=(a(6) AND b(0))XOR(a(5) AND b(1))XOR(a(4) AND b(2))XOR(a(3) AND b(3))XOR(a(2) AND b(4))XOR(a(1) AND b(5))XOR(a(0) AND b(6));
ly(5)<=(a(5) AND b(0))XOR(a(4) AND b(1))XOR(a(3) AND b(2))XOR(a(2) AND b(3))XOR(a(1) AND b(4))XOR(a(0) AND b(5));
ly(4)<=(a(4) AND b(0))XOR(a(3) AND b(1))XOR(a(2) AND b(2))XOR(a(1) AND b(3))XOR(a(0) AND b(4));
ly(3)<=(a(3) AND b(0))XOR(a(2) AND b(1))XOR(a(1) AND b(2))XOR(a(0) AND b(3));
ly(2)<=(a(2) AND b(0))XOR(a(1) AND b(1))XOR(a(0) AND b(2));
ly(1)<=(a(1) AND b(0))XOR(a(0) AND b(1));
ly(0)<=(a(0) AND b(0));
lym(7)<=ly(7);
lym(6)<=ly(6);
lym(5)<=ly(5);
lym(4)<=ly(4);
lym(3)<=ly(3);
lym(2)<=ly(2);
lym(1)<=ly(1);
lym(0)<=ly(0);
c<=lym;
END rtl;
```

2. 正规基乘法

Massey-Omura 乘法器是基于正规基的乘法器，是由 Massey 和 Omura 在 1984 年中的专利中首次提出。以正规基表示的有限域 $GF(2^n)$ 中的元素的平方运算，只需要进行循环移位操作就能完成。基于这个性质，Massey-Omura 乘法器可以抽象出一个基本运算单元——乘积项函数。下面我们介绍正规基乘法。

有限域 $GF(2^n)$ 中的正规基表示为 $\beta = \left\{ x^{2^0}, x^{2^1}, \ldots, x^{2^{n-1}} \right\}$，则有限域 $GF(2^n)$ 中任意一个元素可以表示为 $\alpha = \alpha_0 x + \alpha_1 x^2 + \ldots + \alpha_{n-1} x^{2^{n-1}}$，其中 $\alpha_i \in \{0,1\}$。

以正规基表示有限域 $GF(2^n)$ 中的元素，则域中的元素满足以下性质：

1) 对有限域 $GF(2^n)$ 中的任意两个元素 a 和 b，满足等式 $(a+b)^2 = a^2 + b^2$；
2) 对有限域 $GF(2^n)$ 中的任意一个元素 a，满足等式 $a = a^{2^n}$。

令 a 和 b 为有限域 $GF(2^n)$ 中的元素，表示成向量的形式如下：

$$a = (a_0, a_1, \ldots, a_{n-1})$$
$$b = (b_0, b_1, \ldots, b_{n-1})$$

最终根据以上性质可以得到如下的表达式：

$$c_{n-1} = f(a_0, a_1, \ldots, a_{n-1}, b_0, b_1, \ldots, b_{n-1})$$
$$\ldots$$
$$c_0 = f(a_1, a_2, \ldots, a_{n-1}, b_1, b_2, \ldots, b_{n-1})$$

其中 f 为乘积项函数。

3. Systolic 乘法

基于 Systolic 的乘法器分为 Systolic 阵列和半 Systolic 阵列乘法器。

Systolic 的硬件结构类似于流水线，它是由一组结构简单的基本单元组合而成。这些基本单元执行简单的操作，基本单元只与相邻的单元相连，信号路径较短。Systolic 结构只利用阵列边界的基本单元进行输入和输出操作，使得电路具有良好的可扩展性。其硬件处理速度介于全并行硬件和串行电路之间，面积也介于全并行硬件和串行电路之间。

Wang 等人在 1991 年首次提出了基于 Systolic 的乘法器实现算法，其硬件结构模块性强，数据的吞吐率非常高。

4. 查找表乘法

除了利用数学运算求取乘法结果，计算乘法的方法还有一种，就是查找表乘法，它并不直接计算乘法。查找表乘法的算法描述如下：

首先我们假定 α 是 $GF(2^n)$ 的基元，$f(x)$ 是 $GF(2^n)$ 的不可约多项式，所有非零的有限域元素都可以被表示成它的乘方。对于 $i=0,1,\ldots,n-2$，我们计算 α 的乘方结果的过程如下：

$$k_i(x) = \alpha^i \bmod f(x)$$

这里 i 是指数，$k_i(x)$ 是乘方结果。如表 1-1 所示，我们将 $k_i(x)$ 存储到查找表的第 i 个地址，则 $\alpha^0, \alpha^1, \ldots, \alpha^{n-2}$ 组成了 $GF(2^n)$ 的所有非零元素的有限集合。

表 1-1　　　　　　　　　　$GF(2^n)$ 的乘法查找表

地址（乘方的指数）	域元素（乘方的结果）
$(000\ldots000)_2$	α^0
$(000\ldots001)_2$	α^1
$(000\ldots010)_2$	α^2
$(000\ldots011)_2$	α^3
$(000\ldots101)_2$	α^4
$(000\ldots110)_2$	α^5
…	…
$(111\ldots110)_2$	α^{n-2}

我们下面举例说明如何计算乘法，其中 a 和 b 是 $GF(2^n)$ 的元素。

例 1-1　计算 $a \times b$。我们假定 $a = \alpha^i \bmod f(x)$，$b = \alpha^j \bmod f(x)$，则 $a \times b = (\alpha^i \times \alpha^j) \bmod f(x) = \alpha^{(i+j) \bmod (n-1)}$。若 $c = \alpha^{(i+j) \bmod (n-1)}$，则 $a \times b = c$。

查表法是比较简单的乘法器实现方法，其主要思想是通过空间来换取时间复杂度。它将有限域 $GF(2^n)$ 中所有的元素对相乘的结果预先计算出来，保存在 ROM 中，然后直接通过两个乘数来查表找到结果。其优点是速度快，逻辑简单，可以在很短的时间内查到结果。但是其缺点也非常明显，空间复杂度以 n^2 的速度增长，所以只适合在 n 比较小的情况下使用，在 n 较大的情况下不太实际。

1.4.3　有限域平方

有限域的平方是单操作数的运算，我们下面以 $GF(2^n)$ 的多项式形式为例说明平方的运算。我们假定 $a(x)$ 是有限域 $GF(2^n)$ 的元素，则 $a(x)$ 的有限域平方的计算如下。

$$b(x) = a(x)^2 \bmod f(x)$$

这里 $f(x)$ 是我们选定的 $GF(2^n)$ 的不可约多项式，$b(x)$ 是平方运算的结果。平方一般可以采用乘法的算法来实现。

1.4.4 有限域乘方

有限域乘方是单操作数的运算，我们下面以 $GF(2^n)$ 的多项式形式为例说明乘方的运算。我们假定 $a(x)$ 是有限域 $GF(2^n)$ 的元素，则 $a(x)$ 的有限域乘方的计算如下。

$$b(x) = a(x)^i \bmod f(x)$$

这里 i 是给定的指数，$f(x)$ 是我们选定的 $GF(2^n)$ 的不可约多项式，$b(x)$ 是乘方运算的结果。乘方一般可以采用乘法的算法来实现。

1.4.5 有限域求逆

有限域求逆是一种计算相对复杂的有限域运算，它在很多时候需要的运算时间比有限域乘法还多。我们下面以 $GF(2^n)$ 的多项式形式为例说明求逆运算。我们假定 $a(x)$ 是有限域 $GF(2^n)$ 的元素，则 $a(x)$ 的有限域求逆可以被表示为 $b(x) = a(x)^{-1} \bmod f(x)$。这里 $f(x)$ 是我们选定的 $GF(2^n)$ 的不可约多项式。

1. 费马小定理求逆

有限域求逆的算法很多，我们下面介绍一种常见的费马小定理求逆算法。假定 $a(x)$ 是有限域 $GF(2^n)$ 的元素，$f(x)$ 是选定的 $GF(2^n)$ 的不可约多项式。根据费马小定理，我们可以得到如下计算：

$$a^{-1} = a^{2^n - 2}$$

$$2^n - 2 = \sum_{i=1}^{n-1} 2^i$$

$$a^{-1} = \prod_{i=1}^{n-1} a^{2^i}$$

所以计算求逆 a^{-1} 可以被分解成 $n-1$ 元的乘法，即 $a^2, a^{2^2}, \ldots, a^{2^{n-1}}$。
对于 $i = 1, 2, \ldots, n-1$，我们计算 a^{2^i} 的过程如下：

$$a^{2^i} \bmod f(x) = \sum_{j=0}^{n-1} k_{ij} x^j$$

这样我们就可以分别求出 $a^2, a^{2^2}, \ldots, a^{2^{n-1}}$ 的结果，然后再将它们相乘，乘法结果是 a 的逆元。

以下是有限域 $GF(2^8)$ 的求逆 VHDL 代码示例，其中，inB 是待求逆的有限域元素，

1.4 有限域基础运算

Binvert 是求逆运算结果，gf100101011mult 是乘法部件。

```
LIBRARY IEEE;
USE IEEE.STD_LOGIC_1164.ALL;

ENTITY gf100101011Invert2 IS
      PORT (inB:IN BIT_VECTOR(7 DOWNTO 0);
            BInvert: OUT BIT_VECTOR(7 DOWNTO 0));
END gf100101011Invert2;

ARCHITECTURE behav OF gf100101011Invert2 IS

COMPONENT gf100101011mult
PORT(a,b: IN BIT_VECTOR(7 DOWNTO 0);
     c: OUT BIT_VECTOR(7 DOWNTO 0));
END COMPONENT;

SIGNAL B2,B4,B8,B16,B32,B64,B128,B2B4,B8B16,B32B64,B2B4B8B16,B32B64B128: BIT_VECTOR(7 DOWNTO 0);

BEGIN
B2(0)<= inB(0) XOR inB(4) XOR inB(6) XOR inB(7);
B2(1)<= inB(7);
B2(2)<= inB(1) XOR inB(4) XOR inB(5) XOR inB(6);
B2(3)<= inB(4) XOR inB(6);
B2(4)<= inB(2) XOR inB(4) XOR inB(5) XOR inB(7);
B2(5)<= inB(5);
B2(6)<= inB(3) XOR inB(5) XOR inB(6);
B2(7)<= inB(6);
B4(0)<= inB(0) XOR inB(2) XOR inB(3) XOR inB(6);
B4(1)<= inB(6);
B4(2)<= inB(2) XOR inB(3) XOR inB(4) XOR inB(5) XOR inB(6);
B4(3)<= inB(2) XOR inB(3) XOR inB(4) XOR inB(6) XOR inB(7);
B4(4)<= inB(1) XOR inB(2) XOR inB(5) XOR inB(7);
B4(5)<= inB(5);
B4(6)<= inB(3) XOR inB(4);
B4(7)<= inB(3) XOR inB(5) XOR inB(6);
B8(0)<= inB(0) XOR inB(1) XOR inB(3) XOR inB(4) XOR inB(7);
B8(1)<= inB(3) XOR inB(5) XOR inB(6);
B8(2)<= inB(1) XOR inB(2) XOR inB(3) XOR inB(4) XOR inB(6) XOR inB(7);
B8(3)<= inB(1) XOR inB(2) XOR inB(3) XOR inB(4) XOR inB(5) XOR inB(7);
B8(4)<= inB(1) XOR inB(4) XOR inB(7);
B8(5)<= inB(5);
B8(6)<= inB(2) XOR inB(5) XOR inB(6) XOR inB(7);
B8(7)<= inB(3) XOR inB(4);
B16(0)<= inB(0) XOR inB(2) XOR inB(4) XOR inB(5) XOR inB(6) XOR inB(7);
B16(1)<= inB(3) XOR inB(4);
```

```
B16(2)<= inB(1) XOR inB(2) XOR inB(3) XOR inB(4) XOR inB(5);
B16(3)<= inB(1) XOR inB(2) XOR inB(4) XOR inB(5) XOR inB(6);
B16(4)<= inB(2) XOR inB(4) XOR inB(5) XOR inB(6);
B16(5)<= inB(5);
B16(6)<= inB(1) XOR inB(3) XOR inB(4) XOR inB(5) XOR inB(6);
B16(7)<= inB(2) XOR inB(5) XOR inB(6) XOR inB(7);
B32(0)<= inB(0) XOR inB(1) XOR inB(2) XOR inB(3) XOR inB(4);
B32(1)<= inB(2) XOR inB(5) XOR inB(6) XOR inB(7);
B32(2)<= inB(1) XOR inB(2) XOR inB(4) XOR inB(5);
B32(3)<= inB(1) XOR inB(2) XOR inB(3);
B32(4)<= inB(1) XOR inB(2) XOR inB(3) XOR inB(7);
B32(5)<= inB(5);
B32(6)<= inB(2) XOR inB(3) XOR inB(5);
B32(7)<= inB(1) XOR inB(3) XOR inB(4) XOR inB(5) XOR inB(6);
B64(0)<= inB(0) XOR inB(1) XOR inB(2) XOR inB(6) XOR inB(7);
B64(1)<= inB(1) XOR inB(3) XOR inB(4) XOR inB(5) XOR inB(6);
B64(2)<= inB(1) XOR inB(2) XOR inB(5) XOR inB(6);
B64(3)<= inB(1) XOR inB(5) XOR inB(7);
B64(4)<= inB(1) XOR inB(5) XOR inB(6) XOR inB(7);
B64(5)<= inB(5);
B64(6)<= inB(1);
B64(7)<= inB(2) XOR inB(3) XOR inB(5);
B128(0)<= inB(0) XOR inB(1) XOR inB(3);
B128(1)<= inB(2) XOR inB(3) XOR inB(5);
B128(2)<= inB(1) XOR inB(3) XOR inB(4) XOR inB(5) XOR inB(7);
B128(3)<= inB(5) XOR inB(6) XOR inB(7);
B128(4)<= inB(3) XOR inB(7);
B128(5)<= inB(5);
B128(6)<= inB(7);
B128(7)<= inB(1);

        U0:gf100101011mult PORT MAP (B2,B4,B2B4);
        U1:gf100101011mult PORT MAP (B8,B16,B8B16);
        U2:gf100101011mult PORT MAP (B32,B64,B32B64);
        U3:gf100101011mult PORT MAP (B2B4,B8B16,B2B4B8B16);
        U4:gf100101011mult PORT MAP (B32B64,B128,B32B64B128);
        U5:gf100101011mult PORT MAP (B2B4B8B16,B32B64B128,BInvert);

END behav;
```

2. 查找表求逆

除了利用数学运算求取求逆结果，计算求逆的方法还有一种，就是查找表求逆，它并不直接计算求逆。

如表 1-2 所示，对于 $GF(2^n)$ 和 $GF(p)$ 上的所有非零而且非一的元素分别构建求逆查找表，表中的每一行由地址和地址数值的逆元组成。i_j 是它的地址的逆元。

1）对于 $GF(2^n)$，j 的取值范围是 $j=0,1,2,3,4,5,\ldots,n-6,n-5,n-4,n-3,n-2,n-1$；
2）对于 $GF(p)$，j 的取值范围是 $j=0,1,2,3,4,5,\ldots,p-6,p-5,p-4,p-3,p-2,p-1$。

起始地址是 $(000\ldots010)_2$，这是因为我们不需要求 $(000\ldots000)_2$ 和 $(000\ldots001)_2$ 的逆。结束地址分别是 $GF(2^n)$ 和 $GF(p)$ 的最后一个元素。利用构建的求逆查找表，我们可以很容易通过搜索地址找到要计算的逆元。

表 1-2　　　　$GF(2^n)$ 的求逆查找表和 $GF(p)$ 的求逆查找表

$GF(2^n)$ 的求逆查找表		$GF(p)$ 的求逆查找表	
地址	逆元	地址	逆元
$(000\ldots010)_2$	i_2	$(000\ldots010)_2$	i_2
$(000\ldots011)_2$	i_3	$(000\ldots011)_2$	i_3
$(000\ldots100)_2$	i_4	$(000\ldots100)_2$	i_4
$(000\ldots101)_2$	i_5	$(000\ldots101)_2$	i_5
$(000\ldots110)_2$	i_6	$(000\ldots110)_2$	i_6
$(000\ldots111)_2$	i_7	$(000\ldots111)_2$	i_7
…	…	…	…
$(111\ldots010)_2$	i_{n-6}	$p-6$	i_{p-6}
$(111\ldots011)_2$	i_{n-5}	$p-5$	i_{p-5}
$(111\ldots100)_2$	i_{n-4}	$p-4$	i_{p-4}
$(111\ldots101)_2$	i_{n-3}	$p-3$	i_{p-3}
$(111\ldots110)_2$	i_{n-2}	$p-2$	i_{p-2}
$(111\ldots111)_2$	i_{n-1}	$p-1$	i_{p-1}

1.4.6　有限域除法

有限域的除法是两个操作数的运算，我们下面以 $GF(2^n)$ 的多项式形式为例说明除法的运算。我们假定 $a(x)$、$b(x)$ 是有限域 $GF(2^n)$ 的元素，则 $a(x)$、$b(x)$ 的有限域除法的计算如下：

$$c(x) = a(x) \div b(x) \bmod f(x)$$

这里 $f(x)$ 是我们选定的 $GF(2^n)$ 的不可约多项式，$c(x)$ 是除法运算的结果。除法一般可以采用乘法和求逆的算法来实现，例如：

$$c(x) = a(x) \times b(x)^{-1} \bmod f(x)$$

1.4.7 求解线性方程组

求解有限域的线性方程组的方法主要有高斯消元法（Gaussian Elimination）和高斯约当消元法（Gauss-Jordan Elimination）。这里以求解一个给定的线性方程组 $Ax=b$ 为例，我们分别介绍高斯消元法和高斯约当消元法，其中 A 是一个给定的规模为 $m\times m$ 的系数矩阵，b 是 m 维的向量。

高斯消元法首先把 $Ax=b$ 转换成等效的 $A'x=b'$ 形式，其中 A' 是一个规模为 $m\times m$ 的上三角矩阵，b' 是 m 维向量。在这个过程中我们依次对系数矩阵的每一列做迭代，总共包括 m 次迭代，每次迭代由找主元、归一和消元组成。

1）找主元是在迭代当前列的所有元素中找到一个非零元作为主元。

2）归一是对迭代当前行的所有元素做归一操作，即当前行的每个元素乘以主元的逆元。

3）消元是对迭代当前行下面所有的元素做消元操作，即每个元素加上它所在行在当前列的元素和它所在列在当前行的元素的乘积。

$A'x=b'$ 形式还需经过回溯替代变成等效的 $A''x=b''$，才能求出线性方程组的解，其中 A'' 是一个规模为 $m\times m$ 的一条对角线全为 1、其他元素全为 0 的矩阵，b'' 是 m 维向量，即线性方程组的解。

高斯约当消元法直接把 $Ax=b$ 转换成等效的 $A''x=b''$，b'' 是 m 维向量，即线性方程组的解。它与高斯消元法的不同点在消元部分，它对除了迭代当前行之外所有的元素进行消元，所以迭代结束后 A'' 是一个规模为 $m\times m$ 的一条对角线全为 1、其他元素全为 0 的矩阵，b'' 是线性方程组的解。

下面重点介绍高斯消元法。高斯消元法是求解线性方程组的常用方法，可用于任何域中的线性方程组的求解。对于如下的线性方程组：

$$\begin{cases} a_{11}x_1 + a_{12}x_2 + \ldots + a_{1n}x_n = b_1 \\ a_{21}x_1 + a_{22}x_2 + \ldots + a_{2n}x_n = b_2 \\ \vdots \\ a_{n1}x_1 + a_{n2}x_2 + \ldots + a_{nn}x_n = b_n \end{cases}$$

可以得到方程组中的系数增广矩阵：

$$\begin{bmatrix} a_{11} & a_{12} & \ldots & a_{1n} & b_1 \\ a_{21} & a_{22} & \ldots & a_{2n} & b_2 \\ \vdots & \vdots & \vdots & \vdots \\ a_{n1} & a_{n2} & \ldots & a_{nn} & b_n \end{bmatrix}$$

高斯消元算法对该增广矩阵进行线性变换，将前 n 列转化为上三角矩阵形式，则最后一列为满足线性方程组的解，算法 1-1 描述了整个过程，其算法复杂度在最差的情况

下为 $O(n^3)$。

算法 1-1　高斯消元

输入：M 为 n 个 $n+1$ 维向量组成的矩阵

输出：M_n 为 M 的转秩的第 n 个行向量，即为解

```
FOR i IN 0 TO n-1:
    FOR j IN i TO n-1:
        IF M_{i0} ≠ 0 THEN
            exchange(M_i, M_j);
                Break;
            END IF;
    END FOR;
    FOR j IN i+1 TO n-1:
        M_{ij} := M_{ij} ⊗ M_{ii}^{-1};
    END FOR;
    FOR j IN i+1 TO n-1:
        IF (M_{ji} ≠ 0) THEN
            FOR k IN i+1 TO n:
                M_{jk} := M_{jk} ⊕ M_{ik} ⊗ M_{ji};
            END FOR;
        END IF;
    END FOR;
END FOR;
```

下面分别列出求解线性方程组的软件和硬件实现代码。

以下是求解线性方程组的 C#代码示例。

```csharp
using System;
using System.Collections.Generic;
using System.Linq;
using System.Text;

namespace Gauss
{
    class Program
    {
        public static void Main(string[] args)            // 主函数
        {                                                  // 主函数开始
            // 为了简化程序，本例只考虑方程组有唯一解的情况，不对其他情况进行判断
            //n是线性方程组的个数,数组a是增广矩阵,为了方便调试,在这里直接给n和数组a赋值,
            在实际使用过程中要通过键盘读入它们的值
            int n = 3;
            int[,] a = { { 3, 1, 3, 8 }, { 4, 2, 5, 4 }, { 1, 2, 0, 7 } };
            int[] x = new int[n];
            Gauss(n, a, x);           // 输出方程组的解
            Console.WriteLine("方程组的解为: ");
            for (int i = 0; i < n; i++)
```

```csharp
            Console.Write("x({0})={1,8:F3} ", i, x[i]);
        Console.WriteLine();
        Console.In.Read();
    }

    // 利用高斯消元法求线性方程组的解
    public static void Gauss(int n, int[,] a, int[] x)
    {
        int d;
        Console.WriteLine("高斯消去法解方程组的中间过程");
        Console.WriteLine("============================");
        Console.WriteLine("中间过程");
        Console.WriteLine("增广矩阵：");
        printArray(n, a); //打印增广矩阵
        Console.WriteLine();
            // 消元
        for (int k = 0; k < n; k++)
        {
            Console.WriteLine("第{0}步", k + 1);
            Console.WriteLine("初始矩阵：");
            printArray(n, a);
            Console.WriteLine();
            selectMainElement(n, k, a); // 选择主元素
            Console.WriteLine("选择主元素后的矩阵：");
            printArray(n, a);
            Console.WriteLine();
            d = a[k, k];//主元素
            for (int j = k; j <= n; j++)
                a[k, j] = a[k, j] / d;
            //这里要做除法，既是乘以主元素的乘法逆元
            //首先要算主元素的乘法逆元
            //其次做 n-k 次乘法
            //总计 n 次求逆，n(n+1)/2 次乘法
            Console.WriteLine("将第{0}行中a[{0},{0}]化为1后的矩阵：", k + 1);
            printArray(n, a);
            Console.WriteLine();
            // 高斯消去法与约当消去法的主要区别就是在这一步，高斯消去法是从k+1
            // 到n循环，而约当消去法是从1到n循环，中间跳过第k行
            for (int i = k + 1; i < n; i++)
            {
                d = a[i, k];
                // 这里使用变量d将a[i,k]的值保存下来的原理与上面注释中说明的一样
                for (int j = k; j <= n; j++)
                    a[i, j] = a[i, j] - d * a[k, j];
            //1.减法（异或）
            //2.乘法（做n-k次）
            }
```

1.4 有限域基础运算

```csharp
        Console.WriteLine("消元后的矩阵：");
        printArray(n, a);
        Console.WriteLine();
        //此时已化成上三角
    }
            // 回代
    x[n - 1] = a[n - 1, n];//最后一列
    for (int i = n - 1; i >= 0; i--)
    {
        x[i] = a[i, n];
        for (int j = i + 1; j < n; j++) x[i] = x[i] - a[i, j] * x[j];
    }
}

// 选择主元素  寻找第 K 列的主元素，确定行号
public static void selectMainElement(int n, int k, int[,] a)
{
    // 寻找第 k 列的主元素以及它所在的行号
    int t, mainElement;
    // mainElement 用于保存主元素的值
    int l;
    // 用于保存主元素所在的行号
    // 从第 k 行到第 n 行寻找第 k 列的主元素，记下主元素 mainElement 和所在的行号 l
    mainElement = a[k, k];
    l = k;
    //找到第 k 列最大元素
    for (int i = k + 1; i < n; i++)
    {
        if (mainElement <a[i, k])
        {
            mainElement = a[i, k];
            l = i;
            // 记下主元素所在的行号
        }
    }
            // l 是主元素所在的行。将 l 行与 k 行交换，每行前面的 k 个元素都是 0，不必交换
    if (l != k)
    {
        for (int j = k; j <= n; j++)
        {
            t = a[k, j];
            a[k, j] = a[l, j];
            a[l, j] = t;
        }
    }
}
```

```csharp
            // 打印矩阵
            public static void printArray(int n, int[,] a)
            {
                for (int i = 0; i < n; i++)
                {
                    for (int j = 0; j <= n; j++)
                      // Console.Write("{0,10:F6} ", a[i, j]);
                        Console.Write( a[i, j]);
                    Console.WriteLine();
                }
            }
        }
```

以下是求解线性方程组的 VHDL 代码示例。

```vhdl
LIBRARY IEEE;
USE IEEE.STD_LOGIC_1164.ALL;
USE WORK.RAINBOW_COMPONENTS.ALL;
USE WORK.packgauss.ALL;

ENTITY gauss IS
        PORT (clk:IN std_logic;
                    mtable:IN matrixtable;--要计算的矩阵
                    finish:OUT std_logic;--结束信号为1
                    gaussresult:OUT matrixn1;--方程的解
                    n:IN integer
                    );
END gauss;

ARCHITECTURE behav OF gauss IS

--采用状态机控制
   SIGNAL  present_state: statetype;--定义当前状态

--乘法部件
COMPONENT GF_mult
PORT(a,b: IN STD_LOGIC_VECTOR(7 DOWNTO 0);
         c: OUT STD_LOGIC_VECTOR(7 DOWNTO 0));
END COMPONENT;

--求逆部件
COMPONENT GF_inverse
      PORT (inB:IN STD_LOGIC_VECTOR(7 DOWNTO 0);
      BInvert: OUT STD_LOGIC_VECTOR(7 DOWNTO 0));
END COMPONENT;

SIGNAL mtablebak:matrixtable;--缓存输入矩阵
```

```
signal col:INTEGER:=0;--主元所在行
SIGNAL mainvalue:STD_LOGIC_VECTOR(7 DOWNTO 0);--主元
SIGNAL inversevalue:STD_LOGIC_VECTOR(7 DOWNTO 0);--逆元
SIGNAL multresult:STD_LOGIC_VECTOR(7 DOWNTO 0);--乘法结果
SIGNAL multiplicand:STD_LOGIC_VECTOR(7 DOWNTO 0);--被乘数
SIGNAL multiplier:STD_LOGIC_VECTOR(7 DOWNTO 0);--乘数
SIGNAL midint:integer;--存找到非零主元行
SIGNAL midvector:STD_LOGIC_VECTOR(7 DOWNTO 0);--临时使用

signal gi:integer;
signal oi:integer;
signal ei:integer;
signal ej:integer;

BEGIN

I0:GF_inverse PORT MAP (mainvalue,inversevalue);--求逆元

--乘法
M0:GF_mult PORT MAP (multiplicand,multiplier,multresult);

p1:PROCESS(clk)
BEGIN
      if(clk'event AND clk='1')then
CASE present_state IS

          -- 将输入的矩阵缓存
          WHEN init =>
          col<=0;
          finish<='0';
          L4:FOR i IN 0 TO nmax-1 LOOP
                if(i<=n-1) then
                    L5:FOR j IN 0 TO nmax LOOP
                      if(j<=n) then
                          mtablebak(i,j)<=mtable(i,j);
                      end if;
                      end loop L5;
                  end if;
              end loop L4;
              midvector<=mtable(0,0);--主元
              midint<=0;
              gi<=0;
              oi<=0;
              ei<=0;
              ej<=0;

            mainvalue<=mtable(0,0);
```

```
                present_state <= normalizing1;

       --空闲
       WHEN idle =>
       present_state<=idle;

       ---代入求解
       WHEN outputresult2 =>
        --获得乘法结果
           mtablebak(oi,n)<=multresult XOR mtablebak(oi,n);
        --一列代入结束
        if(oi=0) then
              oi<=col-1;
              gaussresult(col-1)<=mtablebak(col-1,n);
           --结束代入
              if(col=1) then
                 gaussresult(0)<=multresult XOR mtablebak(0,n);
                 finish<='1';
                 present_state<=idle;
              else
                 col<=col-1;
                 --送乘数和被乘数
                 multiplicand<=mtablebak(col-2,col-1);
                 multiplier<=mtablebak(col-1,n);
                 present_state<=outputresult2;
              end if;
           --当列继续代入
           else
              multiplicand<=mtablebak(oi-1,col);
              present_state<=outputresult2;
              oi<=oi-1;
        end if;

              --换行
           WHEN findmainelement2 =>

           --归一
       WHEN normalizing1 =>
          if ((gi>=col) and (gi<=n)) then
                multiplicand<=mtablebak(col,gi);
          end if;
          multiplier<=inversevalue;
          present_state <= normalizing2;

           --归一
       WHEN normalizing2 =>
           if ((gi>=col) and (gi<=n)) then
```

1.4 有限域基础运算

```
              mtablebak(col,gi)<=multresult;
       end if;
       gi<=gi+1;
       ei<=col+1;
       ej<=col+1;
        --继续归一
       if(gi/=n) then
           if ((gi+1>=col) and (gi+1<=n)) then
               multiplicand<=mtablebak(col,gi+1);
           end if;
           present_state <= normalizing2;

           --本行归一结束
       elsif (col/=n-1) then
        multiplier<=mtablebak(col+1,col);
        multiplicand<=mtablebak(col,col+1);
        present_state<=elimiresult;

        --最后一行归一结束，不用消元
       else
              --送乘数和被乘数
             multiplicand<=mtablebak(col-1,col);
             multiplier<=multresult;
             present_state<=outputresult2;
             gaussresult(n-1)<=multresult;
            oi<=col-1;
       end if;

   ---消元
   WHEN  elimiresult=>
        mtablebak(ei,ej)<=multresult XOR mtablebak(ei,ej);
        if ((ej=col+1) and (ei=col+1)) then
             mainvalue<=multresult XOR mtablebak(ei,ej);
        end if;

        --本行消元结束，开始下一行
        if(ej=n) then
            mtablebak(ei,col)<="00000000";
           ej<=col+1;
           ei<=ei+1;
           present_state<=elimiresult;
           multiplier<=mtablebak(ei+1,col);
           multiplicand<=mtablebak(col,col+1);
           end if;

          --结束消元
          if((ei=n-1) and (ej=n)) then
```

```
                            col<=col+1;
                            midvector<=mtablebak(col+1,col+1);--主元
                            midint<=col+1;
                            gi<=0;
                            oi<=0;
                            ei<=0;
                            ej<=0;
                              multiplicand<=mtablebak(col+1,col+2);
                              multiplier<=inversevalue;
                              present_state <= normalizing2;
                        end if;

                        --继续本行的消元
                        if(ej/=n) then
                          ej<=ej+1;
                          present_state<=elimiresult;
                          multiplier<=mtablebak(ei,col);
                          multiplicand<=mtablebak(col,ej+1);
                        end if;
                    when others =>
                        present_state <= idle;
            END CASE;
    END IF;
    END PROCESS;
    END behav;
```

1.5 其他数学问题

1.5.1 MQ 问题

多变量公钥密码体制的安全性基础是 MQ 问题，Garey 等人在 1979 年证明了 MQ 问题在有限域 $GF(2)$ 上是 NP 完全问题，而 Patarin 等人在 1997 年证明了 MQ 问题在任意有限域上都是 NP 完全问题。MQ 问题的定义如下。

定义 1.11 MQ 问题：给定有限域 $GF(q)$ 上的一组多变量非线性多项式方程组 Q，其中 $Q = \{Q_1(x), Q_2(x)..., Q_m(x)\}$，$Q_i(x) = 0$，则求解 $x = \{x_1, x_2..., x_n\} \in GF(q)$ 的问题为 MQ 问题。

MQ 问题中的方程组可以表示为：

$$Q_1(x_1, x_2 \ldots, x_n) = \sum_{1 \leq i \leq j \leq n} \alpha_{1ij} x_i x_j + \sum_{1 \leq i \leq n} \beta_{1i} x_i + \lambda_1$$

…

$$Q_n(x_1, x_2 \ldots, x_n) = \sum_{1 \leq i \leq j \leq n} \alpha_{nij} x_i x_j + \sum_{1 \leq i \leq n} \beta_{ni} x_i + \lambda_n$$

求解 MQ 问题的方法有 Buchberger 算法，首先要构造 Gröbner 基。目前效率最高的求 Gröbner 基的算法是由 Faugère 在 1999 年提出的 F_4 算法和 2002 年提出的 F_5 算法。Faugère 利用 F_5 算法破解了多变量公钥密码方案 HFE。

MQ 问题是多变量公钥密码体制的理论基础，其最基本的形式非常简单。设 L_1 和 L_2 为有限域 $GF(q)$ 上的可逆线性变换，F 为有限域 $GF(q)$ 上二次非线性映射，也称之为中心映射，可以得出基于 MQ 问题的多变量公钥密码体制的公钥为 $L_1 \circ F \circ L_2$，其私钥为 L_1^{-1}、L_2^{-1} 和 L^{-1}。F 的作用在于构造陷门函数，如何构造陷门是整个公钥密码体制的核心所在，例如 RSA 和 ECC 均包含良好密码性质的陷门。目前根据中心映射 F 构造的不同，多变量公钥密码主要分为 MI 公钥密码体制、HFE 公钥密码体制、非平衡油醋公钥密码体制和三角形公钥密码体制等。

1.5.2 线性仿射变换

有限域可逆仿射变换是密码学中非常重要的一个数学运算，它可以被表示为一个线性变换和一个平移。线性仿射变换又被称为线性仿射映射。线性仿射变换是可逆的，在多变量公钥密码算法的构造中有重要作用。其定义如下。

定义 1.12 线性仿射变换：设 $n \times n$ 矩阵 A 中的元素属于有限域 $GF(q)$，x 和 b 为 n 维向量，其中向量中的元素属于有限域 $GF(q)$，则线性仿射变换可以表示为 $F(x) = Ax + b$。

线性仿射变换在多变量公钥密码中，主要用于隐藏中心映射 F 的结构。中心映射的左右有线性仿射变换 L_1 和 L_2，其逆元 L_1^{-1} 和 L_2^{-1} 则被当作私钥的一部分来保存。有限域的可逆仿射变换过程通常由一个矩阵-向量乘法和一个向量之间的加法等运算组成。

算法 1-2 描述了线性仿射变换算法，其中输入 X 为线性仿射变换的参数，为 n 维向量；输入 M 为线性仿射矩阵，由 m 个 n 维向量组成；输入 Y 为线性仿射的偏移量，也是 n 维向量；输出 Z 为线性仿射变换后的结果。向量中的元素属于有限域 $GF(2^l)$，$\oplus$ 为有限域 $GF(2^l)$ 中的加法操作，$\otimes$ 为有限域 $GF(2^l)$ 中的乘法操作。

算法 1-2 线性仿射变换
输入：X, Y, M
输出：Z

```
FOR i IN 0 TO n-1:
    Y_i := Y_i ⊕ X_i ;
END FOR;
FOR i IN 0 TO m-1:
    FOR j IN 0 TO n-1:
Z_i := Z_i ⊕ (Y_j ⊗ M_{ij}) ;
END FOR;
END FOR;
```

1.5.3 随机数发生器

伪随机基于许多数学方法，有线性同余法、非线性同余法、Fibonacci 系列、Tausworthe 系列、进位加一借位减发生器等产生随机数的方法。基于数学算法的随机数发生器由真随机的种子和伪随机网络构成。一旦真随机的种子被暴露，伪随机数发生器就能确定真随机来源于真实的随机物理过程，因而彻底地消除了伪随机数的周期性问题。只有真随机数发生器才能提供真正的、永不重复的随机数序列。

主要的随机数发生器有如下 4 种。

1）LFSR 随机数：线性反馈移位寄存器（Linear Feedback Shift Register, LFSR）是一个反馈移位寄存器。其反馈函数是寄存器中某些位的简单异或。

2）Fibonacci 型将所有输出进行异或操作后送入输入端，Galois 型采用逐级异或方式。两种实现形式实际上是将系数刚好倒过来，性质都是一致的。

3）PUF 随机数：利用制造芯片过程中不可避免的差异性，差异会对 IC 的导线和门电路的传输延迟产生随机的影响，从而提取出不同的传输延迟差异，来保证不可克隆性。

4）基于振荡器采样的真随机数发生器：低频时钟作为采样时钟，高频振荡器的输出作为输入，在低频时钟的上升沿采样。

以下是随机数发生器 VHDL 代码示例。

```
Library IEEE ;
use IEEE.std_logic_1164.all ;
use IEEE.std_logic_arith.all ;
USE IEEE.STD_LOGIC_1164.ALL;
use IEEE.numeric_std.all;
USE IEEE.STD_LOGIC_UNSIGNED.ALL;

entity random is
    port (
        clk     : in std_logic ;
        reset   : in std_logic ;
        get     : in std_logic ;
```

```vhdl
        rand    : out STD_LOGIC_VECTOR(7 DOWNTO 0)
        );
end random ;

architecture rtl of random is
  SIGNAL count:STD_LOGIC_VECTOR(7 DOWNTO 0);
  signal feedback : std_logic ;
  signal lfsr_reg:STD_LOGIC_VECTOR(7 DOWNTO 0);

  begin
   feedback <= lfsr_reg(7) xor lfsr_reg(6) xor lfsr_reg(5) xor lfsr_reg(0) ;
      rand <= lfsr_reg;
      GenerateRandom:process(clk,reset,get)
  begin
          if (reset = '1') then
                  lfsr_reg <= (others => '0') ;
          elsif (clk = '1' and clk'event) then
                  lfsr_reg <= lfsr_reg(lfsr_reg'high - 1 downto 0) & feedback ;
          end if;

          if (get='1')then
                  lfsr_reg <= count ;
          end if;
      end process ;

    Counter:process(clk)
       BEGIN
           IF(clk'EVENT AND clk='1')THEN
                   count<=count+"1";
           END IF;
       END process;

end rtl ;
```

1.6 本章小结

本章从介绍集合、群、环、域的概念开始，展开讲解了多变量公钥密码使用的有限域的概念、域元素和域运算，重点阐述了有限域的乘法和求逆，另外也对多变量公钥密码的基础 MQ 问题、线性仿射变换、随机数发生器等数学问题进行了概要性描述。

1.7 本章参考文献

[1] 陈恭亮. 信息安全数学基础. 北京：清华大学出版社，2004.

[2] 徐常青. 杜先能. 高等代数方法与应用. 合肥：安徽大学出版社，2002.

[3] A. Reyhani-Masoleh, M. Anwar Hasan. Low Complexity Word-Level Sequential Normal Basis Multipliers [J]. IEEE transactions on computers, vol. 54, no. 2, February 2005,pp.98-110.

[4] Paar C. A New Architecture for a Parallel Finite Field Multiplier with Low Complexity Based on Composite Fields [J]. IEEE Trans. Computers, 1996, 45: 856-861.

[5] Oh S, Kim C H, Lim J, et al. Efficient Normal Basis Multipliers in Composite Fields[J]. IEEE Trans. Computers, 2000, 49: 1133-1138.

[6] Orlando G, Paar C. A Super-Serial Galois Fields Multiplier for FPGAs and its Application to PublicKey Algorithms [A]. In: Proceedings of the Seventh Annual IEEE Symposium on Field-Programmable Custom Computing Machines[C]. Washington, DC, USA: IEEE Computer Society. 1999. FCCM '99.

[7] Imana J. Reconfigurable implementation of bit-parallel multipliers over $GF(2^m)$ for two classes of finite fields[A]. In: 2004 IEEE International Conference on Field- Programmable Technology, 2004. Proceedings. [C], 2004: 287 - 290.

[8] Namin A, Wu H, Ahmadi M. A New Finite-Field Multiplier Using Redundant Representation [J]. IEEE Trans. Computers, 2008, 57(5): 716 -720.

[9] Namin A, Wu H, Ahmadi M. A Word-Level Finite Field Multiplier Using Normal Basis [J]. IEEE Trans. Computers, 2011, 60(6): 890 -895.

[10] Hsu I, Truong T, Deutsch L, et al. A comparison of VLSI architecture of finite field multipliers using dual, normal, or standard bases[J]. IEEE Trans. Computers, 1988, 37(6): 735 -739.

[11] Zhou G, Michalik H, Hinsenkamp L. Complexity Analysis and Efficient Implementations of Bit Parallel Finite Field Multipliers Based on Karatsuba-Ofman Algorithm on FPGAs[J]. IEEE Trans. Very Large Scale Integration (VLSI) Systems, 2010, 18(7): 1057 -1066.

[12] Rahaman H, Mathew J, Pradhan D. Test Generation in Systolic Architecture for Multiplication Over $GF(2^m)$ [J]. IEEE Trans. Very Large Scale Integration (VLSI) Systems, 2010, 18(9): 1366 -1371.

[13] Hariri A, Reyhani-Masoleh A. Digit-Level Semi-Systolic and Systolic Structures for the Shifted Polynomial Basis Multiplication Over Binary Extension Fields[J]. IEEE Trans. Very

Large Scale Integration (VLSI) Systems, 2011, 19(11): 2125 -2129.

[14] Wang Z, Fan S. Efficient Montgomery-Based Semi-Systolic Multiplier for Even-Type GNB of $GF(2^m)$ [J]. IEEE Trans. Computers, 2012, 61(3): 415 -419.

[15] Imana J. Low Latency $GF(2^m)$ Polynomial Basis Multiplier[J]. IEEE Trans. Circuits and Systems I: Regular Papers, 2011, 58(5): 935 -946.

[16] Bajard J C, Negre C, Plantard T. Subquadratic Space Complexity Binary Field Multiplier Using Double Polynomial Representation[J]. IEEE Trans. Computers, 2010, 59(12): 1585 -1597.

[17] Xie J, Meher P K, He J. Low-Complexity Multiplier for $GF(2^m)$ Based on All-One Polynomials[J]. IEEE Trans. Very Large Scale Integration (VLSI) Systems, 2013, 21(1): 168 -173.

[18] Hasan M, Namin A, Negre C. Toeplitz Matrix Approach for Binary Field Multiplication Using Quadrinomials[J]. IEEE Trans. Very Large Scale Integration (VLSI) Systems, 2012, 20(3): 449 -458.

[19] Hosseinzadeh Namin A, Wu H, Ahmadi M. High-Speed Architectures for Multiplication Using Reordered Normal Basis[J]. IEEE Trans. Computers, 2012, 61(2): 164 -172.

[20] Park S M, Chang K Y. Fast Bit-Parallel Shifted Polynomial Basis Multiplier Using Weakly Dual Basis Over $GF(2^m)$ [J]. IEEE Trans. Very Large Scale Integration (VLSI) Systems, 2011, 19(12): 2317-2321.

[21] Talapatra S, Rahaman H, Mathew J. Low Complexity Digit Serial Systolic Montgomery Multipliers for Special Class of $GF(2^m)$ [J]. IEEE Trans. Very Large Scale Integration (VLSI) Systems, 2010, 18(5): 847 -852.

[22] Cho Y I, Chang N S, Kim C H, et al. New Bit Parallel Multiplier With Low Space Complexity for All Irreducible Trinomials Over $GF(2^n)$ [J]. IEEE Trans. Very Large Scale Integration (VLSI) Systems, 2012, 20(10): 1903 -1908.

[23] Hasan M, Meloni N, Namin A, et al. Block Recombination Approach for Subquadratic Space Complexity Binary Field Multiplication Based on Toeplitz Matrix-Vector Product[J]. IEEE Trans. Computers, 2012, 61(2): 151 -163.

[24] Imana J. Efficient Polynomial Basis Multipliers for Type-II Irreducible Pentanomials[J]. IEEE Trans. Circuits and Systems II: Express Briefs, 2012, 59(11): 795 -799.

[25] Morales-Sandoval M. Bit-serial and digit-serial $GF(2^n)$ Montgomery multipliers using linear feedback shift registers[J]. IET Computers & Digital Techniques, 2011, 5: 86-94(8).

[26] Mathew J. Single error correctable bit parallel multipliers over $GF(2^n)$ [J]. IET Computers & Digital Techniques, 2009, 3: 281-288(7).

[27] Rahaman H. Simplified bit parallel systolic multipliers for special class of Galois field $GF(2^n)$ with testability[J]. IET Computers & Digital Techniques, 2010, 4: 428-437(9).

[28] Lee C Y. Low complexity bit-parallel systolic multiplier over $GF(2^n)$ using irreducible trinomials[J]. IEE Proceedings - Computers and Digital Techniques, 2003, 150: 39-42(3).

[29] Tsai W. Two systolic architectures for multiplication in $GF(2^m)$ [J]. IEE Proceedings - Computers and Digital Techniques, 2000, 147: 375-382(7).

[30] Furness R. $GF(2^m)$ multiplication over triangular basis for design of Reed-Solomon codes[J]. IEE Proceedings - Computers and Digital Techniques, 1998, 145: 437-443(6).

[31] Wozniak J. Systolic dual basis serial multiplier[J]. IEE Proceedings - Computers and Digital Techniques, 1998, 145: 237-241(4).

[32] Guo J H. Digit-serial systolic multiplier for finite fields $GF(2^m)$ [J]. IEE Proceedings - Computers and Digital Techniques, 1998, 145: 143-148(5).

[33] Fenn S. Bit-serial multiplication in $GF(2^m)$ using irreducible all-one polynomials[J]. IEE Proceedings- Computers and Digital Techniques, 1997, 144: 391-393(2).

[34] Fenn S. Dual basis systolic multipliers for $GF(2^m)$ [J]. IEE Proceedings - Computers and Digital Techniques, 1997, 144: 43-46(3).

[35] Parker M. VLSI structures for bit-serial modular multiplication using basis conversion[J]. IEE Proceedings- Computers and Digital Techniques, 1994, 141: 381-390(9).

[36] Hasan M, Bhargava V. Bit-serial systolic divider and multiplier for finite fields $GF(2^m)$ [J]. IEEE Transactions on Computers, 1992, 41(8): 972-980.

[37] Wei S W. VLSI architectures for computing exponentiations, multiplicative inverses, and divisions in $GF(2^m)$ [J]. IEEE Transactions on Circuits and Systems II: Analog and Digital Signal Processing, 1997, 44(10): 847-855.

[38] Azarderakhsh R, Jarvinen K, Dimitrov V. Fast Inversion in $GF(2^m)$ with Normal Basis Using HybridDouble Multipliers[J]. IEEE Transactions on Computers, 2012, PP(99): 1-1.

[39] Parrilla L, Lloris A, Castillo E, et al. Minimum-clock-cycle Itoh-Tsujii algorithm hardware implementation for cryptography applications over $GF(2^m)$ fields[J]. Electronics Letters, 2012, 48(18): 1126-1128.

[40] Wang C C, Troung T, Shao H M, et al. VLSI Architectures for Computing Multiplications and Inverses in $GF(2^m)$ [J]. IEEE Transactions on Computers, 1985, 100(8): 709-717.

[41] Dinh A, Bolton R, Mason R. A low latency architecture for computing multiplicative inverses and divisions in $GF(2^m)$ [J]. IEEE Transactions on Circuits and Systems II: Analog and Digital Signal Processing, 2001, 48(8): 789-793.

[42] Jing M H, Chen J H, Chen Z H, et al. Low Complexity Architecture for Multiplicative Inversion in $GF(2^m)$ [A]. In: IEEE Asia Pacific Conference on Circuits and Systems 2006, APCCAS 2006[C], 2006: 1492-1495.

[43] Dinh A V, Palmer R, Bolton R J, et al. A low latency architecture for computing

multiplicative inverses and divisions in $GF(2^m)$ [A]. In: 2000 Canadian Conference on Electrical and Computer Engineering[C], 2000. 1:43-47 vol.1.

[44] Fenn S, Benaissa M, Taylor D. Finite field inversion over the dual basis[J]. IEEE Transactions on Very Large Scale Integration (VLSI) Systems, 1996, 4(1): 134 -137.

[45] Gutub A A A, Tenca A F, Savaş E, et al. Scalable and unified hardware to compute Montgomery inverse in $CF(p)$ and $GF(2^n)$ [A]. In: Cryptographic Hardware and Embedded Systems-CHES 2002[C]. Springer, 2003: 484-499.

[46] Savas E. A carry-free architecture for Montgomery inversion[J]. IEEE Transactions on Computers, 2005, 54(12): 1508 - 1519.

[47] Daneshbeh A, Hasan M. A class of unidirectional bit serial systolic architectures for multiplicative inversion and division over $GF(2^m)$ [J]. IEEE Transactions on Computers, 2005, 54(3): 370 - 380.

[48] Fenn S, Benaissa M, Taylor D. Fast normal basis inversion in $GF(2^m)$ [J]. Electronics Letters, 1996, 32(17): 1566 -1567.

[49] Rebeiro C, Roy S, Reddy D, et al. Revisiting the Itoh-Tsujii Inversion Algorithm for FPGA Platforms[J]. IEEE Transactions on Very Large Scale Integration (VLSI) Systems, 2011, 19(8): 1508-1512.

[50] Kaliski J, B.S. The Montgomery inverse and its applications[J]. IEEE Transactions on Computers, 1995, 44(8): 1064-1065.

[51] Bajard J, Imbert L, Negre C. Arithmetic Operations in Finite Fields of Medium Prime Characteristic Using the Lagrange Representation[J]. IEEE Transactions on Computers, 2006, 55(9): 1167-1177.

[52] McIvor C, McLoone M, McCanny J. Improved Montgomery modular inverse algorithm[J]. Electronics Letters, 2004, 40(18): 1110-1112.

[53] Yan Z, Sarwate D. New systolic architectures for inversion and division in $GF(2^m)$ [J]. IEEE Transactions on Computers, 2003, 52(11): 1514-1519.

[54] Huang C T, Wu C W. High-speed easily testable Galois-field inverter[J]. IEEE Transactions on Circuits and Systems II: Analog and Digital Signal Processing, 2000, 47(9): 909-918.

[55] Thomas J, Keller J M, Larsen G. The Calcualtion of Multiplicative Inverses Over $GF(p)$ Efficiently Where p is a Mersenne Prime[J]. IEEE Transactions on Computers, 1986, C-35(5): 478-482.

[56] Hu J, Guo W, Wei J, et al. Fast and Generic Inversion Architectures Over $GF(2^m)$ Using Modified Itoh-Tsujii Algorithms[J]. IEEE Transactions on Circuits and Systems II: Express Briefs, 2015, 62(4): 367-371.

[57] Kim S, Ko K, Chung S Y. Incremental Gaussian elimination decoding of raptor codes over BEC[J]. IEEE communications letters, 2008, 12(4): 307-309.

[58] Bioglio V, Grangetto M, Gaeta R, et al. On the fly Gaussian elimination for LT codes[J]. IEEE communications Letters, 2009, 13(12): 953-955.

[59] Liu C W, Lu C C. A View of Gaussian Elimination Applied to Early-Stopped Berlekamp-Massey Algorithm[J]. IEEE Transactions on Communications, 2007, 55(6): 1131-1143.

[60] Zhou Y, Lai X, Li Y, et al. Ant colony optimization with combining Gaussian eliminations for matrix multiplication[J]. IEEE Transactions on Cybernetics, 2013, 43(1): 347-357.

[61] Trouborst P, Jess J A. The identification of essential variables of systems of algebraic equations applying Gauss-Jordan elimination[J]. IEEE Transactions on Circuits and Systems, 1981, 28(9): 867-876.

[62] Hsieh H. Fill-in comparisons between Gauss-Jordan and Gaussian eliminations[J]. IEEE Transactions on Circuits and Systems, 1974, 21(2): 230-233.

[63] Bryant G, Yeung L. New sequential design procedures for multivariable systems based on GaussJordan factorisation[J]. IEE Proceedings-Control Theory and Applications, 1994, 141(6): 427-436.

[64] Crout P D. A short method for evaluating determinants and solving systems of linear equations with real or complex coefficients[J]. Electrical Engineering, 1941, 60(12): 1235-1240.

[65] Kurzak J, Buttari A, Dongarra J. Solving systems of linear equations on the CELL processor using Cholesky factorization[J]. IEEE Transactions on Parallel and Distributed Systems, 2008, 19(9): 1175-1186.

[66] Ferreira L V, Kaszkurewicz E, Bhaya A. Solving systems of linear equations via gradient systems with discontinuous righthand sides: application to LS-SVM[J]. IEEE Transactions on Neural Networks, 2005, 16(2): 501-505.

[67] Cichocki A, Unbehauen R. Neural networks for solving systems of linear equations and related problems[J]. IEEE Transactions on Circuits and Systems I: Fundamental Theory and Applications, 1992, 39(2): 124-138.

[68] Krishna H, Morgera S. The Levinson recurrence and fast algorithms for solving Toeplitz systems of linear equations[J]. IEEE Transactions on Acoustics, Speech and Signal Processing, 1987, 35(6): 839-848.

[69] Wang J. Recurrent neural networks for solving systems of complex-valued linear equations[J]. Electronics letters, 1992, 28(18): 1751-1753.

[70] Haibin D, Daobo W, Jiaqiang Z. Novel method based on ant colony optimization for solving illconditioned linear systems of equations[J]. Journal of Systems Engineering and Electronics, 2005, 16(3): 606-610.

[71] Paige C C, Saunders M A. Solution of sparse indefinite systems of linear equations[J]. SIAM Journal on Numerical Analysis, 1975, 12(4): 617-629.

[72] Adams W., Loustaunau P. An Introduction to Gröbner Bases [J]. Graduate Texts in Mathematics, vol.3, AMS, 1994.

[73] Courtois N., Klimov A., Patarin J.. Efficient algorithms for solving overdefined systems of multivariate polynomial equations[A]. In Eurocrypt2000, LNCS 1807[C]: Springer-Verlag, 2000: 392-407.

[74] Massey J., Omura J. Computation Method and Apparatus for Finite Field Arithmetic. In US Patent No.4587627,1984.

[75] Wang C., Lin J.. Systolic array implementation of multipilers for finite field $GF(2^m)$. In IEEE Trans on Circuits and Systems. 1991(38):796-800.

[76] Wang C.,Truong T., Shao H., Deutsch L., Omura J., Reed I.. VLSI architecture for computing multiplication and inverses in $GF(2^m)$. In IEEE Trans on Computers. 1985(34): 709-716.

[77] Takagi N., Yoshiki J., Takagi K.. A fast algorithm for multiplicative inversion in $GF(2^m)$ using normal basis. In IEEE Trans on Computers. 2001(50):394-398.

[78] Guo J., Wang C.. New systolic arrays for $C+AB^2$, inversion and division in $GF(2^m)$. In IEEE Trans on Computers. 2000(49):1120-1125.

[79] Brunner H.. On Computing Multiplicative Inverses in $GF(2^m)$ [J]. In IEEE Trans on Computers, 1993(42): 1010-1015.

[80] Araki K.. Fast Inverters over Finite Field Based on Euclid's Algorithm[J]. In IEEE Trans on IEICE, 1989(72): 1230- 1234.

[81] Großschädl J.. High-Speed RSA Hardware Based on Barret's Modular Reduction Method. In CHES2000, LNCS1965. Springer, 2000:95-136.

[82] ZHANG B, GU G, SUN L, et al. Floating-Point FPGA Gaussian Elimination in Reconfigurable Computing System[J]. Chinese Journal of Electronics, 2011, 20(1).

[83] Arias-Garcia J, Pezzuol Jacobi R, Llanos C, et al. A suitable FPGA implementation of floating-point matrix inversion based on Gauss-Jordan elimination[A]. In: 2011 VII Southern Conference on Programmable Logic (SPL)[C], 2011: 263 -268.

[84] Duarte R, Neto H, Vestias M. Double-precision Gauss-Jordan Algorithm with Partial Pivoting on FPGAs[A]. In: 12th Euromicro Conference on Digital System Design, Architectures, Methods and Tools (DSD)[C], 2009: 273-280.

[85] Yoon Kah L, Akoglu A, Guven I, et al. High performance linear equation solver using NVIDIA GPUs[A]. In: 2011 NASA/ESA Conference on Adaptive Hardware and Systems (AHS)[C], 2011: 367-374.

第 2 章 密码学基础

2.1 密码和密码学

2.1.1 密码

密码是一种用来混淆的技术，它希望将可识别的信息转变为无法识别的信息。当然，对一小部分人来说，这种无法识别的信息是可以经过再加工并被恢复的。登录网站、电子邮箱和在银行取款时输入的"密码"其实严格来讲应该被称作"口令"，因为它不是本来意义上的"加密代码"，但是也可以被称为"秘密的号码"。密码按特定法则编成，是对通信双方的信息进行明密变换的符号。换而言之，密码是隐蔽了真实内容的符号序列。密码通过一种变换手段，把用公开的、标准的信息编码表示的信息变为除通信双方以外其他人不能读懂的信息编码。

在公元前，秘密书信已用于战争之中。古希腊"史学之父"希罗多德（Herodotus）的《历史》当中记载了一些最早的秘密书信故事。公元前 5 世纪，希腊城邦为对抗奴役和侵略，与波斯发生多次冲突和战争。公元前 480 年，波斯秘密集结了强大的军队，准备对雅典（Athens）和斯巴达（Sparta）发动一次突袭。希腊人狄马拉图斯（Demaratus）在波斯的苏萨城（Susa）里看到了这次行动，便利用一层蜡把木板上的字遮盖住，然后将木板送往希腊并告知了希腊人波斯的图谋。最后，波斯海军覆没于雅典附近的沙拉米斯湾（Salamis Bay）。由于古时多数人不识字，由此最早的秘密书写的形式只用到纸笔或等同物品。随着识字率提高，就开始需要真正的密码学了。最古典的两个加密技巧如下。

1）置换（Transposition cipher）：将字母顺序重新排列，例如"help me"变成"ehpl em"。

2）替代（Substitution cipher）：有系统地将一组字母换成其他字母或符号，例如"fly at once"变成"gmz bu podf"（每个字母用下一个字母取代）。

2.1.2 密码学

密码是通信双方按约定的法则进行信息特殊变换的一种重要保密手段。密码的相关知识组成了一门深奥的学科，被称为密码学。密码学是研究如何隐密地传递信息的学科，在现代特别指对信息及其传输的数学性研究，常被认为是数学和计算机科学的分支，同时它也和信息论密切相关。

密码学是在编码与破译的斗争实践中逐步发展起来的，随着先进科学技术的应用，它已成为一门综合性的尖端技术科学。它与语言学、数学、电子学、声学、信息论、计算机科学等有着广泛而密切的联系。它的现实研究成果，特别是各国政府使用的密码编制及破译手段都具有高度的机密性。

密码学有着悠久的历史。密码在古代就被用于信息通信传输和存储过程中的保密。在近代和现代战争中，传递情报和指挥战争均离不开密码学。外交斗争中也离不开密码学。著名的密码学者 Ron Rivest 解释道："密码学是关于如何在敌人存在的环境中通信。"从工程学的角度，这相当于密码学与纯数学的异同。密码学是信息安全等相关议题，如认证、访问控制的核心。密码学的首要目的是隐藏信息的涵义，而不是隐藏信息的存在。

随着计算机和信息技术的发展，密码技术的发展也非常迅速，其应用领域不断扩展。密码除了用于信息加密外，也用于数据信息签名和安全认证。因此，密码的应用也不再只局限于为军事、外交斗争服务，它也被广泛应用在社会和经济活动中。例如：

1）可以将密码技术应用在电子商务中，对网上交易双方的身份和商业信用进行识别，防止网上电子商务中的"黑客"和欺诈行为；

2）应用于增值税发票中，可以防伪、防篡改，杜绝了各种利用增值税发票偷、漏、逃、骗国家税收的行为，并大大方便了税务稽查；

3）应用于银行支票鉴别中，可以大大降低利用假支票进行金融诈骗的犯罪行为；

4）应用于个人移动通信中，大大增强了通信信息的保密性。

密码学也促进了计算机科学，特别是在计算机与网络安全所使用的技术方面，如访问控制与信息的机密性。

现代密码学所涉及的学科包括信息论、概率论、数论、计算复杂性理论、近世代数、离散数学、代数几何学和数字逻辑等。

2.1.3 密码系统

依照密码学中的定义：

1）密钥是参与加密、解密变换的参数，分为加密密钥和解密密钥；

2）明文是没有进行加密，能够直接代表原文含义的信息；

3）经过加密处理之后，密文隐藏原文含义的信息；

4）加密是将明文转换成密文的实施过程；

5）解密是将密文转换成明文的实施过程；

6）密码算法是密码系统采用的加密方法和解密方法，随着基于数学密码技术的发展，加密方法一般称为加密算法，解密方法一般称为解密算法；随着通信技术的发展，对语音、图像、数据等都可实施加、解密变换。

密码算法的基本类型可以分为 4 种。

1）错乱：按照规定的图形和线路，改变明文字母或数码等的位置而使其成为密文。

2）代替：用一个或多个代替表将明文字母或数码等代替为密文。

3）密本：用预先编定的字母或数字密码组代替一定的词组单词，变明文为密文。

4）加乱：用有限元素组成的一串序列作为乱数，按规定的算法，同明文序列相结合而变成密文。

以上 4 种密码体制，既可单独使用，也可混合使用，以编制出各种复杂度很高的实用密码算法。

使用密码算法进行密码运算的系统叫作密码系统。一个密码系统的功能是将给定的明文和密钥加密从而变换为密文；在接收端，利用解密密钥完成解密操作，将密文恢复成原来的明文。

在密码系统中，除合法用户外，还有非法的截收者。他们试图通过各种办法窃取机密或窜改消息，利用文字和密码的规律，在一定条件下，采取各种技术手段，通过对截取密文的分析，以求得明文，还原密码编制，即破译密码。破译不同强度的密码，对条件的要求也不相同，甚至很不相同。所以，一个安全的密码系统应该满足：

1）非法截收者很难从密文中推断出明文；

2）加密和解密算法应该相当简便，而且适用于所有密钥空间；

3）密码的保密强度只依赖于密钥；

4）合法接收者能够检验和证实消息的完整性和真实性；

5）消息的发送者无法否认其所发出的消息，同时也不能伪造别人的合法消息；

6）必要时可由仲裁机构进行公断。

2.2 密码体制

2.2.1 对称密码

区别于古典密码学，一般认为现代密码学包括对称密码学（又被称作单钥密码学）和非对称密码学（又被称作公钥密码学）。由于信息技术和电子工业的发展，现有的密码算法的安全性都面对极大的挑战，不断有旧的密码算法被攻破，也有新的密码算法被推出。现代密码学正因为密码算法的推陈出新而不断发展。同时，根据密码算法设计的密码软硬件的应用也极为广泛，这也推动了密码学的发展。

在公钥密码出现之前，密码学主要以对称密码体制为主，即分别用于加密和解密的密钥是相同的，所以我们把它称为对称密码。在对称密码算法中：

1）通常把没有加密的消息称为明文，把所有可能的明文的有限集合称为明文空间；

2）把加密后的消息称为密文，把所有可能的密文的有限集合称为密文空间；

3）把将明文变换成密文的过程称为加密，将密文恢复成明文的过程称为解密；

4）加密和解密这一对逆变换是在同一组密钥（即对称密钥）的控制下进行的，我们把所有有可能的密钥的有限集合称为密钥空间。

对称密码的加密方式一般有两种：

1）第一种是流密码算法，即按明文的字符顺序逐位地进行加密；

2）第二种是分组密码算法，即将明文进行分组，然后逐组地进行加密。

常用的对称密码算法有 DES（Data Encryption Standard）密码算法和 AES（Advanced Encryption Standard）密码算法。

2.2.2 公钥密码

对称密码在一定程度上解决了保密通信的问题，但随着密码学的发展，它可应用的范围的局限性就显现出来了。公钥密码学的思想最早由 Ralph C.Merkle 在 1974 年提出的。Diffie 和 Hellman 于 1976 年以单向函数和单向暗门函数为基础，为密码通讯双方建立密钥。从那以后公钥密码学一直在密码界扮演着重要的角色。

公钥密码的出现改变了密码学的面貌，是密码学发展的一个重要的里程碑。在公钥密码学中，密钥是公开密钥和私有密钥组成的密钥对，简称公钥和私钥。签名者用私钥对消息进行加密，接收者用公钥进行解密，还原消息。由于从公钥不能推算出私钥，所以公钥不会损害私钥持有者的信息安全。公钥无需保密，可以公开传播，而私钥必须由持有者本人保密。因此如果持有者用其私钥加密消息，并且能够用他的公钥正确解密，就可以肯定该消息是他签的字，这就是数字签名的基本原理。

公钥密码其实是一种陷门的单向函数。假如 F 是单向函数，对于任意的明文 X，很容易求出密文 $Y = F(X)$，而当 Y 已知时则非常难求出 X，但给出一些陷门信息时就比较容易求出 X，一般是以陷门信息作为私钥。公钥密码体制都是基于这样的一个原理设计的，密码算法的安全性主要取决于计算的复杂性。

公钥密码为密码学的发展提供了新的理论和技术基础。首先，它打破了密码算法只使用单个密钥的局限；其次，它的设计基于数学问题，取代了简单的代替和换位方法。公钥密码还能够解决对称密码不能解决的问题，即通信的双方需要在公共信道上约定共享的密钥。在现实的通信网络中，如果使用对称密码，约定用于共享的密钥存在极大的安全问题，而公钥密码则解决了这一难题。

公钥密码学出现了一些经典的算法，同时也方便了私钥的管理，如 RSA、ECC、Diffie-Hellman、ElGamal 等，这些算法都已经非常完善并得到了广泛使用。然而主流的公钥密码也有以下问题。

1）RSA、ECC 等是基于大整数分解或椭圆曲线离散对数问题的非对称密码算法，其特点是利用了大量的模指数运算。这类算法的复杂度太高，无法在资源受限的环境下运行。如果需要提高安全性，就需要更多的密钥，比如 RSA。随着安全级别的提高，密钥长度的增长速度要更快，且大整数分解会影响加密算法的速度。因此这些算法不太适合于无线传感器网络、低廉智能卡等低端设备领域。

2）目前，被大家公认的安全性比较高的密码算法都是基于大整数分解或椭圆曲线离散对数问题，其他的公钥密码算法基本被攻破。大整数分解和离散对数是两个非常接近的问题，假设有一天其中之一被攻破了，那么，另一个的安全性问题也会遭到威胁，因此需要更多的备选公钥密码算法。

2.2.3 数字签名

数字签名（又称公钥数字签名、电子签章）是一种类似写在纸上的普通的物理签名，它使用公钥加密领域的技术，用于鉴别数字信息的方法。一套数字签名通常定义两种互补的运算；一种用于签名，另一种用于验证。只有信息的发送者才能产生别人无法伪造的一段数字串，这段数字串同时是对信息的发送者发送信息真实性的一个有效证明。数字签名文件的完整性是很容易验证的，而且数字签名具有不可抵赖性。

简单地说，数字签名就是附加在数据单元上的一些数据，或是对数据单元所做的密码变换。这种数据或变换允许数据单元的接收者用以确认数据单元的来源和数据单元的完整性并保护数据，防止被人伪造。它是对电子形式的消息进行签名的一种方法。一个签名消息能在一个通信网络中传输。

数字签名的过程是将摘要信息用发送者的私钥加密，与原文一起传送给接收者；接收者只有用发送者的公钥才能解密被加密的摘要信息，然后用 HASH 函数对收到的原文产生一个摘要信息，与解密的摘要信息对比。如果相同，则说明收到的信息是完整的，在传输过程中没有被修改，否则说明信息被修改过。所以，数字签名具有保证信息传输的完整性、认证发送者的身份、防止交易中的抵赖发生等特点。

公钥数字签名体制包括普通数字签名和特殊数字签名。普通数字签名算法有 RSA、ElGamal、Fiat-Shamir、Guillou-Quisquarter、Schnorr、Ong-Schnorr-Shamir、DES/DSA、椭圆曲线数字签名算法和有限自动机数字签名算法等；特殊数字签名有盲签名、代理签名、群签名、环签名、不可否认签名、公平盲签名、门限签名、具有消息恢复功能的签名等，它与具体应用环境密切相关。

1. 盲签名

盲签名于 1982 年提出因为具有盲性这一特点，盲签名可以有效保护所签署消息的具体内容,所以在电子商务和电子选举等领域有着广泛的应用。

盲签名允许消息者先将消息盲化,然后让签名者对盲化的消息进行签名,最后消息拥有者将签字的盲因子除去,得到签名者关于原消息的签名。盲签名是接收者在不让签名者获取所签署消息具体内容的情况下所采取的一种特殊的数字签名技术,它除了满足一般的数字签名条件外,还必须满足下面的两条性质:

1)签名者对其所签署的消息是不可见的,即签名者不知道他所签署消息的具体内容。

2)签名消息不可追踪,即当签名消息被公布后,签名者无法知道这是他哪次签署的。

关于盲签名,有一个非常直观的说明:就是先将隐蔽的文件放进信封里,而除去盲因子的过程就是打开这个信封,当文件在一个信封中时,任何人都不能读它。对文件签名就是通过在信封里放一张复写纸,签名者在信封上签名时,他的签名透过复写纸签到文件上。

一般来说,一个好的盲签名应该具有以下性质。

1)不可伪造性:除了签名者本人外,任何人都不能以他的名义生成有效的盲签名。这是一条最基本的性质。

2)不可抵赖性:签名者一旦签署了某个消息,他无法否认自己对消息的签名。

3)盲性:签名者虽然对某个消息进行了签名,但他不可能得到消息的具体内容。

4)不可跟踪性:一旦消息的签名公开,签名者就不能确定自己何时签署了这条消息。

满足上面 4 条性质的盲签名,被认为是安全的。这 4 条性质既是我们设计盲签名所应遵循的标准,又是我们判断盲签名性能优劣的根据。

另外,方案的可操作性和实现的效率也是我们设计盲签名时必须考虑的重要因素。一个盲签名的可操作性和实现速度取决于以下 3 个方面:

1)密钥的长度;

2)盲签名的长度;

3)盲签名的算法和验证算法。

2. 代理签名

代理签名方案是 1996 年被提出来的,由于这种签名机制在许多领域都有着重要的应用,因此引起了人们的极大兴趣。代理签名是指当某签名人因公务或身体健康等原因不能行使签名权力时,将签名权委派给其他人。在一个代理签名方案中,如果假设授权人 A 打算委托代理人 B 进行代理签名,则此签名方案应满足以下 3 个最基本的条件:

1)签名接收方能够像验证 A 的签名那样验证 B 的签名;

2)A 的签名和 B 的签名应该完全不同,并且容易区分;

3)A 和 B 对签名事实不可否认。

一个好的代理签名体制应当具有以下基本性质。

1)不可伪造性(Unforgeability):除了原始签名者,只有指定的代理签名者才能够代表原始签名者产生有效代理签名。

2）可验证性（Verifiability）：从代理签名中，验证者能够相信原始签名者认同了这份签名消息。

3）不可否认性（Undeniability）：一旦代理签名者代替原始签名者产生了有效的代理签名，他就不能向原始签名者否认他所签的有效代理签名。

4）可区分性（Distinguishability）：任何人都可区分代理签名者和原始签名者的签名。

5）代理签名者的不符合性（Proxy signer's deviation）：代理签名者必须创建一个能检测到是代理签名的有效代理签名。

6）可识别性（Identifiability）：原始签名者能够从代理签名中确定代理签名者的身份。

3. 群签名

群签名方案是 Chaum 和 Heyst 在 1991 年首次提出来的。群签名方案允许用户组中合法用户以用户组的名义签名，具有签名者匿名、只有权威才能辨认签名者等多种特点，因此群签名可以被广泛地应用于企业管理、电子商务、电子政务、军事等领域。一般的群签名方案由组、组成员（签名者）、消息接收者（签名验证者）和权威（Authority）或者 GC（Group Center）组成，具有如下特点：

1）只有组中成员才能为消息签名，签名为群签名；

2）消息接收者可以验证群签名的有效性，但不能辨别签名者；

3）一旦发生争论，从消息的群签名权威（Authority）或者 GC（Group Center）处可以辨别签名者。

一个好的群签名方案应满足以下的安全性要求。

1）匿名性：给定一个群签名后，对除了唯一的群管理人之外的任何人来说，确定签名人的身份在计算上是不可行的。

2）不关联性：在不打开签名的情况下，确定两个不同的签名是否为同一个群成员所做在计算上是困难的。

3）防伪造性：只有群成员才能产生有效的群签名。

4）可跟踪性：群管理人在必要时可以打开一个签名以确定签名人的身份，而且签名人不能阻止一个合法签名的打开。

5）防陷害攻击：包括群管理人在内的任何人都不能以其他群成员的名义产生合法的群签名。

6）抗联合攻击：即使一些群成员串通在一起，也不能产生一个合法的不能被跟踪的群签名。

4. 环签名

2001 年，Rivest.Shamir 和 Tauman 三位密码学家首次提出了环签名（Ring Silage）。环签名是一种简化的群签名，环签名中只有环成员没有管理者，不需要环成员间的合作。

环签名同群签名一样也是一种签名者模糊的签名方案。

在环签名中不需要创建、改变或者删除环，也不需要分配指定的密钥，无法撤销签名者的匿名性，除非签名者自己想暴露身份。环签名方案中签名者首先选定一个临时的签名者集合，集合中包括签名者。然后签名者利用自己的私钥和签名集合中其他人的公钥就可以独立地产生签名，而无需他人的帮助。签名者集合中的成员可能并不知道自己被包含在其中。环签名没有可信中心，没有群的建立过程，对于验证者来说，签名人是完全正确匿名的。环签名提供了一种匿名泄露秘密的巧妙方法。环签名的这种无条件匿名性在对信息需要长期保护的一些特殊环境中非常有用。例如，即使 RSA 被攻破也必须保护匿名性的场合。

一个好的环签名必须满足以下的安全性要求。

1）无条件匿名性：攻击者即使非法获取了所有可能签名者的私钥，他能确定真正的签名者的概率也不超过 $1/n$，这里 n 为环成员（可能签名者）的个数。

2）不可伪造性：外部攻击者在不知道任何成员私钥的情况下，即使能够从一个产生环签名的随机预言者那里得到任何消息 m 的签名，他成功伪造一个合法签名的概率也是可以忽略的。

3）环签名具有良好的特性：可以实现签名者的无条件匿名；签名者可以自由指定自己的匿名范围；构成优美的环形逻辑结构；可以实现群签名的主要功能但无需可信第三方或群管理员等。

2.3 常用的密码算法

2.3.1 DES

DES（Data Encryption Standard）密码算法，即数据加密标准，是 1972 年由美国 IBM 公司研制的对称密码算法。

在 1976 年，DES 被美国联邦政府的国家标准局确定为联邦资料的处理标准。DES 密码算法是一种使用密钥加密的块算法，它的明文按 64 位进行分组，参与运算的密钥长度为 56 位（不包括 8 位校验位），分组后的明文组和密钥以按位替代或交换的方法形成密文组。DES 密码算法现在已经不是一种安全的加密方法，主要因为它使用的 56 位密钥过短。对于一切密码而言，最基本的攻击方法是暴力破解法，即我们依次尝试所有可能的密钥。密钥长度决定了可能的密钥数量，因此也决定了这种方法的可行性。不幸的是 DES 密码算法正是因为密钥过短而被破解。1999 年 1 月，distributed.net 与电子前哨基金会合作，在 22 小时 15 分钟内公开破解了一个 DES 密钥。

为了应对 DES 的安全问题，3DES 密码算法作为 DES 密码算法的替代者出现了。事实上，3DES 密码算法是 DES 的派生算法，也存在理论上的攻破方法。它并没有成为替代者，取代 DES 密码算法的是 AES 密码算法。

DES 使用 56 位密钥对 64 位的明文加密得到 64 位密文，其加密过程如下。

1) 对输入的明文 m，经过一个初始变换 IP 得到 m_0，记 $m_0=IP(m)=L_0R_0$，其中 L_0 是 m_0 的左 32 位，R_0 是 m_0 的右 32 位。

2) 然后，进行 16 轮过程相同的数据与轮密钥相结合的运算（$1 \leqslant i \leqslant 16$）：

$$L_i \leqslant R_{i-1},$$
$$R_i \leqslant L_{i-1} \oplus f(R_{i-1}, k_i)。$$

其中，函数 $f(A,J)$ 的计算过程如下：

首先，将 A 根据一个固定的扩展函数 E 把 32 位扩展成 48 位 $E(A)$；

然后，计算 $E(A) \oplus J$，并把结果分成 8 组 6 位的比特串；

接着，把上述 8 组比特串分别作为 8 个 S 盒的输入，产生 8 组共 32 位的输出；

最后，对 S 盒的输出进行 P 转换。

3) 对比特串 $R_{16}L_{16}$ 进行初始变换 IP 的逆变换 IP^{-1}，得到密文 c，即 $c=IP^{-1}(R_{16}L_{16})$。

上述第 2) 步中的 k_i 是轮密钥，每一轮中使用的密钥都是从初始密钥 k 导出的 48 位比特串，其过程如下：

给定 64 比特的密钥 k，去掉 8 个校验位并利用一个固定的转换 $PC1$ 置换剩下的 56 位，记为 $PC1(k)=C_0D_0$，C_0 是 $PC1(k)$ 的前 28 位，D_0 是后 28 位；

对每个 i，$1 \leqslant i \leqslant 16$，计算 $C_i=LS_i(C_{i-1})$，$D_i=LS_i(D_{i-1})$，$k_i=PC2(C_iD_i)$。其中 LS_i 表示当 i=1,2,9,16 时左循环移 1 位，否则移 2 位。$PC2$ 是另一个置换。

DES 的加密的数据路径如图 2-1 所示，其解密采用同一算法实现，把密文 c 作为输入，倒过来使用密钥方案即可实现解密。

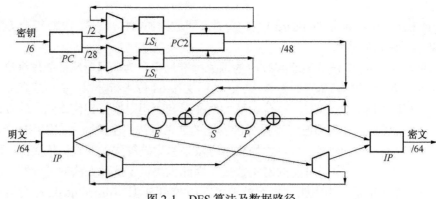

图 2-1　DES 算法及数据路径

用 $e_k(m)$ 表示 DES 加密，$d_k(c)$ 表示解密，则 3DES 就是采用 3 个密钥对明文进行处理得到密文 c，即 $c = e_{k3}\big(d_{k2}\big(e_{k1}(m)\big)\big)$。

2.3.2 AES

AES 密码算法，即高级加密标准，又被称作 Rijndael 加密算法，是美国联邦政府采用的一种区块加密算法。AES 密码算法由两位比利时密码学家 Joan Daemen 和 Vincent Rijmen 设计，他们用自己的姓氏构成组合，以 Rijndael 为名投稿至美国联邦政府的高级加密标准的甄选流程。最终它由美国国家标准与技术研究院于 2001 年 11 月 26 日发布，并在 2002 年 5 月 26 日正式成为有效的加密标准，是 DES 密码算法的替代者。AES 密码算法的区块长度固定为 128 bits，密钥长度的选择则可以是 128bits、192 bits 或 256 bits，就密钥长度而言，AES 密码算法就比 DES 密码算法的安全等级要高得多。

AES 算法也是一个迭代分组密码，其分组长度和密钥长度都可变。分组长度和密钥长度可以独立地指定为 128 bits、192 bits 和 256 bits，根据不同的分组和密钥长度，对应的迭代轮数分别是 10 轮、12 轮和 14 轮。除最后一轮外，每一轮都产生中间状态数据，其过程都是一致的，最后一轮和其他轮略有不同，它删除了其他轮的一个操作（列变换）。一个独立的密钥产生模块通过初始密钥为每一轮产生不同的轮密钥。以下的讨论以 AES128（密钥和分组长度都是 128 位）为例。

轮加密可以分成 4 个操作：

1）位变换（ByteSubstitution）；
2）行移位（ShiftRow）；
3）列变换（MixColumn）；
4）轮密钥加（AddRoundKey）。

它们都是对状态（State）进行操作（AES128 把 State 看成以字节为单位的 4 行 4 列的矩阵）。位变换就是对状态上的每一个字节通过 S 盒进行变换。行移位是对状态矩阵按字节进行移位操作，第一行不移，第二、三、四行分别循环右移动 1、2、3 字节。列变换把第一列乘上一个多项式，可以用矩阵乘法表示：

$$\begin{pmatrix} b_0 \\ b_1 \\ b_2 \\ b_3 \end{pmatrix} = \begin{pmatrix} 02 & 03 & 01 & 01 \\ 01 & 02 & 03 & 01 \\ 01 & 01 & 02 & 03 \\ 03 & 01 & 01 & 02 \end{pmatrix} \begin{pmatrix} a_0 \\ a_1 \\ a_2 \\ a_3 \end{pmatrix}$$

轮密钥加就是把轮密钥和轮数据进行异或操作。AES 的加密过程如图 2-2（a）所示，每一轮操作的数据路径如图 2-2（b）所示。

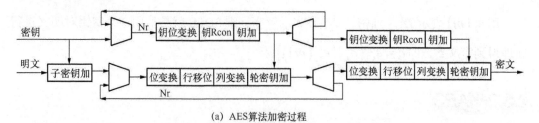

(a) AES算法加密过程

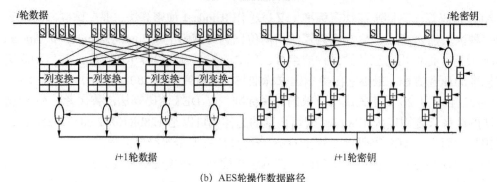

(b) AES轮操作数据路径

图 2-2　AES 算法及数据路径

2.3.3　RSA

RSA 密码算法是目前应用与研究最广泛的公钥密码算法之一。

在 1978 年，3 位美国密码学家 Ron Rivest、Adi Shamir 和 Leonard Adleman 共同提出了以 3 人姓氏首字母命名的 RSA 密码算法。它的设计基于一个十分简单的数学困难问题，即将两个大素数相乘的运算十分容易，但对它们的乘积进行因式分解的运算却极其困难。因此，我们可以将两个大素数的乘积公开作为 RSA 密码的加密密钥。这表明，RSA 密码算法的安全性完全依赖于大整数的因子分解问题的困难性。随着现代计算机的计算能力的不断提升和对大整数的因子的分解算法的进一步改进，为了保证 RSA 密码的安全性，密码算法的设计者就必须增加 RSA 的模数的长度。目前公认的相对安全的 RSA 的模数是 2 048bit。所以，对于计算能力和资源受限的设备，RSA 密码算法并不是一个很好的选择。

2.3.4　ECC

椭圆曲线密码算法 ECC 也是目前最流行的公钥密码算法之一。

在 1985 年，美国密码学家 Neal Koblitz 和 Victor Miller 分别独立提出了椭圆曲线公钥密码。它的设计基于椭圆曲线数学的一个困难问题，即椭圆曲线离散对数问题。它的主要优势是能够在某些情况下使用更小的密钥并且能够保证它的安全性。椭圆曲线密码算法一般使用有限域 $GF(p)$ 和有限域 $GF(2^n)$，其中 p 是素数，所以它容易在通用处理器和硬件上实现，这使得它的应用和研究的热度并不亚于 RSA 密码算法。

2.4 互联网与信息安全

2.4.1 信息安全

信息安全是指信息系统（包括硬件、软件、数据、人、物理环境及其基础设施）受到保护，不因偶然的或者恶意的原因而遭到破坏、更改、泄露，系统可以连续、可靠、正常地运行，信息服务不中断，最终实现业务连续性。

信息安全主要包括 5 方面的内容，即需保证信息的保密性、真实性、完整性、未授权拷贝和所寄生系统的安全性。信息安全本身包括的范围很大，其中包括如何防范商业企业机密泄露、防范青少年对不良信息的浏览、个人信息的泄露等。网络环境下的信息安全体系是保证信息安全的关键，包括计算机安全操作系统、各种安全协议、安全机制（数字签名、消息认证、数据加密等），直至安全系统，如 UniNAC、DLP 等，只要存在安全漏洞，便可以威胁全局安全。

信息安全学科可分为狭义安全与广义安全两个方面：

1）狭义的安全建立在以密码论为基础的计算机安全领域，早期中国信息安全专业通常以此为基准，辅以计算机技术、通信网络技术与编程等方面的内容；

2）广义的信息安全是一门综合性学科，从传统的计算机安全到信息安全，不但是名称的变更，也是对安全发展的延伸。安全不在是单纯的技术问题，而是将管理、技术、法律等问题相结合的产物。

2.4.2 信息安全产业

在我国的信息化建设进程中，信息安全是一个新兴的产业，伴随着我国的信息化建设而生。中国的信息安全产业发端于 21 世纪之初，经过 10 多年的发展，目前该行业已经初具规模，产品线日益丰富，并形成 3 大类、约 20 小类、近百种产品的行业格局，细分程度非常高，并广泛地服务于政府、企事业单位和普通网民。

中国 IT 市场规模在 2014 年达 2 048 亿美元，全球第二。但信息安全市场规模仅为 22.4 亿美元，占 IT 市场的 1.1%，而欧美等发达国家占比约为 8%~12%，中国信息安全市场不论规模还是比例均大幅落后于发达国家。因此，即使不考虑中国整个信息服务业产值的增长，而仅考虑渗透率的提升，未来中国信息安全产业也将具有近 10 倍的成长空间。

随着网络安全形势越发紧迫，我国未来必将持续加大信息安全产业的投入。"十三五"规划将信息安全产品单独列为目标，这就是释放的信号之一。同时，随着个人、企业等网络安全意识的觉醒，市场需求有望进一步释放，给信息安全产业发展提供了源源不断的动力。

综上所述，信息安全关乎国家安全及国计民生，其重要性日益凸显。在政策支持下，信息安全将迎来史无前例的发展契机，前景一片大好。

2.4.3 云计算安全

经过近 10 年的发展，云计算已从概念导入进入广泛普及的新阶段，并成为提升信息化发展水平、打造数字经济新动能的重要支撑。

1）产业规模迅速扩大：据统计，2015 年，我国云计算产业规模约 1 500 亿元，同比增长超过 30%。2016 年，云计算骨干企业收入均实现翻番。SaaS、PaaS 占比不断增加，产业结构持续优化，产业链条趋于完整。

2）关键技术实现突破：云计算骨干企业在大规模并发处理、海量数据存储、数据中心节能等关键领域取得突破，部分指标已达到国际先进水平。在主流开源社区和国际标准化组织中的作用日益重要。

3）骨干企业加速形成：云计算骨干企业加快战略布局，加快丰富业务种类，围绕咨询设计、应用开发、运维服务、人才培训等环节培育合作伙伴，构建生态体系。

4）应用范畴不断拓展：大型企业、政府机构、金融机构不断加快云计算应用步伐，大量中小微企业已应用云服务。云计算正从游戏、电商、视频向制造、政务、金融、教育、医疗等领域延伸、拓展。

5）支撑"双创"快速发展：云计算降低了创业创新门槛，汇聚了数以百万计的开发者，催生了平台经济、分享经济等新模式，进一步丰富了数字经济的内涵。

在云计算的架构下，云计算开放网络和业务共享场景更加复杂多变，安全性方面的挑战更加严峻，一些新型的安全问题变得比较突出，如多个虚拟机租户间并行业务的安全运行，公有云中海量数据的安全存储等。云计算的安全问题涉及内容广泛，主要包括以下 3 个方面。

1）用户身份安全问题：云计算通过网络提供弹性可变的 IT 服务，用户需要登录到云端来使用应用与服务，系统需要确保使用者身份的合法性，才能为其提供服务。如果非法用户取得了用户身份，则会危及合法用户的数据和业务。

2）共享业务安全问题：云计算的底层架构（IaaS 和 PaaS 层）通过虚拟化技术实现资源共享调用，优点是资源利用率高，但是共享会引入新的安全问题，一方面需要保证用户资源间的隔离，另一方面需要面向虚拟机、虚拟交换机、虚拟存储等虚拟对象的安全提供保护策略，这与传统的硬件上的安全策略完全不同。

3）用户数据安全问题：数据的安全性是用户最为关注的问题。广义的数据不仅包括用户的业务数据，还包括用户的应用程序和用户的整个业务系统。数据安全问题包括数据丢失、泄漏、篡改等。传统的 IT 架构中，数据是离用户很"近"的，数据离用户越"近"则越安全。而云计算架构下数据常常存储在离用户很"远"的数据中心中，需要对数据采用有效的保护措施，如多份复制、数据存储加密，以确保数据的安全。

2.4.4 公钥基础设施

公钥基础设施（Public Key Infrastructure，PKI）是由公开密钥密码技术、数字证书、证书认证中心（Certification Authority，CA）和关于公开密钥的安全策略等基本成分共同组成的，管理密钥和证书的系统或平台。通过采用 PKI 框架管理密钥和证书可以建立一个安全的网络环境。

一个典型、完整、有效的 PKI 应用系统至少应具有以下 5 个部分。

1）证书认证中心（CA）：CA 是 PKI 的核心，负责管理 PKI 结构下的所有用户（包括各种应用程序）的证书，把用户的公钥和用户的其他信息捆绑在一起，在网上验证用户的身份。CA 还负责用户证书的黑名单登记和黑名单发布。

2）目录服务器 X.500：目录服务器用于发布用户的证书和黑名单信息。用户可通过标准的 LDAP 协议查询自己或其他人的证书和下载黑名单信息。

3）具有高强度密码算法和安全协议：SSL 协议最初由 Netscape 公司研发，现已成为互联网用来鉴别网站和网页浏览者身份，以及在浏览器使用者及网页服务器之间进行加密通讯的全球化标准。

4）Web 安全通信平台：Web 安全通信平台包括 Web Client 端和 Web Server 端两部分，分别安装在用户端和服务器端，通过具有高强度密码算法的 SSL 协议保证用户端和服务器端数据的机密性、完整性、身份验证。

5）自开发安全应用系统：自开发安全应用系统是指各行业自开发的各种具体应用系统，例如银行、证券的应用系统等。

完整的 PKI 应该还包括认证政策的制定（包括遵循的技术标准、各 CA 之间的上下级或同级关系、安全策略、安全程度、服务对象、管理原则和框架等）、认证规则、运作制度的制定、所涉及的各方法律关系内容以及技术的实现等。

CA 作为 PKI 的核心部分，实现了 PKI 中的重要的功能。概括地说，CA 的功能有：证书发放、证书更新、证书撤销和证书验证。CA 的核心功能就是发放和管理数字证书，具体描述如下：

1）接收验证最终用户数字证书的申请；

2）确定是否接受最终用户数字证书的申请—证书的审批；

3）向申请者颁发、拒绝颁发数字证书—证书；

4）接收、处理最终用户数字证书的请求——证书的更新；

5）接收最终用户数字证书的查询、撤销；

6）产生和发布证书废止列表（CRL）；

7）数字证书的归档；

8）密钥的归档；

9）历史数据的归档。

CA 为了实现其功能，主要由以下 3 部分组成。

1）注册服务器：通过 Web Server 建立的站点，可为用户提供不间断的服务。用户在网上提出证书申请和填写相应的证书申请表。

2）证书申请受理和审核机构：负责证书的申请和审核。它的主要功能是接受用户证书的申请并进行审核。

3）认证中心服务器：是数字证书生成、发放的运行实体，同时提供发放证书的管理、证书废止列表（CRL）的生成和处理等服务。

2.4.5 身份与访问管理

身份与访问管理（Identity and Access Management，IAM）具有单点登录、强大的认证管理、基于策略的集中式授权和审计、动态授权、企业可管理性等功能。IAM 是一套全面的建立和维护数字身份，并提供有效的、安全的 IT 资源访问的业务流程和管理手段，从而实现组织信息资产统一的身份认证、授权和身份数据集中管理与审计。身份和访问管理是一套业务处理流程，也是一个用于创建、维护和使用数字身份的支持基础结构。通俗地讲，IAM 是让合适的自然人在恰当的时间通过统一的方式访问授权的信息资产，提供集中式的数字身份管理、认证、授权、审计的模式和平台。

IAM 包括以下功能。

1）单点登录（SSO）：通过对跨多种不同 Web 应用程序、门户和安全域的无缝访问允许单点登录，还支持对企业应用程序（例如，SAP、Siebel、PeopleSoft 以及 Oracle 应用程序）的无缝访问。

2）强大的认证管理：提供了统一的认证策略，确保 Internet 和局域网应用程序中的安全级别都正确。这使高安全级别的应用程序受到更强的认证方法保护，而低安全级别应用程序可以只用较简单的用户名/密码方法保护。为许多认证系统（包括密码、令牌、X.509 证书、智能卡、定制表单和生物识别）及多种认证方法组合提供了访问管理支持。

3）基于策略的集中式授权和审计：将一个企业 Web 应用程序中的用户、合作伙伴和员工的访问管理都集起来。因此，不需要冗余、特定于应用程序的安全逻辑。可以按用户属性、角色、组和动态组对访问权进行限制，并按位置和时间确定访问权。授权可以在文件、页面或对象级别上进行。此外，受控制的"模拟"（在此情形中，诸如用户服务代表的某个授权用户，可以访问其他用户可以访问的资源）也由策略定义。

4）动态授权：从不同本地或外部源（包括 Web 服务和数据库）实时触发评估数据的安全策略，从而确定进行访问授权或拒绝访问。通过环境相关的评估，可获得更加细化的授权。例如，限制满足特定条件（最小账户余额）的用户对特定应用程序（特定银行服务）的访问权。授权策略还可以与外部系统（例如，基于风险的安全系统）结合应用。

5）企业可管理性：提供了企业级系统管理工具，使安全人员可以更有效地监控、管理和维护多种环境（包括管理开发、测试和生产环境）。

2.4.6 后量子密码

虽然以 RSA 和椭圆曲线密码算法为主的现代密码学在飞速地发展，但是随着计算机科学的不断进步，量子计算机逐渐成为密码学领域潜在的巨大威胁。贝尔实验室的密码学家 Peter Shor 在 1994 年提出了利用量子计算机在多项式时间内解决大整数因子分解和离散对数问题的方法，这对基于这两种困难问题的密码算法是毁灭性的打击。这是因为半导体只能记录 0 与 1，而量子能够同时表示多种状态，它一次的运算可以处理多种不同状况。所以，假设现在有一个 40 量子位的量子计算机，它能够在很短时间内解决一般计算机花上数十年才能解决的问题。

虽然目前的量子计算机的研究还不是很成熟，但是在 2011 年 5 月 11 日，加拿大的 D-Wave System Inc 公司发布了全球第一款商用型量子计算机 D-Wave One。两年后的 2013 年 5 月，D-Wave System Inc 公司再次与美国国家航空航天局（National Aeronautics and Space Administration，NASA）和 Google 公司共同发布了采用 512 量子位的 D-Wave Two 量子计算机。在国内，量子计算机的研究也如火如荼。2007 年，中国科学技术大学潘建伟院士领衔的量子光学和量子信息团队的陆朝阳、刘乃乐研究小组与英国的牛津大学的研究人员合作，在国际上首次利用光量子计算机演示了 Peter Shor 的分解大整数因子的算法。虽然他们仅仅实现的是15 = 3×5 这一非常简单的质因子分解，但这表明，一旦第一台实用量子计算机问世，基于大整数因子分解或者基于离散对数问题的公钥密码算法就将受到严重的威胁。

尽管目前第一台真正意义上能够破解现代密码体制的量子计算机还没有被研制出来，但是量子计算机带来的潜在威胁给现代密码学提出了一个新的问题，那就是在量子计算机面世以后，我们有哪些密码能够抵御量子计算机的攻击。所幸的是，在现有的密码学算法中，尚有几种不能被量子计算机征服的公钥密码算法，它们是基于 Hash 函数的密码、基于编码的密码、基于格的密码和多变量公钥密码。

1）基于 Hash 函数的密码：基于 Hash 函数的密码主要基于 Merkle 的思想是用 Hash 树将多个一次性验证密钥（Hash 树的叶子）的有效性降低到一个公钥（Hash 树的根）的有效性。他最初的构造与 RSA 等方案相比并不够有效，然而由于 Merkle 的安全性是基于 Hash 函数的抗强碰撞性，并且理论计算表明最先进的 Hash 函数能确保 Merkle 的高安全性级别，抵御量子计算的攻击。因此，基于 Hash 函数的密码仍然受到了学者们的青睐。

2）基于编码的密码：基于编码的密码体制算法核心是应用了一种刻错码，主要特征是编码时添加一定数量的错误码字或根据码的检验矩阵计算伴随式。基于纠错码的第一个密码体制是 1978 年的 McEliece 提出的公钥加密方案。该方案安全性高且加密运算快，但它的公钥规模和签名代价太大。1986 年 Niederreiter 提出了背包型基于编码的公钥密码体制。1990 年王新梅提出了第一个基于编码的数字签名方案——Xinmei 方案。1991 年李元兴构造了一类同时具有签名、加密和纠错能力的公钥体制。由于基于数论的公钥密码体制容易受到量子计算的攻击，基于编码的公钥密码体制已然成为基于数论的公钥密码体制的一个很好的替代。

3）基于格的密码：基于格的密码体制基础是格上面的一些困难问题，如最短向量问题（SVP）、最近向量问题（CVP）、最小基问题（SBP）等。在量子计算飞速发展的时代背景下，学者们对格密码对抗量子计算寄予厚望。自 1996 年 Ajtai 首次在格难题基础上提出一个具有里程碑意义的密码体制后，格理论的相关研究不断取得突破，已经逐步成为抵抗量子计算攻击的公钥密码体制理论的核心研究内容。基于格的公钥密码算法以其安全性高和效率高的优势受到研究者的持续关注。1996 年 Hoffstein、Pipher 和 SiLverman 提出 3NTRV（Number Theory Research Vnit）公钥密码体制，该体制具有抗量子攻击、安全性强、运动速度快、密钥生成快、所需内存小等优势。2000 年，Hoffstein 等人利用 NTRU 提出了 NSS 签名体制。

4）多变量公钥密码：它的设计基于求解有限域的多变量多次多项式方程组的数学困难问题，其中方程组中的多项式大多是二次多项式，而这一问题已经被证明是 NP-Hard 问题。此外，由于多变量公钥密码一般是在小的有限域进行运算，因此要比在大的有限域的运算的密码更加高效。

2.4.7 散列

哈希（Hash）也叫作散列，它把任意长度的输入，通过散列算法，转换成固定长度的输出，该输出就是散列值。这种转换是一种压缩映射，也就是说，散列值的空间通常远小于输入的空间，不同的输入可能会散列成相同的输出，但不可能从散列值来唯一地确定输入值。简单地说就是一种将任意长度的消息压缩到某一固定长度的消息摘要的函数。

Hash 主要用于信息安全领域中的加密算法，它把一些不同长度的信息转化成杂乱的 128 位的编码里，叫作 Hash 值。也可以说，Hash 就是找到一种数据内容和数据存放地址之间的映射关系。MD5 和 SHA1 是目前应用最广泛的 Hash 算法，而它们都是以 MD4 为基础设计的。

MD4 是 MIT Ronald L. Rivest 在 1990 年设计的，MD 是 Message Digest 的缩写。它适合在 32 位字长的处理器上用高速软件实现，是基于 32 位操作数的位操作来实现的。

MD5 是 Rivest 于 1991 年对 MD4 的改进版本。它对输入仍以 512 位分组，其输出是 4 个 32 位字的级联，与 MD4 相同。虽然 MD5 比 MD4 复杂，并且速度较之要慢一点，但更安全，在抗分析和抗差分方面表现更好。由于计算机运算水平的发展，通过大型计算机的运算采用暴力破解方式可以在秒级内破解 MD5。

SHA1 是由 NIST NSA 设计，同 DSA 一起使用的，它对长度小于 264bit 的输入，会产生长度为 160bit 的散列值，因此抗穷举性更好。SHA1 设计时的原理和 MD4 相同原理，并且模仿了该算法。

Hash 算法在信息安全方面的应用主要体现在以下的 3 个方面。

1）文件校验：校验算法主要有奇偶校验和 CRC 校验，这两种校验并没有抗数据篡改的能力，它们一定程度上能检测并纠正数据传输中的信道误码，但不能防止对数据的恶意破坏。MD5 Hash 算法的"数字指纹"特性，使它成为目前应用最广泛的一种文件完整性校验和算法，不少 UNIX 系统提供计算 md5 checksum 的命令。MD5-Hash 文件的数字文摘通过 Hash 函数计算得到。不管文件长度如何，它的 Hash 函数计算结果都是一个固定长度的数字。与加密算法不同，这一个 Hash 算法是一个不可逆的单向函数。采用安全性高的 Hash 算法，如 MD5、SHA 时，两个不同的文件几乎不可能得到相同的 Hash 结果。因此，一旦文件被修改，就可检测出来。

2）数字签名：Hash 算法也是现代密码体系中的一个重要组成部分。由于非对称算法的运算速度较慢，所以在数字签名协议中，单向散列函数扮演了一个重要的角色。对 Hash 值进行数字签名，在统计上可以认为与对文件本身进行数字签名等效。

3）鉴权协议：鉴权协议又被称作"挑战——认证"模式，在传输信道是可被侦听，但不可被篡改的情况下，这是一种简单而安全的方法。

2.5 本章参考代码

2.5.1 DES

DES 的 VHDL 代码示例如下。

```
library IEEE;
use IEEE.std_logic_1164.all;
use IEEE.std_logic_unsigned.all;
use IEEE.std_logic_signed.all;
use IEEE.numeric_std.all;

entity TripleDESsmall is port
(
                k: in std_logic_vector(162 downto 0);
                x1,y1:in std_logic_vector(162 downto 0);
                clk,pmulstart,reset: in std_logic;
                mdone,mbusy: out std_logic;
```

```vhdl
                    x2,y2: out std_logic_vector(162 downto 0)
    );
    end TripleDESsmall;

    architecture structural of TripleDESsmall is
            type ctl_state is (
    idle,init,highbit,highbitzero,highbitjmp,sm1,sm1w,sm2,sm2w,sm3,sm3a,sm3b,sm3w,
sm4,sm4w,sm4wa,sm4put,sm5,sm5a,sm5b,sm5w,sm5put,sm6,sm6w,sm7,sm7w,sm8,sm8w,sm9,sm9w
,sm10,sm10w,sm11,sm11w,sm12,sm12w,sm13,sm13w,sm14,sm14w,sm15,sm15w,sm16,sm16w,sm17,
sm17w,sm18,sm18w,sm19,sm19w,sm20,sm20w,sm21,sm21w,sm21wa,sm22,sm22w,sm23,sm23w,sm23
wa,sm24,sm24w,smend,smout
    );
    signal mstate:ctl_state;
            signal
mulina,mulinb,mulresult,r0,r1,r2,r3,r4,r5,r6,r7,r8,r9,r10:std_logic_vector(162
downto 0);
            signal mulstart,muldone,mulbusy:std_logic;

    component multiply IS
    PORT(
                clk   : IN std_logic;
                reset : IN std_logic;
                load  : IN std_logic;

            a_in  : IN std_logic_vector(162 downto 0);
                b_in  : IN std_logic_vector(162 downto 0);
                result: OUT std_logic_vector(162 downto 0);

                done  : out std_logic;
                busy  : out std_logic
    );
    END component   ;

    begin

     multiplier: multiply
     port map(
                    clk=>clk,
                    reset=>'0',
                    load=>mulstart,
                a_in=>mulina,
                b_in=>mulinb,
                result=>mulresult,
                    done=>muldone,
                    busy=>mulbusy);

    eccpmachine : process ( clk )
        variable m:integer range 0 to 162;
        begin
            if(rising_edge(CLK))then
```

```vhdl
                    case mstate is
when idle =>
 //r0:k  r1:x  r2:y  r3:X1  r4:Z1  r5:X2  r6:Z2  r7:T1  r8:T2  r9:T3  r10:T4
                    if (pmulstart = '1') then
                            r1<=x1;
                            r2<=y1;
                            r0<=k;
                                mbusy<='1';
                                mdone<='0';
                                mstate<=init;
                        else
                                mstate <= idle;
                                mdone<='0';
                                mbusy<='0';
                        end if;
                when init =>
  //3.r3:X1=x    r4:Z1=1     r5:X2=x^4+b    r6:Z2=x^2
                    r3<=r1;
                    r4<=x"ffffffffffffffffffffffffffffffffffffffffff"&"111";
                    r5<=not(x1(160 downto 0)&x1(162 downto 161));
                    r6<=x1(161 downto 0) & x1(162);
                    m:=162;
                    mstate<=highbit;
    when highbit=>
            if r0(m)='1' then
                        mstate<=highbitjmp;
            else
                        mstate<=highbitzero;
            end if;
     when highbitzero=>
             if m=0 then
                        r5<=x"000000000000000000000000000000000000000000"&"000";
                        mstate<=smout;
                else
                        m:=m-1;
                        mstate<=highbit;
                end if;
when highbitjmp=>
        if m=0 then
                mstate<=sm6;
            else
                m:=m-1;
                mstate<=sm1;
            end if;
     when sm1 =>
            mulina<=r3;
            mulinb<=r6;
            mulstart<='1';
            mstate<=sm1w;
```

```
        when sm1w =>
               mulstart<='0';
               if muldone='1' then
                      mstate<=sm2;
               else
                       mstate<=sm1w;
               end if;
    when sm2  =>
               r7<=mulresult;--上次乘法结果放入目的寄存器中
               mulina<=r5;
               mulinb<=r4;
               mulstart<='1';
               mstate<=sm2w;
       when sm2w =>
               mulstart<='0';
               if muldone='1' then
                      mstate<=sm3;
               else
                       mstate<=sm2w;
               end if;
     when sm3 =>
               r8<=mulresult;
               mulina<=r7;
               mulinb<=mulresult;
               mulstart<='1';
               mstate<=sm3a;
    when sm3a =>
               mulstart<='0';
               r10<=r7 xor r8;
               mstate<=sm3b;
    when sm3b =>
               r10<=r10(161 downto 0) & r10(162);
               mstate<=sm3w;
     when sm3w =>
               if muldone='1' then
                      mstate<=sm4;
               else
                       mstate<=sm3w;
               end if;
     when sm4  =>
               r9<=mulresult;
               mulina<=r1;
               mulinb<=r10;
               mulstart<='1';
               mstate<=sm4w;
       when sm4w =>
               mulstart<='0';
               if muldone='1' then
                      mstate<=sm4wa;
```

```
            else
                    mstate<=sm4w;
            end if;
      when sm4wa =>
            r7<=mulresult;
            r8<=mulresult xor r9;
            mstate<=sm4put;
when sm4put =>
            if r0(m)='1' then
                    r3<=r8;
                    r4<=r10;
                    r7<=r5(161 downto 0) & r5(162);
                    r9<=r6(161 downto 0) & r6(162);
            else
                    r5<=r8;
                    r6<=r10;
                    r7<=r3(161 downto 0) & r3(162);
                    r9<=r4(161 downto 0) & r4(162);
            end if;
            mstate<=sm5;
      when sm5 =>
            mulina<=r7;
            mulinb<=r9;
            mulstart<='1';
            mstate<=sm5a;
      when sm5a =>
            mulstart<='0';
            r7<=r7(161 downto 0) & r7(162);
            r9<=r9(161 downto 0) & r9(162);
            mstate<=sm5b;
      when sm5b =>
            r10<=r7 xor r9;
            mstate<=sm5w;
      when sm5w =>
            if muldone='1' then
                    mstate<=sm5put;
            else
                    mstate<=sm5w;
            end if;
when sm5put=>
            r8<=mulresult;
            if r0(m)='1' then
                    r6<=mulresult;
                    r5<=r10;
            else
                    r4<=mulresult;
                    r3<=r10;
            end if;
            if m=0 then
```

```
                        mstate<=sm6;
            else
                    m:=m-1;
                    mstate<=sm1;
            end if;
    when sm6=>
            mulina<=r4;
            mulinb<=r6;
            r7<=r1;
            r8<=r2;
            mulstart<='1';
            mstate<=sm6w;
    when sm6w=>
            mulstart<='0';
            if muldone='1' then
                    mstate<=sm7;
            else
                    mstate<=sm6w;
            end if;
    when sm7=>
            r9<=mulresult;
            mulina<=r4;
            mulinb<=r7;
            mulstart<='1';
            mstate<=sm7w;
    when sm7w=>
            mulstart<='0';
            if muldone='1' then
                    mstate<=sm8;
            else
                    mstate<=sm7w;
            end if;
    when sm8=>
            r4<=mulresult xor r3;
            mulina<=r6;
            mulinb<=r7;
            mulstart<='1';
            mstate<=sm8w;
    when sm8w=>
            mulstart<='0';
            if muldone='1' then
                    mstate<=sm9;
            else
                    mstate<=sm8w;
            end if;
    when sm9=>
            r6<=mulresult;
            mulina<=mulresult;
            mulinb<=r3;
```

```
                mulstart<='1';
                mstate<=sm9w;
        when sm9w=>
                mulstart<='0';
                if muldone='1' then
                        r6<=r6 xor r5;
                        mstate<=sm10;
                else
                        mstate<=sm9w;
                end if;
        when sm10=>
                r3<=mulresult;
                mulina<=r6;
                mulinb<=r4;
                mulstart<='1';
                r10<=r7(161 downto 0) & r7(162);
                mstate<=sm10w;
        when sm10w=>
                mulstart<='0';
                if muldone='1' then
                        r10<=r10 xor r8;
                        mstate<=sm11;
                else
                        mstate<=sm10w;
                end if;
        when sm11=>
                r6<=mulresult;
                mulina<=r10;
                mulinb<=r9;
                mulstart<='1';
                mstate<=sm11w;
        when sm11w=>
                mulstart<='0';
                if muldone='1' then
                        mstate<=sm12;
                else
                        mstate<=sm11w;
                end if;
        when sm12=>
                r10<=mulresult xor r6;
                mulina<=r9;
                mulinb<=r7;
                mulstart<='1';
                mstate<=sm12w;
        when sm12w=>
                mulstart<='0';
                if muldone='1' then
                        mstate<=sm13;
                else
```

```vhdl
                    mstate<=sm12w;
            end if;
    when sm13=>
            r9<=mulresult;
            mulina<=mulresult;
            mulinb<=mulresult(161 downto 0)&mulresult(162);
            mulstart<='1';
            mstate<=sm13w;
    when sm13w=>
            mulstart<='0';
            if muldone='1' then
                    mstate<=sm14;
            else
                    mstate<=sm13w;
            end if;
    when sm14=>
            r1<=mulresult;
            mulina<=mulresult;
            mulinb<=mulresult(160 downto 0)&mulresult(162 downto 161);
            mulstart<='1';
            mstate<=sm14w;
    when sm14w=>
            mulstart<='0';
            if muldone='1' then
                    mstate<=sm15;
            else
                    mstate<=sm14w;
            end if;
    when sm15=>
            r2<=mulresult;
            mulina<=r9;
            mulinb<=mulresult(161 downto 0)&mulresult(162);
            mulstart<='1';
            mstate<=sm15w;
    when sm15w=>
            mulstart<='0';
            if muldone='1' then
                    mstate<=sm16;
            else
                    mstate<=sm15w;
            end if;
    when sm16=>
            r1<=mulresult;
            mulina<=mulresult;
            mulinb<=mulresult(157 downto 0)&mulresult(162 downto 158);
            mulstart<='1';
            mstate<=sm16w;
    when sm16w=>
            mulstart<='0';
```

```vhdl
            if muldone='1' then
                    mstate<=sm17;
            else
                    mstate<=sm16w;
            end if;
when sm17=>
        r2<=mulresult;
        mulina<=mulresult;
        mulinb<=mulresult(152 downto 0)&mulresult(162 downto 153);
        mulstart<='1';
        mstate<=sm17w;
when sm17w=>
        mulstart<='0';
        if muldone='1' then
                mstate<=sm18;
        else
                mstate<=sm17w;
        end if;
when sm18=>
        r1<=mulresult;
        mulina<=mulresult;
        mulinb<=mulresult(142 downto 0)&mulresult(162 downto 143);
        mulstart<='1';
        mstate<=sm18w;
when sm18w=>
        mulstart<='0';
        if muldone='1' then
                mstate<=sm19;
        else
                mstate<=sm18w;
        end if;
when sm19=>
        r2<=mulresult;
        mulina<=mulresult;
        mulinb<=mulresult(122 downto 0)&mulresult(162 downto 123);
        mulstart<='1';
        mstate<=sm19w;
when sm19w=>
        mulstart<='0';
        if muldone='1' then
                mstate<=sm20;
        else
                mstate<=sm19w;
        end if;
when sm20=>
        r1<=mulresult;
        mulina<=r9;
        mulinb<=mulresult(161 downto 0)&mulresult(162);
        mulstart<='1';
```

```vhdl
                mstate<=sm20w;
        when sm20w=>
                mulstart<='0';
                if muldone='1' then
                        mstate<=sm21;
                else
                        mstate<=sm20w;
                end if;
        when sm21=>
                r2<=mulresult;
                mulina<=mulresult;
                mulinb<=mulresult(81 downto 0)&mulresult(162 downto 82);
                mulstart<='1';
                mstate<=sm21w;
        when sm21w=>
                mulstart<='0';
                if muldone='1' then
                        mstate<=sm21wa;
                else
                        mstate<=sm21w;
                end if;
        when sm21wa=>
                r1<=mulresult;
                r9<=mulresult(161 downto 0) & mulresult(162);
                mstate<=sm22;
        when sm22=>
                mulina<=r9;
                mulinb<=r10;
                mulstart<='1';
                mstate<=sm22w;
        when sm22w=>
                mulstart<='0';
                if muldone='1' then
                        mstate<=sm23;
                else
                        mstate<=sm22w;
                end if;
        when sm23=>
                r10<=mulresult;
                mulina<=r3;
                mulinb<=r9;
                mulstart<='1';
                mstate<=sm23w;
        when sm23w=>
                mulstart<='0';
                if muldone='1' then
                        mstate<=sm23wa;
                else
                        mstate<=sm23w;
                end if;
--   //r0:k   r1:x   r2:y   r3:X1   r4:Z1   r5:X2   r6:Z2   r7:T1   r8:T2   r9:T3   r10:T4
```

```vhdl
            when sm23wa=>
                 r5<=mulresult;
                 r6<=mulresult xor r7;
                 mstate<=sm24;
            when sm24 =>
                 mulina<=r6;
                 mulinb<=r10;
                 mulstart<='1';
                 mstate<=sm24w;
            when sm24w=>
                 mulstart<='0';
                 if muldone='1' then
                      mstate<=smend;
                 else
                      mstate<=sm24w;
                 end if;
            when smend=>
                 r6<=mulresult xor r8;
                 mdone<='1';
                 mstate<=smout;

            when smout=>
                 x2<=r5;
                 y2<=r6;
                 mdone<='0';
                 mbusy<='0';
                 mstate<=idle;

            when others =>
                 mstate <= idle;
            end case;
       end if;
end process;

end structural;
```

2.5.2 AES

1. VHDL 代码（AES）

以下是 AES 的 VHDL 代码示例。

```vhdl
--********************************************************************
-- Project     : AES128
--
-- Block Name  : aes128_fast.vhd
     ********************************************************************
library ieee;
```

```vhdl
    use ieee.std_logic_1164.all;
    use ieee.std_logic_unsigned.all;
    use work.aes_package.all;

    entity aes128_fast is
    port(
        clk        in std_logic;
        reset      in std_logic;
        start      in std_logic; -- to initiate the encryption/decryption process after loading
        mode       in std_logic; -- to select encryption or decryption
        load       in std_logic; -- to load the input and keys.has to
        key        in std_logic_vector(63 downto 0);
        data_in    in std_logic_vector(63 downto 0);
        data_out   out std_logic_vector(127 downto 0);
        done       out std_logic
        );

    end aes128_fast;

    architecture mapping of aes128_fast is

    component key_expander
    port(
        clk       in std_logic;
        reset     in std_logic;
        key_in_c0: in state_array_type;
        key_in_c1: in state_array_type;
        key_in_c2: in state_array_type;
        key_in_c3: in state_array_type;
        count     in integer;
        mode      in std_logic;
        keyout_c0: out state_array_type;
        keyout_c1: out state_array_type;
        keyout_c2: out state_array_type;
        keyout_c3: out state_array_type
        );
    end component;

    signal data_in_reg0: state_array_type;
    signal data_in_reg1: state_array_type;
    signal data_in_reg2: state_array_type;
    signal data_in_reg3: state_array_type;
    signal key_reg0: state_array_type;
    signal key_reg1: state_array_type;
    signal key_reg2: state_array_type;
    signal key_reg3: state_array_type;
    signal s0       state_array_type;
    signal s1       state_array_type;
    signal s2       state_array_type;
```

```
signal s3        state_array_type;
signal s_00      state_array_type;
signal s_01      state_array_type;
signal s_02      state_array_type;
signal s_03      state_array_type;
signal r_00      state_array_type;
signal r_01      state_array_type;
signal r_02      state_array_type;
signal r_03      state_array_type;
signal load_d1 : std_logic;
signal start_d1: std_logic;
signal start_d2: std_logic;
signal round_cnt: integer range 0 to 15;
signal flag_cnt: std_logic;
signal done_d1 : std_logic;
signal done_d2 : std_logic;

signal mixcol_0: state_array_type;
signal mixcol_1: state_array_type;
signal mixcol_2: state_array_type;
signal mixcol_3: state_array_type;

signal new_key0: state_array_type;
signal new_key1: state_array_type;
signal new_key2: state_array_type;
signal new_key3: state_array_type;
signal new_key0_d1: state_array_type;
signal new_key1_d1: state_array_type;
signal new_key2_d1: state_array_type;
signal new_key3_d1: state_array_type;

signal s0_buf    state_array_type;
signal s1_buf    state_array_type;
signal s2_buf    state_array_type;
signal s3_buf    state_array_type;

signal next_round_data_0: state_array_type;
signal next_round_data_1: state_array_type;
signal next_round_data_2: state_array_type;
signal next_round_data_3: state_array_type;

signal pr_data_0: state_array_type;
signal pr_data_1: state_array_type;
signal pr_data_2: state_array_type;
signal pr_data_3: state_array_type;

signal mix_col_array    std_logic_vector(0 to 127);
signal mixcol_key_array: std_logic_vector(0 to 127);
signal mixcol_key_0     state_array_type;
signal mixcol_key_1     state_array_type;
```

```vhdl
    signal mixcol_key_2       state_array_type;
    signal mixcol_key_3       state_array_type;
    signal key_select_0       state_array_type;
    signal key_select_1       state_array_type;
    signal key_select_2       state_array_type;
    signal key_select_3       state_array_type;
begin

    Loading the data and keys
process(clk,reset)
begin
 if(reset = '1') then
    key_reg0 <= (others =>(others => '0'));
    key_reg1 <= (others =>(others => '0'));
    key_reg2 <= (others =>(others => '0'));
    key_reg3 <= (others =>(others => '0'));
    data_in_reg0 <= (others =>(others => '0'));
    data_in_reg1 <= (others =>(others => '0'));
    data_in_reg2 <= (others =>(others => '0'));
    data_in_reg3 <= (others =>(others => '0'));
   elsif rising_edge(clk) then
     if(load = '1' and load_d1 = '0') then
       key_reg0 <= (key(63 downto 56),key(55 downto 48),key(47 downto 40),key(39 downto 32));
       key_reg1 <= (key(31 downto 24),key(23 downto 16),key(15 downto 8),key(7 downto 0));
       data_in_reg0 <= (data_in(63 downto 56),data_in(55 downto 48),data_in(47 downto 40),data_in(39 downto 32));
       data_in_reg1 <= (data_in(31 downto 24),data_in(23 downto 16),data_in(15 downto 8),data_in(7 downto 0));
     elsif(load_d1 = '1' and load = '0') then
       key_reg2 <= (key(63 downto 56),key(55 downto 48),key(47 downto 40),key(39 downto 32));
       key_reg3 <= (key(31 downto 24),key(23 downto 16),key(15 downto 8),key(7 downto 0));
       data_in_reg2 <= (data_in(63 downto 56),data_in(55 downto 48),data_in(47 downto 40),data_in(39 downto 32));
       data_in_reg3 <= (data_in(31 downto 24),data_in(23 downto 16),data_in(15 downto 8),data_in(7 downto 0));
     end if;
   end if;
end process;

----------STATE MATRIX ROW WORDS ------
-- Given input xored with given key for generating input to the first round
s0(0) <= data_in_reg0(0) xor key_reg0(0);
s0(1) <= data_in_reg0(1) xor key_reg0(1);
s0(2) <= data_in_reg0(2) xor key_reg0(2);
s0(3) <= data_in_reg0(3) xor key_reg0(3);
s1(0) <= data_in_reg1(0) xor key_reg1(0);
s1(1) <= data_in_reg1(1) xor key_reg1(1);
s1(2) <= data_in_reg1(2) xor key_reg1(2);
```

```vhdl
      s1(3) <= data_in_reg1(3) xor key_reg1(3);
      s2(0) <= data_in_reg2(0) xor key_reg2(0);
      s2(1) <= data_in_reg2(1) xor key_reg2(1);
      s2(2) <= data_in_reg2(2) xor key_reg2(2);
      s2(3) <= data_in_reg2(3) xor key_reg2(3);
      s3(0) <= data_in_reg3(0) xor key_reg3(0);
      s3(1) <= data_in_reg3(1) xor key_reg3(1);
      s3(2) <= data_in_reg3(2) xor key_reg3(2);
      s3(3) <= data_in_reg3(3) xor key_reg3(3);

   ----------------SUB BYTES TRANSFORMATION--------------------------------------
   process(s0_buf,s1_buf,s2_buf,s3_buf,mode)
   begin
     if(mode = '1') then
       s_00(0) <= sbox_val(s0_buf(0));
       s_00(1) <= sbox_val(s0_buf(1));
       s_00(2) <= sbox_val(s0_buf(2));
       s_00(3) <= sbox_val(s0_buf(3));

       s_01(0) <= sbox_val(s1_buf(0));
       s_01(1) <= sbox_val(s1_buf(1));
       s_01(2) <= sbox_val(s1_buf(2));
       s_01(3) <= sbox_val(s1_buf(3));

       s_02(0) <= sbox_val(s2_buf(0));
       s_02(1) <= sbox_val(s2_buf(1));
       s_02(2) <= sbox_val(s2_buf(2));
       s_02(3) <= sbox_val(s2_buf(3));

       s_03(0) <= sbox_val(s3_buf(0));
       s_03(1) <= sbox_val(s3_buf(1));
       s_03(2) <= sbox_val(s3_buf(2));
       s_03(3) <= sbox_val(s3_buf(3));
     else
       s_00(0) <= inv_sbox_val(s0_buf(0));
       s_00(1) <= inv_sbox_val(s0_buf(1));
       s_00(2) <= inv_sbox_val(s0_buf(2));
       s_00(3) <= inv_sbox_val(s0_buf(3));

       s_01(0) <= inv_sbox_val(s1_buf(0));
       s_01(1) <= inv_sbox_val(s1_buf(1));
       s_01(2) <= inv_sbox_val(s1_buf(2));
       s_01(3) <= inv_sbox_val(s1_buf(3));

       s_02(0) <= inv_sbox_val(s2_buf(0));
       s_02(1) <= inv_sbox_val(s2_buf(1));
       s_02(2) <= inv_sbox_val(s2_buf(2));
       s_02(3) <= inv_sbox_val(s2_buf(3));
```

```vhdl
            s_03(0) <= inv_sbox_val(s3_buf(0));
            s_03(1) <= inv_sbox_val(s3_buf(1));
            s_03(2) <= inv_sbox_val(s3_buf(2));
            s_03(3) <= inv_sbox_val(s3_buf(3));
        end if;
    end process;

    ----------SHIFT ROWS TRANSFORMATION-------------------------------------
    process(s_00,s_01,s_02,s_03,mode)
    begin
     if(mode = '1') then
            r_00 <= (s_00(0),s_01(1),s_02(2),s_03(3));
            r_01 <= (s_01(0),s_02(1),s_03(2),s_00(3));
            r_02 <= (s_02(0),s_03(1),s_00(2),s_01(3));
            r_03 <= (s_03(0),s_00(1),s_01(2),s_02(3));
        else
            r_00 <= (s_00(0),s_03(1),s_02(2),s_01(3));
            r_01 <= (s_01(0),s_00(1),s_03(2),s_02(3));
            r_02 <= (s_02(0),s_01(1),s_00(2),s_03(3));
            r_03 <= (s_03(0),s_02(1),s_01(2),s_00(3));
        end if;
    end process;
    ----------MIX COLUMNS TRANSFORMATION------------------------------------
    mix_col_array <= mix_cols_routine(r_00,r_01,r_02,r_03,mode);
        mixcol_0 <= (mix_col_array(0 to 7),mix_col_array(8 to 15),mix_col_array(16 to 23),mix_col_array(24 to 31));
        mixcol_1 <= (mix_col_array(32 to 39),mix_col_array(40 to 47),mix_col_array(48 to 55),mix_col_array(56 to 63));
        mixcol_2 <= (mix_col_array(64 to 71),mix_col_array(72 to 79),mix_col_array(80 to 87),mix_col_array(88 to 95));
        mixcol_3 <= (mix_col_array(96 to 103),mix_col_array(104 to 111),mix_col_array(112 to 119),mix_col_array(120 to 127));

        mixcol_key_array <= mix_cols_routine(new_key0_d1,new_key1_d1,new_key2_d1,new_key3_d1,mode);
        mixcol_key_0 <= (mixcol_key_array(0 to 7),mixcol_key_array(8 to 15),mixcol_key_array(16 to 23),mixcol_key_array(24 to 31));
        mixcol_key_1 <= (mixcol_key_array(32 to 39),mixcol_key_array(40 to 47),mixcol_key_array(48 to 55),mixcol_key_array(56 to 63));
        mixcol_key_2 <= (mixcol_key_array(64 to 71),mixcol_key_array(72 to 79),mixcol_key_array(80 to 87),mixcol_key_array(88 to 95));
        mixcol_key_3 <= (mixcol_key_array(96 to 103),mixcol_key_array(104 to 111),mixcol_key_array(112 to 119),mixcol_key_array(120 to 127));

    ---------ADD ROUND KEY STEP--------------------------------------------
    expand_key:   key_expander
            port map(
                        clk         => clk,
                        reset       => reset,
```

```vhdl
                    key_in_c0 => key_reg0,
                    key_in_c1 => key_reg1,
                    key_in_c2 => key_reg2,
                    key_in_c3 => key_reg3,
                    count     => round_cnt,
                    mode      => mode,
                    keyout_c0 => new_key0,
                    keyout_c1 => new_key1,
                    keyout_c2 => new_key2,
                    keyout_c3 => new_key3
                );

    process(clk,reset)  ---- registered to increase speed
    begin
      if(reset = '1') then
        new_key0_d1 <= (others =>(others => '0'));
        new_key1_d1 <= (others =>(others => '0'));
        new_key2_d1 <= (others =>(others => '0'));
        new_key3_d1 <= (others =>(others => '0'));
      elsif rising_edge(clk) then
        new_key0_d1 <= new_key0;
        new_key1_d1 <= new_key1;
        new_key2_d1 <= new_key2;
        new_key3_d1 <= new_key3;
      end if;
    end process;

    -- Previous round output as input to next round
    next_round_data_0 <=  (pr_data_0(0) xor  key_select_0(0),pr_data_0(1)  xor key_select_0(1),pr_data_0(2) xor key_select_0(2),pr_data_0(3) xor key_select_0(3));
    next_round_data_1 <=  (pr_data_1(0) xor  key_select_1(0),pr_data_1(1)  xor key_select_1(1),pr_data_1(2) xor key_select_1(2),pr_data_1(3) xor key_select_1(3));
    next_round_data_2 <=  (pr_data_2(0) xor  key_select_2(0),pr_data_2(1)  xor key_select_2(1),pr_data_2(2) xor key_select_2(2),pr_data_2(3) xor key_select_2(3));
    next_round_data_3 <=  (pr_data_3(0) xor  key_select_3(0),pr_data_3(1)  xor key_select_3(1),pr_data_3(2) xor key_select_3(2),pr_data_3(3) xor key_select_3(3));

    -- Muxing for choosing data for the last round
    pr_data_0 <= r_00 when round_cnt=11 else
                 mixcol_0;
    pr_data_1 <= r_01 when round_cnt=11 else
                 mixcol_1;
    pr_data_2 <= r_02 when round_cnt=11 else
                 mixcol_2;
    pr_data_3 <= r_03 when round_cnt=11 else
                 mixcol_3;

    key_select_0 <= new_key0_d1 when (mode = '1') else
                    mixcol_key_0 when(mode = '0' and round_cnt < 11) else
```

```vhdl
                new_key0_d1;
  key_select_1 <= new_key1_d1 when (mode = '1') else
                  mixcol_key_1 when(mode = '0' and round_cnt < 11) else
                  new_key1_d1;
  key_select_2 <= new_key2_d1 when (mode = '1') else
                  mixcol_key_2 when(mode = '0' and round_cnt < 11) else
                  new_key2_d1;
  key_select_3 <= new_key3_d1 when (mode = '1') else
                  mixcol_key_3 when(mode = '0' and round_cnt < 11) else
                  new_key3_d1;
  done <= done_d2;

  -- Registering start and load
  process(clk,reset)
  begin
    if(reset = '1') then
      load_d1  <= '0';
      start_d1 <= '0';
      start_d2 <= '0';
    elsif rising_edge(clk) then
      load_d1  <= load;
      start_d1 <= start;
      start_d2 <= start_d1;
    end if;
  end process;

  -- Register outputs at end of each round
  process(clk,reset)
  begin
    if(reset = '1') then
      s0_buf <= (others =>(others => '0'));
      s1_buf <= (others =>(others => '0'));
      s2_buf <= (others =>(others => '0'));
      s3_buf <= (others =>(others => '0'));
    elsif rising_edge(clk) then
      if(round_cnt = 0 or round_cnt = 1) then
        s0_buf <= s0;
        s1_buf <= s1;
        s2_buf <= s2;
        s3_buf <= s3;
      else
        s0_buf <= next_round_data_0;
        s1_buf <= next_round_data_1;
        s2_buf <= next_round_data_2;
        s3_buf <= next_round_data_3;
      end if;
    end if;
  end process;
```

```vhdl
-- Initiator process
process(clk,reset)
begin
  if(reset = '1') then
    round_cnt <= 0;
    flag_cnt <= '0';
  elsif rising_edge(clk) then
    if((start_d2 = '1' and start_d1 = '0') or flag_cnt = '1') then
      if(round_cnt < 11) then
        round_cnt <= round_cnt + 1;
        flag_cnt <= '1';
      else
        round_cnt <= 0;
        flag_cnt <= '0';
      end if;
    end if;
  end if;
end process;

-- Completion signalling process
process(clk,reset)
begin
  if(reset = '1') then
    done_d1 <= '0';
    done_d2 <= '0';
  elsif rising_edge(clk) then
    if(start_d2 = '1' and start_d1 = '0') then
      done_d1 <= '0';
      done_d2 <= '0';
    elsif(round_cnt = 10) then
      done_d1 <= '1';
    end if;
    done_d2 <= done_d1;
  end if;
end process;

-- Output assignment process
process(clk,reset)
begin
  if(reset= '1') then
    data_out <= (others => '0');
  elsif rising_edge(clk) then
    if(done_d1 = '1' and done_d2 = '0') then
      data_out <= (next_round_data_0(0) & next_round_data_0(1) & next_round_data_0(2) & next_round_data_0(3) &
                   next_round_data_1(0) & next_round_data_1(1) & next_round_data_1(2) & next_round_data_1(3) &
                   next_round_data_2(0) & next_round_data_2(1) & next_round_data_2(2) & next_round_data_2(3) &
```

```vhdl
                        next_round_data_3(0) & next_round_data_3(1) & next_round_data_
3(2) & next_round_data_3(3));
            end if;
        end if;
    end process;

end mapping;

    --***********************************************************************
    -- Project    : AES128                                           *
    --                                                               *
    -- Block Name : aes_package.vhd                                   *
    --                                                               *
    --***********************************************************************

    library ieee;
    use ieee.std_logic_1164.all;
    use ieee.std_logic_unsigned.all;

    package aes_package is

    -- This data type is declared to make all operations on a vector of 4 bytes ecach
    -- refer fips-197 doc, sec 3.5
    type state_array_type is array (0 to 3) of std_logic_vector(7 downto 0);

    -- S-Box look up function
    function sbox_val(address: std_logic_vector(7 downto 0)) return std_logic_vector;
    -- Inverse S-Box look up function
    function inv_sbox_val(address: std_logic_vector(7 downto 0)) return std_logic_vector;
    -- column generation fucntion for Mix columns routine
    function col_transform(p: state_array_type) return std_logic_vector;
    -- column generation fucntion for Inverse Mix columns routine
    function col_inv_transform(s: state_array_type) return std_logic_vector;
    -- Mix Columns function
    function mix_cols_routine
        (
            a_r0 : state_array_type;
            a_r1 : state_array_type;
            a_r2 : state_array_type;
            a_r3 : state_array_type;
            mode : std_logic
        )
    return std_logic_vector;

    end aes_package;

    package body aes_package is

    function sbox_val(address: std_logic_vector(7 downto 0)) return std_logic_vector is
```

```vhdl
        variable data: bit_vector(7 downto 0);
        variable data_stdlogic: std_logic_vector(7 downto 0);
        begin
        case address is

          when "00000000" => data := X"63";
          when "00000001" => data := X"7C";
          when "00000010" => data := X"77";
          when "00000011" => data := X"7B";
          when "00000100" => data := X"F2";
          when "00000101" => data := X"6B";
          when "00000110" => data := X"6F";
          when "00000111" => data := X"C5";
          when "00001000" => data := X"30";
          when "00001001" => data := X"01";
          when "00001010" => data := X"67";
          when "00001011" => data := X"2B";
          when "00001100" => data := X"FE";
          when "00001101" => data := X"D7";
          when "00001110" => data := X"AB";
          when "00001111" => data := X"76";
          when "00010000" => data := X"CA";
          when "00010001" => data := X"82";
          when "00010010" => data := X"C9";
          when "00010011" => data := X"7D";
          when "00010100" => data := X"FA";
          when "00010101" => data := X"59";
          when "00010110" => data := X"47";
          when "00010111" => data := X"F0";
          when "00011000" => data := X"AD";
          when "00011001" => data := X"D4";
          when "00011010" => data := X"A2";
          when "00011011" => data := X"AF";
          when "00011100" => data := X"9C";
          when "00011101" => data := X"A4";
          when "00011110" => data := X"72";
          when "00011111" => data := X"C0";
          when "00100000" => data := X"B7";
          when "00100001" => data := X"FD";
          when "00100010" => data := X"93";
          when "00100011" => data := X"26";
          when "00100100" => data := X"36";
          when "00100101" => data := X"3F";
          when "00100110" => data := X"F7";
          when "00100111" => data := X"CC";
          when "00101000" => data := X"34";
          when "00101001" => data := X"A5";
          when "00101010" => data := X"E5";
          when "00101011" => data := X"F1";
```

```vhdl
            when "00101100" => data := X"71";
            when "00101101" => data := X"D8";
            when "00101110" => data := X"31";
            when "00101111" => data := X"15";
            when "00110000" => data := X"04";
            when "00110001" => data := X"C7";
            when "00110010" => data := X"23";
            when "00110011" => data := X"C3";
            when "00110100" => data := X"18";
            when "00110101" => data := X"96";
            when "00110110" => data := X"05";
            when "00110111" => data := X"9A";
            when "00111000" => data := X"07";
            when "00111001" => data := X"12";
            when "00111010" => data := X"80";
            when "00111011" => data := X"E2";
            when "00111100" => data := X"EB";
            when "00111101" => data := X"27";
            when "00111110" => data := X"B2";
            when "00111111" => data := X"75";
            when "01000000" => data := X"09";
            when "01000001" => data := X"83";
            when "01000010" => data := X"2C";
            when "01000011" => data := X"1A";
            when "01000100" => data := X"1B";
            when "01000101" => data := X"6E";
            when "01000110" => data := X"5A";
            when "01000111" => data := X"A0";
            when "01001000" => data := X"52";
            when "01001001" => data := X"3B";
            when "01001010" => data := X"D6";
            when "01001011" => data := X"B3";
            when "01001100" => data := X"29";
            when "01001101" => data := X"E3";
            when "01001110" => data := X"2F";
            when "01001111" => data := X"84";
            when "01010000" => data := X"53";
            when "01010001" => data := X"D1";
            when "01010010" => data := X"00";
            when "01010011" => data := X"ED";
            when "01010100" => data := X"20";
            when "01010101" => data := X"FC";
            when "01010110" => data := X"B1";
            when "01010111" => data := X"5B";
            when "01011000" => data := X"6A";
            when "01011001" => data := X"CB";
            when "01011010" => data := X"BE";
            when "01011011" => data := X"39";
            when "01011100" => data := X"4A";
```

```vhdl
when "01011101" => data := X"4C";
when "01011110" => data := X"58";
when "01011111" => data := X"CF";
when "01100000" => data := X"D0";
when "01100001" => data := X"EF";
when "01100010" => data := X"AA";
when "01100011" => data := X"FB";
when "01100100" => data := X"43";
when "01100101" => data := X"4D";
when "01100110" => data := X"33";
when "01100111" => data := X"85";
when "01101000" => data := X"45";
when "01101001" => data := X"F9";
when "01101010" => data := X"02";
when "01101011" => data := X"7F";
when "01101100" => data := X"50";
when "01101101" => data := X"3C";
when "01101110" => data := X"9F";
when "01101111" => data := X"A8";
when "01110000" => data := X"51";
when "01110001" => data := X"A3";
when "01110010" => data := X"40";
when "01110011" => data := X"8F";
when "01110100" => data := X"92";
when "01110101" => data := X"9D";
when "01110110" => data := X"38";
when "01110111" => data := X"F5";
when "01111000" => data := X"BC";
when "01111001" => data := X"B6";
when "01111010" => data := X"DA";
when "01111011" => data := X"21";
when "01111100" => data := X"10";
when "01111101" => data := X"FF";
when "01111110" => data := X"F3";
when "01111111" => data := X"D2";
when "10000000" => data := X"CD";
when "10000001" => data := X"0C";
when "10000010" => data := X"13";
when "10000011" => data := X"EC";
when "10000100" => data := X"5F";
when "10000101" => data := X"97";
when "10000110" => data := X"44";
when "10000111" => data := X"17";
when "10001000" => data := X"C4";
when "10001001" => data := X"A7";
when "10001010" => data := X"7E";
when "10001011" => data := X"3D";
when "10001100" => data := X"64";
when "10001101" => data := X"5D";
```

```vhdl
            when "10001110" => data := X"19";
            when "10001111" => data := X"73";
            when "10010000" => data := X"60";
            when "10010001" => data := X"81";
            when "10010010" => data := X"4F";
            when "10010011" => data := X"DC";
            when "10010100" => data := X"22";
            when "10010101" => data := X"2A";
            when "10010110" => data := X"90";
            when "10010111" => data := X"88";
            when "10011000" => data := X"46";
            when "10011001" => data := X"EE";
            when "10011010" => data := X"B8";
            when "10011011" => data := X"14";
            when "10011100" => data := X"DE";
            when "10011101" => data := X"5E";
            when "10011110" => data := X"0B";
            when "10011111" => data := X"DB";
            when "10100000" => data := X"E0";
            when "10100001" => data := X"32";
            when "10100010" => data := X"3A";
            when "10100011" => data := X"0A";
            when "10100100" => data := X"49";
            when "10100101" => data := X"06";
            when "10100110" => data := X"24";
            when "10100111" => data := X"5C";
            when "10101000" => data := X"C2";
            when "10101001" => data := X"D3";
            when "10101010" => data := X"AC";
            when "10101011" => data := X"62";
            when "10101100" => data := X"91";
            when "10101101" => data := X"95";
            when "10101110" => data := X"E4";
            when "10101111" => data := X"79";
            when "10110000" => data := X"E7";
            when "10110001" => data := X"C8";
            when "10110010" => data := X"37";
            when "10110011" => data := X"6D";
            when "10110100" => data := X"8D";
            when "10110101" => data := X"D5";
            when "10110110" => data := X"4E";
            when "10110111" => data := X"A9";
            when "10111000" => data := X"6C";
            when "10111001" => data := X"56";
            when "10111010" => data := X"F4";
            when "10111011" => data := X"EA";
            when "10111100" => data := X"65";
            when "10111101" => data := X"7A";
            when "10111110" => data := X"AE";
```

```vhdl
when "10111111" => data := X"08";
when "11000000" => data := X"BA";
when "11000001" => data := X"78";
when "11000010" => data := X"25";
when "11000011" => data := X"2E";
when "11000100" => data := X"1C";
when "11000101" => data := X"A6";
when "11000110" => data := X"B4";
when "11000111" => data := X"C6";
when "11001000" => data := X"E8";
when "11001001" => data := X"DD";
when "11001010" => data := X"74";
when "11001011" => data := X"1F";
when "11001100" => data := X"4B";
when "11001101" => data := X"BD";
when "11001110" => data := X"8B";
when "11001111" => data := X"8A";
when "11010000" => data := X"70";
when "11010001" => data := X"3E";
when "11010010" => data := X"B5";
when "11010011" => data := X"66";
when "11010100" => data := X"48";
when "11010101" => data := X"03";
when "11010110" => data := X"F6";
when "11010111" => data := X"0E";
when "11011000" => data := X"61";
when "11011001" => data := X"35";
when "11011010" => data := X"57";
when "11011011" => data := X"B9";
when "11011100" => data := X"86";
when "11011101" => data := X"C1";
when "11011110" => data := X"1D";
when "11011111" => data := X"9E";
when "11100000" => data := X"E1";
when "11100001" => data := X"F8";
when "11100010" => data := X"98";
when "11100011" => data := X"11";
when "11100100" => data := X"69";
when "11100101" => data := X"D9";
when "11100110" => data := X"8E";
when "11100111" => data := X"94";
when "11101000" => data := X"9B";
when "11101001" => data := X"1E";
when "11101010" => data := X"87";
when "11101011" => data := X"E9";
when "11101100" => data := X"CE";
when "11101101" => data := X"55";
when "11101110" => data := X"28";
when "11101111" => data := X"DF";
```

```vhdl
      when "11110000" => data := X"8C";
      when "11110001" => data := X"A1";
      when "11110010" => data := X"89";
      when "11110011" => data := X"0D";
      when "11110100" => data := X"BF";
      when "11110101" => data := X"E6";
      when "11110110" => data := X"42";
      when "11110111" => data := X"68";
      when "11111000" => data := X"41";
      when "11111001" => data := X"99";
      when "11111010" => data := X"2D";
      when "11111011" => data := X"0F";
      when "11111100" => data := X"B0";
      when "11111101" => data := X"54";
      when "11111110" => data := X"BB";
      when "11111111" => data := X"16";
      when others => null;
end case;
data_stdlogic := to_StdLogicVector(data);
return data_stdlogic;
end function sbox_val;

function inv_sbox_val(address: std_logic_vector(7 downto 0)) return std_logic_vector is
variable inv_data: bit_vector(7 downto 0);
variable inv_data_stdlogic: std_logic_vector(7 downto 0);
begin
case address is

  when "00000000" => inv_data := X"52";
  when "00000001" => inv_data := X"09";
  when "00000010" => inv_data := X"6a";
  when "00000011" => inv_data := X"d5";
  when "00000100" => inv_data := X"30";
  when "00000101" => inv_data := X"36";
  when "00000110" => inv_data := X"a5";
  when "00000111" => inv_data := X"38";
  when "00001000" => inv_data := X"bf";
  when "00001001" => inv_data := X"40";
  when "00001010" => inv_data := X"a3";
  when "00001011" => inv_data := X"9e";
  when "00001100" => inv_data := X"81";
  when "00001101" => inv_data := X"f3";
  when "00001110" => inv_data := X"d7";
  when "00001111" => inv_data := X"fb";
  when "00010000" => inv_data := X"7c";
  when "00010001" => inv_data := X"e3";
  when "00010010" => inv_data := X"39";
  when "00010011" => inv_data := X"82";
  when "00010100" => inv_data := X"9b";
```

```vhdl
when "00010101" => inv_data := X"2f";
when "00010110" => inv_data := X"ff";
when "00010111" => inv_data := X"87";
when "00011000" => inv_data := X"34";
when "00011001" => inv_data := X"8e";
when "00011010" => inv_data := X"43";
when "00011011" => inv_data := X"44";
when "00011100" => inv_data := X"c4";
when "00011101" => inv_data := X"de";
when "00011110" => inv_data := X"e9";
when "00011111" => inv_data := X"cb";
when "00100000" => inv_data := X"54";
when "00100001" => inv_data := X"7b";
when "00100010" => inv_data := X"94";
when "00100011" => inv_data := X"32";
when "00100100" => inv_data := X"a6";
when "00100101" => inv_data := X"c2";
when "00100110" => inv_data := X"23";
when "00100111" => inv_data := X"3d";
when "00101000" => inv_data := X"ee";
when "00101001" => inv_data := X"4c";
when "00101010" => inv_data := X"95";
when "00101011" => inv_data := X"0b";
when "00101100" => inv_data := X"42";
when "00101101" => inv_data := X"fa";
when "00101110" => inv_data := X"c3";
when "00101111" => inv_data := X"4e";
when "00110000" => inv_data := X"08";
when "00110001" => inv_data := X"2e";
when "00110010" => inv_data := X"a1";
when "00110011" => inv_data := X"66";
when "00110100" => inv_data := X"28";
when "00110101" => inv_data := X"d9";
when "00110110" => inv_data := X"24";
when "00110111" => inv_data := X"b2";
when "00111000" => inv_data := X"76";
when "00111001" => inv_data := X"5b";
when "00111010" => inv_data := X"a2";
when "00111011" => inv_data := X"49";
when "00111100" => inv_data := X"6d";
when "00111101" => inv_data := X"8b";
when "00111110" => inv_data := X"d1";
when "00111111" => inv_data := X"25";
when "01000000" => inv_data := X"72";
when "01000001" => inv_data := X"f8";
when "01000010" => inv_data := X"f6";
when "01000011" => inv_data := X"64";
when "01000100" => inv_data := X"86";
when "01000101" => inv_data := X"68";
```

```
            when "01000110" => inv_data := X"98";
            when "01000111" => inv_data := X"16";
            when "01001000" => inv_data := X"d4";
            when "01001001" => inv_data := X"a4";
            when "01001010" => inv_data := X"5c";
            when "01001011" => inv_data := X"cc";
            when "01001100" => inv_data := X"5d";
            when "01001101" => inv_data := X"65";
            when "01001110" => inv_data := X"b6";
            when "01001111" => inv_data := X"92";
            when "01010000" => inv_data := X"6c";
            when "01010001" => inv_data := X"70";
            when "01010010" => inv_data := X"48";
            when "01010011" => inv_data := X"50";
            when "01010100" => inv_data := X"fd";
            when "01010101" => inv_data := X"ed";
            when "01010110" => inv_data := X"b9";
            when "01010111" => inv_data := X"da";
            when "01011000" => inv_data := X"5e";
            when "01011001" => inv_data := X"15";
            when "01011010" => inv_data := X"46";
            when "01011011" => inv_data := X"57";
            when "01011100" => inv_data := X"a7";
            when "01011101" => inv_data := X"8d";
            when "01011110" => inv_data := X"9d";
            when "01011111" => inv_data := X"84";
            when "01100000" => inv_data := X"90";
            when "01100001" => inv_data := X"d8";
            when "01100010" => inv_data := X"ab";
            when "01100011" => inv_data := X"00";
            when "01100100" => inv_data := X"8c";
            when "01100101" => inv_data := X"bc";
            when "01100110" => inv_data := X"d3";
            when "01100111" => inv_data := X"0a";
            when "01101000" => inv_data := X"f7";
            when "01101001" => inv_data := X"e4";
            when "01101010" => inv_data := X"58";
            when "01101011" => inv_data := X"05";
            when "01101100" => inv_data := X"b8";
            when "01101101" => inv_data := X"b3";
            when "01101110" => inv_data := X"45";
            when "01101111" => inv_data := X"06";
            when "01110000" => inv_data := X"d0";
            when "01110001" => inv_data := X"2c";
            when "01110010" => inv_data := X"1e";
            when "01110011" => inv_data := X"8f";
            when "01110100" => inv_data := X"ca";
            when "01110101" => inv_data := X"3f";
            when "01110110" => inv_data := X"0f";
```

```
when "01110111" => inv_data := X"02";
when "01111000" => inv_data := X"c1";
when "01111001" => inv_data := X"af";
when "01111010" => inv_data := X"bd";
when "01111011" => inv_data := X"03";
when "01111100" => inv_data := X"01";
when "01111101" => inv_data := X"13";
when "01111110" => inv_data := X"8a";
when "01111111" => inv_data := X"6b";
when "10000000" => inv_data := X"3a";
when "10000001" => inv_data := X"91";
when "10000010" => inv_data := X"11";
when "10000011" => inv_data := X"41";
when "10000100" => inv_data := X"4f";
when "10000101" => inv_data := X"67";
when "10000110" => inv_data := X"dc";
when "10000111" => inv_data := X"ea";
when "10001000" => inv_data := X"97";
when "10001001" => inv_data := X"f2";
when "10001010" => inv_data := X"cf";
when "10001011" => inv_data := X"ce";
when "10001100" => inv_data := X"f0";
when "10001101" => inv_data := X"b4";
when "10001110" => inv_data := X"e6";
when "10001111" => inv_data := X"73";
when "10010000" => inv_data := X"96";
when "10010001" => inv_data := X"ac";
when "10010010" => inv_data := X"74";
when "10010011" => inv_data := X"22";
when "10010100" => inv_data := X"e7";
when "10010101" => inv_data := X"ad";
when "10010110" => inv_data := X"35";
when "10010111" => inv_data := X"85";
when "10011000" => inv_data := X"e2";
when "10011001" => inv_data := X"f9";
when "10011010" => inv_data := X"37";
when "10011011" => inv_data := X"e8";
when "10011100" => inv_data := X"1c";
when "10011101" => inv_data := X"75";
when "10011110" => inv_data := X"df";
when "10011111" => inv_data := X"6e";
when "10100000" => inv_data := X"47";
when "10100001" => inv_data := X"f1";
when "10100010" => inv_data := X"1a";
when "10100011" => inv_data := X"71";
when "10100100" => inv_data := X"1d";
when "10100101" => inv_data := X"29";
when "10100110" => inv_data := X"c5";
when "10100111" => inv_data := X"89";
```

```vhdl
            when "10101000" => inv_data := X"6f";
            when "10101001" => inv_data := X"b7";
            when "10101010" => inv_data := X"62";
            when "10101011" => inv_data := X"0e";
            when "10101100" => inv_data := X"aa";
            when "10101101" => inv_data := X"18";
            when "10101110" => inv_data := X"be";
            when "10101111" => inv_data := X"1b";
            when "10110000" => inv_data := X"fc";
            when "10110001" => inv_data := X"56";
            when "10110010" => inv_data := X"3e";
            when "10110011" => inv_data := X"4b";
            when "10110100" => inv_data := X"c6";
            when "10110101" => inv_data := X"d2";
            when "10110110" => inv_data := X"79";
            when "10110111" => inv_data := X"20";
            when "10111000" => inv_data := X"9a";
            when "10111001" => inv_data := X"db";
            when "10111010" => inv_data := X"c0";
            when "10111011" => inv_data := X"fe";
            when "10111100" => inv_data := X"78";
            when "10111101" => inv_data := X"cd";
            when "10111110" => inv_data := X"5a";
            when "10111111" => inv_data := X"f4";
            when "11000000" => inv_data := X"1f";
            when "11000001" => inv_data := X"dd";
            when "11000010" => inv_data := X"a8";
            when "11000011" => inv_data := X"33";
            when "11000100" => inv_data := X"88";
            when "11000101" => inv_data := X"07";
            when "11000110" => inv_data := X"c7";
            when "11000111" => inv_data := X"31";
            when "11001000" => inv_data := X"b1";
            when "11001001" => inv_data := X"12";
            when "11001010" => inv_data := X"10";
            when "11001011" => inv_data := X"59";
            when "11001100" => inv_data := X"27";
            when "11001101" => inv_data := X"80";
            when "11001110" => inv_data := X"ec";
            when "11001111" => inv_data := X"5f";
            when "11010000" => inv_data := X"60";
            when "11010001" => inv_data := X"51";
            when "11010010" => inv_data := X"7f";
            when "11010011" => inv_data := X"a9";
            when "11010100" => inv_data := X"19";
            when "11010101" => inv_data := X"b5";
            when "11010110" => inv_data := X"4a";
            when "11010111" => inv_data := X"0d";
            when "11011000" => inv_data := X"2d";
```

```vhdl
      when "11011001" => inv_data := X"e5";
      when "11011010" => inv_data := X"7a";
      when "11011011" => inv_data := X"9f";
      when "11011100" => inv_data := X"93";
      when "11011101" => inv_data := X"c9";
      when "11011110" => inv_data := X"9c";
      when "11011111" => inv_data := X"ef";
      when "11100000" => inv_data := X"a0";
      when "11100001" => inv_data := X"e0";
      when "11100010" => inv_data := X"3b";
      when "11100011" => inv_data := X"4d";
      when "11100100" => inv_data := X"ae";
      when "11100101" => inv_data := X"2a";
      when "11100110" => inv_data := X"f5";
      when "11100111" => inv_data := X"b0";
      when "11101000" => inv_data := X"c8";
      when "11101001" => inv_data := X"eb";
      when "11101010" => inv_data := X"bb";
      when "11101011" => inv_data := X"3c";
      when "11101100" => inv_data := X"83";
      when "11101101" => inv_data := X"53";
      when "11101110" => inv_data := X"99";
      when "11101111" => inv_data := X"61";
      when "11110000" => inv_data := X"17";
      when "11110001" => inv_data := X"2b";
      when "11110010" => inv_data := X"04";
      when "11110011" => inv_data := X"7e";
      when "11110100" => inv_data := X"ba";
      when "11110101" => inv_data := X"77";
      when "11110110" => inv_data := X"d6";
      when "11110111" => inv_data := X"26";
      when "11111000" => inv_data := X"e1";
      when "11111001" => inv_data := X"69";
      when "11111010" => inv_data := X"14";
      when "11111011" => inv_data := X"63";
      when "11111100" => inv_data := X"55";
      when "11111101" => inv_data := X"21";
      when "11111110" => inv_data := X"0c";
      when "11111111" => inv_data := X"7d";
      when others => null;
    end case;
    inv_data_stdlogic := to_StdLogicVector(inv_data);
    return inv_data_stdlogic;
    end function inv_sbox_val;

    function col_transform(p: state_array_type) return std_logic_vector is
     variable result: std_logic_vector(7 downto 0);
     variable m,n: std_logic_vector(7 downto 0);
     begin
```

```vhdl
      if(p(0)(7) = '1') then
        m := (p(0)(6 downto 0) & '0') xor "00011011";
      else
        m := (p(0)(6 downto 0) & '0');
      end if;
      if(p(1)(7) = '1') then
        n := (p(1)(6 downto 0) & '0') xor "00011011" xor p(1);
      else
        n := (p(1)(6 downto 0) & '0') xor p(1);
      end if;
      result := m xor n xor p(2) xor p(3);
        return result;
    end function col_transform;

    function col_inv_transform(s: state_array_type) return std_logic_vector is
    variable result: std_logic_vector(7 downto 0);
    variable sub0,sub1,sub2,sub3: std_logic_vector(7 downto 0);
    variable x0,y0,z0: std_logic_vector(7 downto 0);
    variable x1,y1,z1: std_logic_vector(7 downto 0);
    variable x2,y2,z2: std_logic_vector(7 downto 0);
    variable x3,y3,z3: std_logic_vector(7 downto 0);
    begin
      if(s(0)(7) = '1') then
        x0 := (s(0)(6 downto 0) & '0') xor "00011011";
      else
        x0 := (s(0)(6 downto 0) & '0');
      end if;
      if(x0(7) = '1') then
        y0 := (x0(6 downto 0) & '0') xor "00011011";
      else
        y0 := (x0(6 downto 0) & '0');
      end if;
      if(y0(7) = '1') then
        z0 := (y0(6 downto 0) & '0') xor "00011011";
      else
        z0 := (y0(6 downto 0) & '0');
      end if;
      sub0 := (x0 xor y0 xor z0);----------

      if(s(1)(7) = '1') then
        x1 := (s(1)(6 downto 0) & '0') xor "00011011";
      else
        x1 := (s(1)(6 downto 0) & '0');
      end if;
      if(x1(7) = '1') then
        y1 := (x1(6 downto 0) & '0') xor "00011011";
      else
        y1 := (x1(6 downto 0) & '0');
      end if;
```

```
      if(y1(7) = '1') then
         z1 := (y1(6 downto 0) & '0') xor "00011011";
      else
         z1 := (y1(6 downto 0) & '0');
      end if;
      sub1 := (x1 xor z1 xor s(1));----------

      if(s(2)(7) = '1') then
         x2 := (s(2)(6 downto 0) & '0') xor "00011011";
      else
         x2 := (s(2)(6 downto 0) & '0');
      end if;
      if(x2(7) = '1') then
         y2 := (x2(6 downto 0) & '0') xor "00011011";
      else
         y2 := (x2(6 downto 0) & '0');
      end if;
      if(y2(7) = '1') then
         z2 := (y2(6 downto 0) & '0') xor "00011011";
      else
         z2 := (y2(6 downto 0) & '0');
      end if;
      sub2 := (y2 xor z2 xor s(2));----------

      if(s(3)(7) = '1') then
         x3 := (s(3)(6 downto 0) & '0') xor "00011011";
      else
         x3 := (s(3)(6 downto 0) & '0');
      end if;
      if(x3(7) = '1') then
         y3 := (x3(6 downto 0) & '0') xor "00011011";
      else
         y3 := (x3(6 downto 0) & '0');
      end if;
      if(y3(7) = '1') then
         z3 := (y3(6 downto 0) & '0') xor "00011011";
      else
         z3 := (y3(6 downto 0) & '0');
      end if;
      sub3 := (z3 xor s(3));----------

      result := sub0 xor sub1 xor sub2 xor sub3;
      return result;
end function col_inv_transform;

-- combo logic for mix columns
function mix_cols_routine
     (
        a_r0 :state_array_type;
```

```
        a_r1 :state_array_type;
        a_r2 :state_array_type;
        a_r3 :state_array_type;
        mode :std_logic
    )
    return std_logic_vector is
    variable b        std_logic_vector(0 to 127);
    variable b0       state_array_type;
    variable b1       state_array_type;
    variable b2       state_array_type;
    variable b3       state_array_type;
    ------------------------------------------------
    variable b_0_0 : std_logic_vector(7 downto 0);
    variable s_0_0 : state_array_type;
    ------------------------------------------------
    variable b_0_1 : std_logic_vector(7 downto 0);
    variable s_0_1 : state_array_type;
    ------------------------------------------------
    variable b_0_2 : std_logic_vector(7 downto 0);
    variable s_0_2 : state_array_type;
    ------------------------------------------------
    variable b_0_3 : std_logic_vector(7 downto 0);
    variable s_0_3 : state_array_type;
    ------------------------------------------------
    variable b_1_0 : std_logic_vector(7 downto 0);
    variable s_1_0 : state_array_type;
    ------------------------------------------------
    variable b_1_1 : std_logic_vector(7 downto 0);
    variable s_1_1 : state_array_type;
    ------------------------------------------------
    variable b_1_2 : std_logic_vector(7 downto 0);
    variable s_1_2 : state_array_type;
    ------------------------------------------------
    variable b_1_3 : std_logic_vector(7 downto 0);
    variable s_1_3 : state_array_type;
    ------------------------------------------------
    variable b_2_0 : std_logic_vector(7 downto 0);
    variable s_2_0 : state_array_type;
    ------------------------------------------------
    variable b_2_1 : std_logic_vector(7 downto 0);
    variable s_2_1 : state_array_type;
    ------------------------------------------------
    variable b_2_2 : std_logic_vector(7 downto 0);
    variable s_2_2 : state_array_type;
    ------------------------------------------------
    variable b_2_3 : std_logic_vector(7 downto 0);
    variable s_2_3 : state_array_type;
    ------------------------------------------------
    variable b_3_0 : std_logic_vector(7 downto 0);
```

```vhdl
    variable s_3_0 : state_array_type;
    --------------------------------------------
    variable b_3_1 : std_logic_vector(7 downto 0);
    variable s_3_1 : state_array_type;
    --------------------------------------------
    variable b_3_2 : std_logic_vector(7 downto 0);
    variable s_3_2 : state_array_type;
    --------------------------------------------
    variable b_3_3 : std_logic_vector(7 downto 0);
    variable s_3_3 : state_array_type;
    ----------------------------------------------
begin
if(mode = '1') then
    s_0_0 := a_r0;
    b_0_0 := col_transform(s_0_0);
    -------------------------------------------------
    s_0_1 := a_r1;
    b_0_1 := col_transform(s_0_1);
    -------------------------------------------------
    s_0_2 := a_r2;
    b_0_2 := col_transform(s_0_2);
    -------------------------------------------------
    s_0_3 := a_r3;
    b_0_3 := col_transform(s_0_3);
    --***********************************************
    s_1_0 := (a_r0(1),a_r0(2),a_r0(3),a_r0(0));
    b_1_0 := col_transform(s_1_0);
    -------------------------------------------------
    s_1_1 := (a_r1(1),a_r1(2),a_r1(3),a_r1(0));
    b_1_1 := col_transform(s_1_1);
    -------------------------------------------------
    s_1_2 := (a_r2(1),a_r2(2),a_r2(3),a_r2(0));
    b_1_2 := col_transform(s_1_2);
    -------------------------------------------------
    s_1_3 := (a_r3(1),a_r3(2),a_r3(3),a_r3(0));
    b_1_3 := col_transform(s_1_3);
    --***********************************************
    s_2_0 := (a_r0(2),a_r0(3),a_r0(0),a_r0(1));
    b_2_0 := col_transform(s_2_0);
    -------------------------------------------------
    s_2_1 := (a_r1(2),a_r1(3),a_r1(0),a_r1(1));
    b_2_1 := col_transform(s_2_1);
    -------------------------------------------------
    s_2_2 := (a_r2(2),a_r2(3),a_r2(0),a_r2(1));
    b_2_2 := col_transform(s_2_2);
    -------------------------------------------------
    s_2_3 := (a_r3(2),a_r3(3),a_r3(0),a_r3(1));
    b_2_3 := col_transform(s_2_3);
    --***********************************************
```

```
    s_3_0 := (a_r0(3),a_r0(0),a_r0(1),a_r0(2));
    b_3_0 := col_transform(s_3_0);
--------------------------------------------------
    s_3_1 := (a_r1(3),a_r1(0),a_r1(1),a_r1(2));
    b_3_1 := col_transform(s_3_1);
--------------------------------------------------
    s_3_2 := (a_r2(3),a_r2(0),a_r2(1),a_r2(2));
    b_3_2 := col_transform(s_3_2);
--------------------------------------------------
    s_3_3 := (a_r3(3),a_r3(0),a_r3(1),a_r3(2));
    b_3_3 := col_transform(s_3_3);
--**************************************************************
else
    s_0_0 := a_r0;
    b_0_0 := col_inv_transform(s_0_0);
--------------------------------------------------
    s_0_1 := a_r1;
    b_0_1 := col_inv_transform(s_0_1);
--------------------------------------------------
    s_0_2 := a_r2;
    b_0_2 := col_inv_transform(s_0_2);
--------------------------------------------------
    s_0_3 := a_r3;
    b_0_3 := col_inv_transform(s_0_3);
--**************************************************************
    s_1_0 := (a_r0(1),a_r0(2),a_r0(3),a_r0(0));
    b_1_0 := col_inv_transform(s_1_0);
--------------------------------------------------
    s_1_1 := (a_r1(1),a_r1(2),a_r1(3),a_r1(0));
    b_1_1 := col_inv_transform(s_1_1);
--------------------------------------------------
    s_1_2 := (a_r2(1),a_r2(2),a_r2(3),a_r2(0));
    b_1_2 := col_inv_transform(s_1_2);
--------------------------------------------------
    s_1_3 := (a_r3(1),a_r3(2),a_r3(3),a_r3(0));
    b_1_3 := col_inv_transform(s_1_3);
--**************************************************************
    s_2_0 := (a_r0(2),a_r0(3),a_r0(0),a_r0(1));
    b_2_0 := col_inv_transform(s_2_0);
--------------------------------------------------
    s_2_1 := (a_r1(2),a_r1(3),a_r1(0),a_r1(1));
    b_2_1 := col_inv_transform(s_2_1);
--------------------------------------------------
    s_2_2 := (a_r2(2),a_r2(3),a_r2(0),a_r2(1));
    b_2_2 := col_inv_transform(s_2_2);
--------------------------------------------------
    s_2_3 := (a_r3(2),a_r3(3),a_r3(0),a_r3(1));
    b_2_3 := col_inv_transform(s_2_3);
--**************************************************************
```

```vhdl
        s_3_0 := (a_r0(3),a_r0(0),a_r0(1),a_r0(2));
        b_3_0 := col_inv_transform(s_3_0);
--------------------------------------------------
        s_3_1 := (a_r1(3),a_r1(0),a_r1(1),a_r1(2));
        b_3_1 := col_inv_transform(s_3_1);
--------------------------------------------------
        s_3_2 := (a_r2(3),a_r2(0),a_r2(1),a_r2(2));
        b_3_2 := col_inv_transform(s_3_2);
--------------------------------------------------
        s_3_3 := (a_r3(3),a_r3(0),a_r3(1),a_r3(2));
        b_3_3 := col_inv_transform(s_3_3);
--***********************************************************
end if;
b := (b_0_0 & b_1_0 & b_2_0 & b_3_0 & b_0_1 & b_1_1 & b_2_1 & b_3_1 &
     b_0_2 & b_1_2 & b_2_2 & b_3_2 & b_0_3 & b_1_3 & b_2_3 & b_3_3);
return b;
end function mix_cols_routine;

end package body aes_package;
--****************************************************************************
-- Project    : AES128                                                       *
-- Block Name : key_expander.vhd                                             *
    --****************************************************************************

library ieee;
use ieee.std_logic_1164.all;
use ieee.std_logic_unsigned.all;
use work.aes_package.all;


entity key_expander is
port(
        clk       in std_logic;
        reset     in std_logic;
        key_in_c0: in state_array_type; -- given input keys
        key_in_c1: in state_array_type; -- given input keys
        key_in_c2: in state_array_type; -- given input keys
        key_in_c3: in state_array_type; -- given input keys
        count    : in integer;    -- to synchronise with input transformation rounds
        mode     : in std_logic;       -- high=encrypt, low=decrypt
        keyout_c0: out state_array_type;-- output key value for each round
        keyout_c1: out state_array_type;-- output key value for each round
        keyout_c2: out state_array_type;-- output key value for each round
        keyout_c3: out state_array_type -- output key value for each round
        );
end key_expander;

architecture expansion of key_expander is
signal X0       state_array_type;
```

```vhdl
    signal X1        state_array_type;
    signal X2        state_array_type;
    signal X3        state_array_type;
    signal w_i_nk0 : state_array_type;
    signal w_i_nk1 : state_array_type;
    signal w_i_nk2 : state_array_type;
    signal w_i_nk3 : state_array_type;
    signal temp0     state_array_type;
    signal k_rot     state_array_type;
    signal key_sub : state_array_type;
    signal key_xor_rcon: state_array_type;
    signal rcon: std_logic_vector(7 downto 0);
    begin

    -- transformation of keys
    process(mode,rcon,temp0,k_rot,key_sub,key_xor_rcon,X0,X1,X2,X3,w_i_nk0,w_i_nk1
,w_i_nk2,w_i_nk3)
       begin
         if(mode = '1') then -- if encrypt
           k_rot <= (temp0(1),temp0(2),temp0(3),temp0(0)); -- ROTATE word
           -- SUB word
           key_sub(0) <= sbox_val(k_rot(0));
           key_sub(1) <= sbox_val(k_rot(1));
           key_sub(2) <= sbox_val(k_rot(2));
           key_sub(3) <= sbox_val(k_rot(3));
           -- XOR with rcon
           key_xor_rcon <= ((key_sub(0) xor rcon),key_sub(1),key_sub(2),key_sub(3));

           -- XOR with Wi's
           X0   <=  (  key_xor_rcon(0)   xor   w_i_nk0(0)   ,key_xor_rcon(1)   xor
w_i_nk0(1),key_xor_rcon(2) xor w_i_nk0(2),key_xor_rcon(3) xor w_i_nk0(3));
           X1 <= ((X0(0) xor w_i_nk1(0)) , (X0(1) xor w_i_nk1(1)) , (X0(2) xor w_i_nk1(2)) ,
(X0(3) xor w_i_nk1(3)));
           X2 <= ((X1(0) xor w_i_nk2(0)) , (X1(1) xor w_i_nk2(1)) , (X1(2) xor w_i_nk2(2)) ,
(X1(3) xor w_i_nk2(3)));
           X3 <= ((X2(0) xor w_i_nk3(0)) , (X2(1) xor w_i_nk3(1)) , (X2(2) xor w_i_nk3(2)) ,
(X2(3) xor w_i_nk3(3)));
         else -- if decrypt
           X3 <= (w_i_nk3(0) xor w_i_nk2(0) , w_i_nk3(1) xor w_i_nk2(1) , w_i_nk3(2) xor
w_i_nk2(2) , w_i_nk3(3) xor w_i_nk2(3));
           X2 <= (w_i_nk2(0) xor w_i_nk1(0) , w_i_nk2(1) xor w_i_nk1(1) , w_i_nk2(2) xor
w_i_nk1(2) , w_i_nk2(3) xor w_i_nk1(3));
           X1 <= (w_i_nk1(0) xor w_i_nk0(0) , w_i_nk1(1) xor w_i_nk0(1) , w_i_nk1(2) xor
w_i_nk0(2) , w_i_nk1(3) xor w_i_nk0(3));
           X0   <=  (  key_xor_rcon(0)   xor   w_i_nk0(0)   ,key_xor_rcon(1)   xor
w_i_nk0(1),key_xor_rcon(2) xor w_i_nk0(2),key_xor_rcon(3) xor w_i_nk0(3));

           k_rot <= (X3(1),X3(2),X3(3),X3(0));
           key_sub(0) <= sbox_val(k_rot(0));
```

```vhdl
      key_sub(1) <= sbox_val(k_rot(1));
      key_sub(2) <= sbox_val(k_rot(2));
      key_sub(3) <= sbox_val(k_rot(3));
      key_xor_rcon <= ((key_sub(0) xor rcon),key_sub(1),key_sub(2),key_sub(3));
   end if;
end process;

-- registering key outputs for each round and generating rcon values for each round
process(clk,reset)
begin
  if(reset = '1') then
    temp0  <= (others =>(others => '0'));
    w_i_nk0 <= (others =>(others => '0'));
    w_i_nk1 <= (others =>(others => '0'));
    w_i_nk2 <= (others =>(others => '0'));
    w_i_nk3 <= (others =>(others => '0'));
    rcon    <= (others => '0');
  elsif clk'event and clk = '1' then
    if(count = 0) then
      temp0  <= key_in_c3;
      w_i_nk0 <= key_in_c0;
      w_i_nk1 <= key_in_c1;
      w_i_nk2 <= key_in_c2;
      w_i_nk3 <= key_in_c3;
    else
      temp0  <= X3;
      w_i_nk0 <= X0;
      w_i_nk1 <= X1;
      w_i_nk2 <= X2;
      w_i_nk3 <= X3;
    end if;
    if(mode = '1') then
      case count is
        when 0 => rcon <= "00000001";
        when 1 => rcon <= "00000010";
        when 2 => rcon <= "00000100";
        when 3 => rcon <= "00001000";
        when 4 => rcon <= "00010000";
        when 5 => rcon <= "00100000";
        when 6 => rcon <= "01000000";
        when 7 => rcon <= "10000000";
        when 8 => rcon <= "00011011";
        when 9 => rcon <= "00110110";
        when others => rcon <= "00000000";
      end case;
    else-------------------------->>>>>>>>>>>>>
      case count is
        when 0 => rcon <= "00110110";
        when 1 => rcon <= "00011011";
```

```
        when 2 => rcon <= "10000000";
        when 3 => rcon <= "01000000";
        when 4 => rcon <= "00100000";
        when 5 => rcon <= "00010000";
        when 6 => rcon <= "00001000";
        when 7 => rcon <= "00000100";
        when 8 => rcon <= "00000010";
        when 9 => rcon <= "00000001";
        when others => rcon <= "00000000";
      end case;
    end if;
  end if;
end process;

keyout_c0 <= X0;
keyout_c1 <= X1;
keyout_c2 <= X2;
keyout_c3 <= X3;

end expansion;
```

2. VHDL 代码（TestBench）

以下是 AES 的 TestBench VHDL 代码示例。

```
--**************************************************************************
-- Project      : AES128                                          *
--                                                                *
-- Block Name : aes_fips_mctester.vhd                             *
--                                                          *
--**************************************************************************
library ieee;
use ieee.std_logic_1164.all;
use ieee.std_logic_arith.all;
use ieee.math_real.all;
use std.textio.all;
use ieee.std_logic_textio.all;
use work.aes_tb_package.all;

entity aes_tester is end aes_tester;

architecture behavioral of aes_tester is

component aes128_fast
port(
     clk       in std_logic;
     reset     in std_logic;
     start     in std_logic;
     mode      in std_logic;
```

```vhdl
        load       in std_logic;
        key        in std_logic_vector(63 downto 0);
        data_in    in std_logic_vector(63 downto 0);
        data_out   out std_logic_vector(127 downto 0);
        done       out std_logic
        );
    end component;
    constant total_num_char: integer:= 160;--
    type array1 is array (1 to 300) of std_logic_vector(7 downto 0);
    -- array to hold binary representation of text input
    type array2 is array (1 to (total_num_char/16 + 1)) of std_logic_vector(127 downto 0);
    -- array to hold decrypted binary values
    type array3 is array (1 to 9) of std_logic_vector(127 downto 0);
    --@@@@@@@@@@@@@@@@@@@@@@@@@@@
    signal indicator: integer:=2; -- 1-> text_2_bits; 0-> bits_2_text; 2-> bits_2_bits
    --@@@@@@@@@@@@@@@@@@@@@@@@@@@
    signal clock_tb: std_logic:='0';
    signal reset_tb: std_logic:='0';
    signal load_tb : std_logic:='0';
    signal start_tb: std_logic:='0';
    signal done_tb : std_logic;
    --###############################
    signal mode_tb: std_logic:='0'; -- 1-> encode; 0-> decode
    --###############################
    signal data_in_tb: std_logic_vector(63 downto 0);
    signal data_out_tb: std_logic_vector(127 downto 0);
    signal key_tb: std_logic_vector(63 downto 0);
    constant key_val: std_logic_vector(127 downto 0):=X"000102030405060708090A0B0C0D0E0F";
    signal char_vector: array1:=(others =>(others => '0'));
    signal code_out: array2;

    signal decode_vector: array3;
    signal num_vecs: integer:=0;
    constant key_val_decode:std_logic_vector(127 downto 0):=X"13111D7FE3944A17F307A78B4D2B30C5";
    signal decode_out: array3;
    signal one_block_in: std_logic_vector(127 downto 0);
    signal one_block_out: std_logic_vector(127 downto 0);

    signal chk: character;
    signal chk3: character;
    signal chk1: std_logic:='0';
    signal length_inline: integer:=0;
    signal itr_cnt: integer:=0;
    signal chk2: integer:=0;
    begin
```

```vhdl
aes_i: aes128_fast
    port map(
                clk       => clock_tb,
                reset     => reset_tb,
                start     => start_tb,
                mode      => mode_tb,
                load      => load_tb,
                key       => key_tb,
                data_in   => data_in_tb,
                data_out  => data_out_tb,
                done      => done_tb
            );

process
file infile1: text open read_mode is "text_in.txt";
file infile2: text open read_mode is "encoded_text.txt";
file infile3: text open read_mode is "aes_data_in.txt";
type char_array is array (1 to 6,1 to 40) of character; -- upto 6 lines of 40 characters each for
variable inline: line;                                  -- text_2_bit conversion mode
variable one_char: character;
variable linestr: char_array;
variable outline: line;
variable j: integer:=0;
variable bits_128: std_logic_vector(127 downto 0);
begin
  wait for 1 ns;
  if(indicator = 1) then
    while(not endfile(infile1)) loop
      readline(infile1,inline);
      length_inline <= inline'length + length_inline;
      wait for 1 ns;
      for i in 1 to inline'length loop
        read(inline,one_char);
        char_vector(i+j) <= ascii_2_std_logic_vector(one_char);
        chk <= one_char;
      end loop;
      j := length_inline;
    end loop;
    chk1 <= '1';
  elsif(indicator = 0) then
    while(not endfile(infile2)) loop
    j:= j+1;
     num_vecs <= j;
     readline(infile2,inline);
     read(inline,bits_128);
     decode_vector(j) <= bits_128;
    end loop;
  elsif(indicator = 2) then
```

```vhdl
      while(not endfile(infile3)) loop
        readline(infile3,inline);
        read(inline,bits_128);
        one_block_in <= bits_128;
      end loop;
    end if;
  end process;

  clock_tb <= not clock_tb after 50 ns;
  reset_tb <= '1','0' after 150 ns;

  itr_cnt <= length_inline/16 when ((length_inline rem 16) = 0) else (length_inline/16 +1);
  process
    file outfile1: text open write_mode is "coded_text.txt";
    file outfile2: text open write_mode is "decoded_text.txt";
    file outfile3: text open write_mode is "aes_data_out.txt";
    variable outline: line;
    variable x: integer:=0;
    variable getchar: character;
  begin
    key_tb <= (others => '0');
    data_in_tb <= (others => '0');
    code_out <=(others =>(others => '0'));
    wait for 10 ns;
    wait until (reset_tb = '0');
   if(indicator = 1) then
    wait until(clock_tb'event and clock_tb = '1');
    for i in 1 to itr_cnt loop
       load_tb <= '1';
       key_tb <= key_val(127 downto 64);
       data_in_tb <= (char_vector(1+x) & char_vector(2+x) & char_vector(3+x) & char_vector(4+x) &
                      char_vector(5+x) & char_vector(6+x) & char_vector(7+x) & char_vector(8+x));

       wait until(clock_tb'event and clock_tb = '1');
       load_tb <= '0';
       key_tb <= key_val(63 downto 0);
       data_in_tb <= (char_vector(9+x) & char_vector(10+x) & char_vector(11+x) & char_vector(12+x) &
                      char_vector(13+x) & char_vector(14+x) & char_vector(15+x) & char_vector(16+x));
       wait until(clock_tb'event and clock_tb = '1');
       wait until(clock_tb'event and clock_tb = '1');
       start_tb <= '1';
       wait until(clock_tb'event and clock_tb = '1');
       start_tb <= '0';
       wait until(clock_tb'event and clock_tb = '1');
       wait until done_tb = '1';
```

```vhdl
            code_out(i) <= data_out_tb;
            wait for 1 ns;
            write(outline,code_out(i));
            writeline(outfile1,outline);
            wait until(clock_tb'event and clock_tb = '1');
            chk2 <=i;
            x:= x+16;
            wait until(clock_tb'event and clock_tb = '1');
         end loop;
        wait until(clock_tb'event and clock_tb = '1');
      elsif(indicator = 0) then
        for i in 1 to num_vecs loop
           load_tb <= '1';
           key_tb <= key_val_decode(127 downto 64);
           data_in_tb <= decode_vector(i)(127 downto 64);
           wait until(clock_tb'event and clock_tb = '1');
           load_tb <= '0';
           key_tb <= key_val_decode(63 downto 0);
           data_in_tb <= decode_vector(i)(63 downto 0);
           wait until(clock_tb'event and clock_tb = '1');
           wait until(clock_tb'event and clock_tb = '1');
           start_tb <= '1';
           wait until(clock_tb'event and clock_tb = '1');
           start_tb <= '0';
           wait until(clock_tb'event and clock_tb = '1');
           wait until done_tb = '1';
           decode_out(i) <= data_out_tb;
           wait for 1 ns;
           for k in 0 to 15 loop
             getchar := std_logic_vector_2_ascii(decode_out(i)((127-(8*k)) downto (120-(8*k))));
             chk3 <= getchar;
             chk2 <= k;
             wait for 1 ns;
             if(getchar = '~') then
                writeline(outfile2,outline);
             else
                write(outline,getchar);
             end if;
           end loop;
           wait until(clock_tb'event and clock_tb = '1');
        end loop;
      elsif(indicator = 2)  then
         load_tb <= '1';
         if(mode_tb = '1') then
           key_tb <= key_val(127 downto 64);
         else
           key_tb <= key_val_decode(127 downto 64);
         end if;
         data_in_tb <= one_block_in(127 downto 64);
```

```vhdl
    wait until(clock_tb'event and clock_tb = '1');
    wait until(clock_tb'event and clock_tb = '1');
    wait until(clock_tb'event and clock_tb = '1');
    load_tb <= '0';
    if(mode_tb = '1') then
      key_tb <= key_val(63 downto 0);
    else
      key_tb <= key_val_decode(63 downto 0);
    end if;
    data_in_tb <= one_block_in(63 downto 0);
    wait until(clock_tb'event and clock_tb = '1');
    wait until(clock_tb'event and clock_tb = '1');
    start_tb <= '1';
    wait until(clock_tb'event and clock_tb = '1');
    start_tb <= '0';
    wait until(clock_tb'event and clock_tb = '1');
    wait until done_tb = '1';
    one_block_out <= data_out_tb;
    wait for 1 ns;
    write(outline,one_block_out);
    writeline(outfile3,outline);
    hwrite(outline,one_block_out);
    writeline(outfile3,outline);
    wait until(clock_tb'event and clock_tb = '1');
  end if;

  wait;
end process;
end behavioral;
```

3. Verilog 代码（AES）

以下是 AES 加密、解密的 Verilog 代码示例。

```verilog
///////////////////////////////
//       SubBytes            //
///////////////////////////////
module SubBytes (x, y);
  input  [31:0] x;
  output [31:0] y;
  function [7:0] S;
  input    [7:0] x;
    case (x)
      0:S= 99; 1:S=124; 2:S=119; 3:S=123; 4:S=242; 5:S=107; 6:S=111; 7:S=197;
      8:S= 48; 9:S=  1; 10:S=103; 11:S= 43; 12:S=254; 13:S=215; 14:S=171; 15:S=118;
      16:S=202; 17:S=130; 18:S=201; 19:S=125; 20:S=250; 21:S= 89; 22:S= 71; ; 23:S=240;
      24:S=173; 25:S=212; 26:S=162; 27:S=175; 28:S=156; 29:S=164; 30:S=114; 31:S=192;
      32:S=183; 33:S=253; 34:S=147; 35:S= 38; 36:S= 54; 37:S= 63; 38:S=247; 39:S=204;
      40:S= 52; ; 41:S=165; 42:S=229; 43:S=241; 44:S=113; 45:S=216; 46:S= 49; 47:S= 21;
      48:S=  4; 49:S=199; 50:S= 35; 51:S=195; 52:S= 24; 53:S=150; 54:S=  5; 55:S=154;
```

56:S= 7; 57:S= 18; 58:S=128; 59:S=226; 60:S=235; 61:S= 39; 62:S=178; 63:S=117;
64:S= 9; 65:S=131; 66:S= 44; 67:S= 26; 68:S= 27; 69:S=110; 70:S= 90; 71:S=160;
72:S= 82; 73:S= 59; 74:S=214; 75:S=179; 76:S= 41; 77:S=227; 78:S= 47; 79:S=132;
80:S= 83; 81:S=209; 82:S= 0; 83:S=237; 84:S= 32; 85:S=252; 86:S=177; 87:S= 91;
88:S=106; 89:S=203; 90:S=190; 91:S= 57; 92:S= 74; 93:S= 76; 94:S= 88; 95:S=207;
96:S=208; 97:S=239; 98:S=170; 99:S=251; 100:S= 67; 101:S= 77; 102:S= 51; 103:S=133;
104:S= 69; 105:S=249; 106:S= 2; 107:S=127; 108:S= 80; 109:S= 60; 110:S=159; 111:S=168;
112:S= 81; 113:S=163; 114:S= 64; 115:S=143; 116:S=146; 117:S=157; 118:S= 56; 119:S=245;
120:S=188; 121:S=182; 122:S=218; 123:S= 33; 124:S= 16; 125:S=255; 126:S=243; 127:S=210;
128:S=205; 129:S= 12; 130:S= 19; 131:S=236; 132:S= 95; 133:S=151; 134:S= 68; 135:S= 23;
136:S=196; 137:S=167; 138:S=126; 139:S= 61; 140:S=100; 141:S= 93; 142:S= 25; 143:S=115;
144:S= 96; 145:S=129; 146:S= 79; 147:S=220; 148:S= 34; 149:S= 42; 150:S=144; 151:S=136;
152:S= 70; 153:S=238; 154:S=184; 155:S= 20; 156:S=222; 157:S= 94; 158:S= 11; 159:S=219;
160:S=224; 161:S= 50; 162:S= 58; 163:S= 10; 164:S= 73; 165:S= 6; 166:S= 36; 167:S= 92;
168:S=194; 169:S=211; 170:S=172; 171:S= 98; 172:S=145; 173:S=149; 174:S=228; 175:S=121;
176:S=231; 177:S=200; 178:S= 55; 179:S=109; 180:S=141; 181:S=213; 182:S= 78; 183:S=169;
184:S=108; 185:S= 86; 186:S=244; 187:S=234; 188:S=101; 189:S=122; 190:S=174; 191:S= 8;
192:S=186; 193:S=120; 194:S= 37; 195:S= 46; 196:S= 28; 197:S=166; 198:S=180; 199:S=198;
200:S=232; 201:S=221; 202:S=116; 203:S= 31; 204:S= 75; 205:S=189; 206:S=139; 207:S=138;
208:S=112; 209:S= 62; 210:S=181; 211:S=102; 212:S= 72; 213:S= 3; 214:S=246; 215:S= 14;
216:S= 97; 217:S= 53; 218:S= 87; 219:S=185; 220:S=134; 221:S=193; 222:S= 29; 223:S=158;
224:S=225; 225:S=248; 226:S=152; 227:S= 17; 228:S=105; 229:S=217; 230:S=142; 231:S=148;
232:S=155; 233:S= 30; 234:S=135; 235:S=233; 236:S=206; 237:S= 85; 238:S= 40; 239:S=223;
240:S=140; 241:S=161; 242:S=137; 243:S= 13; 244:S=191; 245:S=230; 246:S= 66; 247:S=104;
248:S= 65; 249:S=153; 250:S= 45; 251:S= 15; 252:S=176; 253:S= 84; 254:S=187; 255:S= 22;
        endcase
    endfunction
    assign y = {S(x[31:24]), S(x[23:16]), S(x[15: 8]), S(x[7: 0])};
```
```

```
endmodule
/////////////////////////////
//      InvSubBytes        //
/////////////////////////////
module InvSubBytes (x, y);
   input  [31:0] x;
   output [31:0] y;
   function [7:0] S;
   input    [7:0] x;
     case (x)
          0:S= 82; 1:S=  9; 2:S=106; 3:S=213; 4:S= 48; 5:S= 54; 6:S=165; 7:S= 56;
          8:S=191; 9:S= 64; 10:S=163; 11:S=158; 12:S=129; 13:S=243; 14:S=215; 15:S=251;
         16:S=124; 17:S=227; 18:S= 57; 19:S=130; 20:S=155; 21:S= 47; 22:S=255; 23:S=135;
         24:S= 52; 25:S=142; 26:S= 67; 27:S= 68; 28:S=196; 29:S=222; 30:S=233; 31:S=203;
         32:S= 84; 33:S=123; 34:S=148; 35:S= 50; 36:S=166; 37:S=194; 38:S= 35; 39:S= 61;
         40:S=238; 41:S= 76; 42:S=149; 43:S= 11; 44:S= 66; 45:S=250; 46:S=195; 47:S= 78;
         48:S=  8; 49:S= 46; 50:S=161; 51:S=102; 52:S= 40; 53:S=217; 54:S= 36;
55:S=178;
         56:S=118; 57:S= 91; 58:S=162; 59:S= 73; 60:S=109; 61:S=139; 62:S=209; 63:S= 37;
         64:S=114; 65:S=248; 66:S=246; 67:S=100; 68:S=134; 69:S=104; 70:S=152; 71:S= 22;
         72:S=212; 73:S=164; 74:S= 92; 75:S=204; 76:S= 93; 77:S=101; 78:S=182; 79:S=146;
         80:S=108; 81:S=112; 82:S= 72; 83:S= 80; 84:S=253; 85:S=237; 86:S=185; 87:S=218;
         88:S= 94; 89:S= 21; 90:S= 70; 91:S= 87; 92:S=167; 93:S=141; 94:S=157;
95:S=132;
         96:S=144; 97:S=216; 98:S=171; 99:S=  0; 100:S=140; 101:S=188; 102:S=211; 103:S= 10;
         104:S=247; 105:S=228; 106:S= 88; 107:S=  5; 108:S=184; 109:S=179; 110:S= 69;
111:S=  6;
         112:S=208; 113:S= 44; 114:S= 30; 115:S=143; 116:S=202; 117:S= 63; 118:S= 15; 119:S=  2;
         120:S=193; 121:S=175; 122:S=189; 123:S=  3; 124:S=  1; 125:S= 19; 126:S=138;
127:S=107;
         128:S= 58; 129:S=145; 130:S= 17; 131:S= 65; 132:S= 79; 133:S=103; 134:S=220;
135:S=234;
         136:S=151; 137:S=242; 138:S=207; 139:S=206; 140:S=240; 141:S=180; 142:S=230;
143:S=115;
         144:S=150; 145:S=172; 146:S=116; 147:S= 34; 148:S=231; 149:S=173; 150:S= 53; 151:S=133;
         152:S=226; 153:S=249; 154:S= 55; 155:S=232; 156:S= 28; 157:S=117; 158:S=223;
159:S=110;
         160:S= 71; 161:S=241; 162:S= 26; 163:S=113; 164:S= 29; 165:S= 41; 166:S=197;
167:S=137;
         168:S=111; 169:S=183; 170:S= 98; 171:S= 14; 172:S=170; 173:S= 24; 174:S=190;
175:S= 27;
         176:S=252; 177:S= 86; 178:S= 62; 179:S= 75; 180:S=198; 181:S=210; 182:S=121;
183:S= 32;
         184:S=154; 185:S=219; 186:S=192; 187:S=254; 188:S=120; 189:S=205; 190:S= 90; 191:S=244;
         192:S= 31; 193:S=221; 194:S=168; 195:S= 51; 196:S=136; 197:S=  7; 198:S=199;
199:S= 49;
```

```
                200:S=177; 201:S= 18; 202:S= 16; 203:S= 89; 204:S= 39; 205:S=128; 206:S=236;
207:S= 95;
                208:S= 96; 209:S= 81; 210:S=127; 211:S=169; 212:S= 25; 213:S=181; 214:S=
74; 215:S= 13;
                216:S= 45; 217:S=229; 218:S=122; 219:S=159; 220:S=147; 221:S=201; 222:S=156;
223:S=239;
                224:S=160; 225:S=224; 226:S= 59; 227:S= 77; 228:S=174; 229:S= 42; 230:S=245;
231:S=176;
                232:S=200; 233:S=235; 234:S=187; 235:S= 60; 236:S=131; 237:S= 83; 238:S=153;
239:S= 97;
                240:S= 23; 241:S= 43; 242:S=  4; 243:S=126; 244:S=186; 245:S=119; 246:S=214;
247:S= 38;
                248:S=225; 249:S=105; 250:S= 20; 251:S= 99; 252:S= 85; 253:S= 33; 254:S=
12; 255:S=125;
        endcase
    endfunction
    assign y = {S(x[31:24]), S(x[23:16]), S(x[15: 8]), S(x[ 7: 0])};
endmodule
////////////////////////////
//      MixColumns        //
////////////////////////////
module MixColumns(x, y);
    input  [31:0]  x;
    output [31:0]  y;
    wire [7:0] a3, a2, a1, a0, b3, b2, b1, b0;
    assign a3 = x[31:24];
    assign a2 = x[23:16];
    assign a1 = x[15: 8];
    assign a0 = x[ 7: 0];
    assign b3 = a3 ^ a2;
    assign b2 = a2 ^ a1;
    assign b1 = a1 ^ a0;
    assign b0 = a0 ^ a3;
    assign y = {a2[7] ^ b1[7] ^ b3[6],
                a2[6] ^ b1[6] ^ b3[5],
                a2[5] ^ b1[5] ^ b3[4],
                a2[4] ^ b1[4] ^ b3[3] ^ b3[7],
                a2[3] ^ b1[3] ^ b3[2] ^ b3[7],
                a2[2] ^ b1[2] ^ b3[1],
                a2[1] ^ b1[1] ^ b3[0] ^ b3[7],
                a2[0] ^ b1[0] ^ b3[7],
                a3[7] ^ b1[7] ^ b2[6],
                a3[6] ^ b1[6] ^ b2[5],
                a3[5] ^ b1[5] ^ b2[4],
                a3[4] ^ b1[4] ^ b2[3] ^ b2[7],
                a3[3] ^ b1[3] ^ b2[2] ^ b2[7],
                a3[2] ^ b1[2] ^ b2[1],
                a3[1] ^ b1[1] ^ b2[0] ^ b2[7],
                a3[0] ^ b1[0] ^ b2[7],
```

```verilog
                a0[7] ^ b3[7] ^ b1[6],
                a0[6] ^ b3[6] ^ b1[5],
                a0[5] ^ b3[5] ^ b1[4],
                a0[4] ^ b3[4] ^ b1[3] ^ b1[7],
                a0[3] ^ b3[3] ^ b1[2] ^ b1[7],
                a0[2] ^ b3[2] ^ b1[1],
                a0[1] ^ b3[1] ^ b1[0] ^ b1[7],
                a0[0] ^ b3[0] ^ b1[7],
                a1[7] ^ b3[7] ^ b0[6],
                a1[6] ^ b3[6] ^ b0[5],
                a1[5] ^ b3[5] ^ b0[4],
                a1[4] ^ b3[4] ^ b0[3] ^ b0[7],
                a1[3] ^ b3[3] ^ b0[2] ^ b0[7],
                a1[2] ^ b3[2] ^ b0[1],
                a1[1] ^ b3[1] ^ b0[0] ^ b0[7],
                a1[0] ^ b3[0] ^ b0[7]};
endmodule
///////////////////////////
//      InvMixColumns    //
///////////////////////////
module InvMixColumns(x, y);
  input  [31:0] x;
  output [31:0] y;
  wire [7:0] a3, a2, a1, a0, b3, b2, b1, b0;
  wire [7:0] c3, c2, c1, c0, d3, d2, d1, d0;
  assign a3 = x[31:24];
  assign a2 = x[23:16];
  assign a1 = x[15: 8];
  assign a0 = x[ 7: 0];
  assign b3 = a3 ^ a2;
  assign b2 = a2 ^ a1;
  assign b1 = a1 ^ a0;
  assign b0 = a0 ^ a3;
  assign c3 = {a2[7] ^ b1[7] ^ b3[6],
               a2[6] ^ b1[6] ^ b3[5],
               a2[5] ^ b1[5] ^ b3[4],
               a2[4] ^ b1[4] ^ b3[3] ^ b3[7],
               a2[3] ^ b1[3] ^ b3[2] ^ b3[7],
               a2[2] ^ b1[2] ^ b3[1],
               a2[1] ^ b1[1] ^ b3[0] ^ b3[7],
               a2[0] ^ b1[0] ^ b3[7]};
  assign c2 = {a3[7] ^ b1[7] ^ b2[6],
               a3[6] ^ b1[6] ^ b2[5],
               a3[5] ^ b1[5] ^ b2[4],
               a3[4] ^ b1[4] ^ b2[3] ^ b2[7],
               a3[3] ^ b1[3] ^ b2[2] ^ b2[7],
               a3[2] ^ b1[2] ^ b2[1],
               a3[1] ^ b1[1] ^ b2[0] ^ b2[7],
               a3[0] ^ b1[0] ^ b2[7]};
```

```verilog
    assign c1 = {a0[7] ^ b3[7] ^ b1[6],
                 a0[6] ^ b3[6] ^ b1[5],
                 a0[5] ^ b3[5] ^ b1[4],
                 a0[4] ^ b3[4] ^ b1[3] ^ b1[7],
                 a0[3] ^ b3[3] ^ b1[2] ^ b1[7],
                 a0[2] ^ b3[2] ^ b1[1],
                 a0[1] ^ b3[1] ^ b1[0] ^ b1[7],
                 a0[0] ^ b3[0] ^ b1[7]};
    assign c0 = {a1[7] ^ b3[7] ^ b0[6],
                 a1[6] ^ b3[6] ^ b0[5],
                 a1[5] ^ b3[5] ^ b0[4],
                 a1[4] ^ b3[4] ^ b0[3] ^ b0[7],
                 a1[3] ^ b3[3] ^ b0[2] ^ b0[7],
                 a1[2] ^ b3[2] ^ b0[1],
                 a1[1] ^ b3[1] ^ b0[0] ^ b0[7],
                 a1[0] ^ b3[0] ^ b0[7]};
    assign d3 = {c3[5], c3[4], c3[3] ^ c3[7], c3[2] ^ c3[7] ^ c3[6],
                 c3[1] ^ c3[6], c3[0] ^ c3[7], c3[7] ^ c3[6], c3[6]};
    assign d2 = {c2[5], c2[4], c2[3] ^ c2[7], c2[2] ^ c2[7] ^ c2[6],
                 c2[1] ^ c2[6], c2[0] ^ c2[7], c2[7] ^ c2[6], c2[6]};
    assign d1 = {c1[5], c1[4], c1[3] ^ c1[7], c1[2] ^ c1[7] ^ c1[6],
                 c1[1] ^ c1[6], c1[0] ^ c1[7], c1[7] ^ c1[6], c1[6]};
    assign d0 = {c0[5], c0[4], c0[3] ^ c0[7], c0[2] ^ c0[7] ^ c0[6],
                 c0[1] ^ c0[6], c0[0] ^ c0[7], c0[7] ^ c0[6], c0[6]};
    assign y  = {d3 ^ d1 ^ c3, d2 ^ d0 ^ c2, d3 ^ d1 ^ c1, d2 ^ d0 ^ c0};
endmodule
///////////////////////////
//     Encryotion Core   //
///////////////////////////
module EncCore(di, ki, Rrg, do, ko);
  input  [127:0] di;
  input  [127:0] ki;
  input  [9:0]   Rrg;
  output [127:0] do;
  output [127:0] ko;
  wire   [127:0] sb, sr, mx;
  wire   [31:0]  so;
SubBytes SB3 (di[127:96], sb[127:96]);
SubBytes SB2 (di[ 95:64], sb[ 95:64]);
SubBytes SB1 (di[ 63:32], sb[ 63:32]);
SubBytes SB0 (di[ 31: 0], sb[ 31: 0]);
  assign sr = {sb[127:120], sb[ 87: 80], sb[ 47: 40], sb[  7:  0],
               sb[ 95: 88], sb[ 55: 48], sb[ 15:  8], sb[103: 96],
               sb[ 63: 56], sb[ 23: 16], sb[111:104], sb[ 71: 64],
               sb[ 31: 24], sb[119:112], sb[ 79: 72], sb[ 39: 32]};

MixColumns MX3 (sr[127:96], mx[127:96]);
MixColumns MX2 (sr[ 95:64], mx[ 95:64]);
MixColumns MX1 (sr[ 63:32], mx[ 63:32]);
```

```verilog
    MixColumns MX0 (sr[ 31: 0], mx[ 31: 0]);
    assign do = ((Rrg[0] == 1)? sr: mx) ^ ki;
    function [7:0] rcon;
    input [9:0] x;
      casex (x)
       10'bxxxxxxxxx1: rcon = 8'h01;
       10'bxxxxxxxx1x: rcon = 8'h02;
       10'bxxxxxxx1xx: rcon = 8'h04;
       10'bxxxxxx1xxx: rcon = 8'h08;
       10'bxxxxx1xxxx: rcon = 8'h10;
       10'bxxxx1xxxxx: rcon = 8'h20;
       10'bxxx1xxxxxx: rcon = 8'h40;
       10'bxx1xxxxxxx: rcon = 8'h80;
       10'bx1xxxxxxxx: rcon = 8'h1b;
       10'b1xxxxxxxxx: rcon = 8'h36;
      endcase
    endfunction
    SubBytes SBK ({ki[23:16], ki[15:8], ki[7:0], ki[31:24]}, so);
    assign ko[127:96] = ki[127:96] ^ {so[31:24] ^ rcon(Rrg), so[23: 0]};
    assign ko[ 95:64] = ki[ 95:64] ^ ko[127:96];
    assign ko[ 63:32] = ki[ 63:32] ^ ko[ 95:64];
    assign ko[ 31: 0] = ki[ 31: 0] ^ ko[ 63:32];
endmodule
///////////////////////////
//     Decryotion Core    //
///////////////////////////
module DecCore(di, ki, Rrg, do, ko);
    input [127:0] di;
    input [127:0] ki;
    input [9:0]   Rrg;
    output [127:0] do;
    output [127:0] ko;
    wire [127:0] sb, sr, mx, dx;
    wire [31:0] so;
    InvMixColumns MX3 (di[127:96], mx[127:96]);
    InvMixColumns MX2 (di[ 95:64], mx[ 95:64]);
    InvMixColumns MX1 (di[ 63:32], mx[ 63:32]);
    InvMixColumns MX0 (di[ 31: 0], mx[ 31: 0]);
    assign dx = (Rrg[8] == 1)? di: mx;
    assign sr = {dx[127:120], dx[ 23: 16], dx[ 47: 40], dx[ 71: 64],
                 dx[ 95: 88], dx[119:112], dx[ 15:  8], dx[ 39: 32],
                 dx[ 63: 56], dx[ 87: 80], dx[111:104], dx[  7:  0],
                 dx[ 31: 24], dx[ 55: 48], dx[ 79: 72], dx[103: 96]};
    InvSubBytes SB3 (sr[127:96], sb[127:96]);
    InvSubBytes SB2 (sr[ 95:64], sb[ 95:64]);
    InvSubBytes SB1 (sr[ 63:32], sb[ 63:32]);
    InvSubBytes SB0 (sr[ 31: 0], sb[ 31: 0]);
    assign do = sb ^ ki;
    function [7:0] rcon;
```

```verilog
    input [9:0] x;
      casex (x)
        10'bxxxxxxxxx1: rcon = 8'h01;
        10'bxxxxxxxx1x: rcon = 8'h02;
        10'bxxxxxxx1xx: rcon = 8'h04;
        10'bxxxxxx1xxx: rcon = 8'h08;
        10'bxxxxx1xxxx: rcon = 8'h10;
        10'bxxxx1xxxxx: rcon = 8'h20;
        10'bxxx1xxxxxx: rcon = 8'h40;
        10'bxx1xxxxxxx: rcon = 8'h80;
        10'bx1xxxxxxxx: rcon = 8'h1b;
        10'b1xxxxxxxxx: rcon = 8'h36;
      endcase
    endfunction
    SubBytes SBK ({ko[23:16], ko[15:8], ko[7:0], ko[31:24]}, so);
    assign ko[127:96] = ki[127:96] ^ {so[31:24] ^ rcon(Rrg), so[23: 0]};
    assign ko[ 95:64] = ki[ 95:64] ^ ki[127:96];
    assign ko[ 63:32] = ki[ 63:32] ^ ki[ 95:64];
    assign ko[ 31: 0] = ki[ 31: 0] ^ ki[ 63:32];
endmodule
////////////////////////////
//    AES for encryption  //
////////////////////////////
module AES_ENC(Din, Key, Dout, Drdy, Krdy, RSTn, EN, CLK, BSY, Dvld);
  input  [127:0] Din;   // Data input
  input  [127:0] Key;   // Key input
  output [127:0] Dout;  // Data output
  input  Drdy;          // Data input ready
  input  Krdy;          // Key input ready
  input  RSTn;          // Reset (Low active)
  input  EN;            // AES circuit enable
  input  CLK;           // System clock
  output BSY;           // Busy signal
  output Dvld;          // Data output valid

  reg [127:0] Drg;   // Data register
  reg [127:0] Krg;   // Key register
  reg [127:0] KrgX;  // Temporary key Register
  reg [9:0]   Rrg;   // Round counter
  reg Dvldrg, BSYrg;
  wire [127:0] Dnext, Knext;
  EncCore EC (Drg, KrgX, Rrg, Dnext, Knext);
  assign Dvld = Dvldrg;
  assign Dout = Drg;
  assign BSY  = BSYrg;
  always @(posedge CLK) begin
    if (RSTn == 0) begin
      Rrg   <= 10'b0000000001;
      Dvldrg <= 0;
```

2.5 本章参考代码

```verilog
            BSYrg <= 0;
         end
      else if (EN == 1) begin
         if (BSYrg == 0) begin
            if (Krdy == 1) begin
               Krg   <= Key;
               KrgX  <= Key;
               Dvldrg <= 0;
            end
            else if (Drdy == 1) begin
               Rrg   <= {Rrg[8:0], Rrg[9]};
               KrgX  <= Knext;
               Drg   <= Din ^ Krg;
               Dvldrg <= 0;
               BSYrg <= 1;
            end
         end
         else begin
            Drg <= Dnext;
            if (Rrg[0] == 1) begin
               KrgX  <= Krg;
               Dvldrg <= 1;
               BSYrg <= 0;
            end
            else begin
               Rrg   <= {Rrg[8:0], Rrg[9]};
               KrgX  <= Knext;
            end
         end
      end
   end
endmodule
////////////////////////////
//  AES for decryption    //
////////////////////////////
module AES_DEC(Din, Key, Dout, Drdy, Krdy, RSTn, EN, CLK, BSY, Dvld);
   input  [127:0] Din;    // Data input
   input  [127:0] Key;    // Key input
   output [127:0] Dout;   // Data output
   input  Drdy;           // Data input ready
   input  Krdy;           // Key input ready
   input  RSTn;           // Reset (Low active)
   input  EN;             // AES circuit enable
   input  CLK;            // System clock
   output BSY;            // Busy signal
   output Dvld;           // Data output valid

   reg [127:0] Drg;       // Data register
   reg [127:0] Krg;       // Key register
```

```verilog
    reg  [127:0] KrgX;        // Temporary key Register
    reg  [9:0]   Rrg;         // Round counter
    reg  Dvldrg, BSYrg;
    wire [127:0] Dnext, Knext;
    DecCore DC (Drg, KrgX, Rrg, Dnext, Knext);
    assign Dvld = Dvldrg;
    assign Dout = Drg;
    assign BSY  = BSYrg;
    always @(posedge CLK) begin
      if (RSTn == 0) begin
        Rrg    <= 10'b1000000000;
        Dvldrg <= 0;
        BSYrg  <= 0;
      end
      else if (EN == 1) begin
        if (BSYrg == 0) begin
          if (Krdy == 1) begin
            Krg   <= Key;
            KrgX  <= Key;
            Dvldrg <= 0;
          end
          else if (Drdy == 1) begin
            KrgX  <= Knext;
            Drg   <= Din ^ Krg;
            Rrg <= {Rrg[0], Rrg[9:1]};
            Dvldrg <= 0;
            BSYrg  <= 1;
          end
        end
        else begin
          Drg <= Dnext;
          if (Rrg[9] == 1) begin
            KrgX   <= Krg;
            Dvldrg <= 1;
            BSYrg  <= 0;
          end
          else begin
            Rrg    <= {Rrg[0], Rrg[9:1]};
            KrgX   <= Knext;
          end
        end
      end
    end
endmodule
```

4. Verilog 代码（TestBench）

以下是 AES 的 TestBench Verilog 代码示例。

//////////////////////////////////////

```verilog
//    AES TestBench encryption    //
////////////////////////////////////
`timescale 1ns/1ns
module AES_TB;
parameter CLOCK = 10;
reg  [127:0] Din;
reg  [127:0] Key;
reg  Drdy;
reg  Krdy;
reg  RSTn;
reg  EN_E, EN_D;
reg  CLK;
wire [127:0] Dout_E, Dout_D;
wire BSY_E, BSY_D;
wire Dvld_E, Dvld_D;
AES_ENC AES_ENC(
   Din(Din),
   Key(Key),
   Dout(Dout_E),
   Drdy(Drdy),
   Krdy(Krdy),
   RSTn(RSTn),
   EN(EN_E),
   CLK(CLK),
   BSY(BSY_E),
   Dvld(Dvld_E)
);
AES_DEC AES_DEC(
   Din(Din),
   Key(Key),
   Dout(Dout_D),
   Drdy(Drdy),
   Krdy(Krdy),
   RSTn(RSTn),
   EN(EN_D),
   CLK(CLK),
   BSY(BSY_D),
   Dvld(Dvld_D)
);
// FIPS-197 Test Vectors
reg [127:0] KE, KD, CT, PT;
initial KE = 128'h000102030405060708090a0b0c0d0e0f; // Encryption Key
initial KD = 128'h13111d7fe3944a17f307a78b4d2b30c5; // Decryption Key (Encryption finlarl round key)
initial PT = 128'h00112233445566778899aabbccddeeff; // Plain Text
initial CT = 128'h69c4e0d86a7b0430d8cdb78070b4c55a; // Cipher Text
initial CLK = 1;
always #(CLOCK/2)
   CLK <= ~CLK;
```

```verilog
initial begin
#(CLOCK/2)
// Reset
  EN_E <= 0;
  EN_D <= 0;
  RSTn <= 0;
  Krdy <= 0;
  Drdy <= 0;
// Eecryption key set
#(CLOCK)
  RSTn <= 1;
  EN_E <= 1;
  Key  <= KE;
  Krdy <= 1;
// Cipher text set
#(CLOCK)
  Krdy <= 0;
  Din  <= PT;
  Drdy <= 1;
#(CLOCK*1)
  Drdy <= 0;
#(CLOCK*10)
// Encryption key set
#(CLOCK)
  EN_E <= 0;
  EN_D <= 1;
  Key  <= KD;
  Krdy <= 1;
// Plain text set
#(CLOCK)
  Krdy <= 0;
  Din  <= CT;
  Drdy <= 1;
#(CLOCK*1)
  Drdy <= 0;
#(CLOCK*10)
  $finish;
end
endmodule
```

2.5.3 RSA

RSA 的 Verilog 代码示例如下。

```
/*
模块名称：基于 Montgomery 算法的 RSA 大数分解乘法模块
*/
`timescale 1 ns / 1 ns
```

2.5 本章参考代码

```verilog
module Montgomery_multiplier_modif(x, y, clk, reset, start, z, done);
  include "montgomery_exponentiator_parameters.v"
  input [K+1:0]   x;
  input [K+1:0]   y;
  input           clk;
  input           reset;
  input           start;
  output [K-1:0]  z;
  output          done;

  wire            done;

  reg             done_reg;

  wire [K:0]      p;
  reg [K+1:0]     pc;
  reg [K+1:0]     ps;
  wire [K+1:0]    y_by_xi;
  wire [K+1:0]    next_pc;
  wire [K+1:0]    next_ps;
  wire [K+1:0]    half_ac;
  wire [K+1:0]    half_as;
  wire [K+1:0]    half_bc;
  wire [K+1:0]    half_bs;
  wire [K:0]      p_minus_m;
  wire [K+2:0]    ac;
  wire [K+2:0]    as;
  wire [K+2:0]    bc;
  wire [K+2:0]    bs;
  wire [K+1:0]    long_m;
  reg [K+1:0]     int_x;
  wire            xi;
  reg             load;
  reg             ce_p;
  wire            equal_zero;
  reg             load_timer;
  wire            time_out;
  reg [0:4]       current_state;
  reg [logK:0]    count;
  reg [logK-1:0]  timer_state;
  reg [K/2:0]     phigh2;
  reg [K/2:0]     phigh;
  reg [K/2:0]     plow;

  assign done = done_reg & (!start);

  generate
    begin : xhdl0
```

```verilog
            genvar            i;
            for (i = 0; i <= K + 1; i = i + 1)
            begin : and_gates
                assign y_by_xi[i] = y[i] & xi;
            end
        end
        endgenerate
        //assign y_by_xi[K] = 1'b0;
        generate
            begin : xhdl1
                genvar            i;
                for (i = 0; i <= K+1; i = i + 1)//第一种情况能整除2
                begin : first_csa
                    assign as[i] = pc[i] ^ ps[i] ^ y_by_xi[i];//下面是进位的。
                    assign ac[i + 1] = (pc[i] & ps[i]) | (pc[i] & y_by_xi[i]) | (ps[i] & y_by_xi[i]);
                end
            end
        endgenerate
        assign ac[0] = 1'b0;
        assign as[K + 2] = 1'b0;
        assign long_m = {2'b00, M};//第二种情况不能整除2,加M
        generate
            begin : xhdl2
                genvar            i;
                for (i = 0; i <= K+1; i = i + 1)
                begin : second_csa
                    assign bs[i] = ac[i] ^ as[i] ^ long_m[i];//下面是进位 < 2M 所以为K+1位
                    assign bc[i + 1] = (ac[i] & as[i]) | (ac[i] & long_m[i]) | (as[i] & long_m[i]);
                end
            end
        endgenerate
        
        assign bc[0] = 1'b0;
        assign bs[K + 2] = ac[K + 2];  //肯定为0
        assign half_as = as[K + 2:1];//相当于除以2了
        assign half_ac = ac[K + 2:1];
        assign half_bs = bs[K + 2:1];
        assign half_bc = bc[K + 2:1];
        
        //是否要加M
        assign next_pc = (as[0] == 1'b0) ? half_ac :
                         half_bc;
        
        assign next_ps = (as[0] == 1'b0) ? half_as :
                         half_bs;
        
        always @(posedge clk)
        begin: parallel_register
```

```verilog
    begin
      if (load == 1'b1)
      begin
        pc <= {K+2{1'b0}};
        ps <= {K+2{1'b0}};
      end
      else if (ce_p == 1'b1)
      begin
        pc <= next_pc;
        ps <= next_ps;
      end
    end
end

assign equal_zero = (count == ZERO) ? 1'b1 :
                    1'b0;
always @(posedge clk)
begin
if(reset)
begin
  plow <= 0;
  phigh <= 0;
  phigh2 <= 0;
end
else
begin
 plow  <= ps[255:0]+pc[255:0];
 phigh <= ps[511:256]+pc[511:256];
 phigh2 <= phigh+plow[256];
end
end
assign p = {phigh2[255:0],plow[255:0]};
//assign p_minus_m = p + minus_M;

//assign z = (p_minus_m[K] == 1'b0) ? p[K - 1:0] :
//           p_minus_m[K - 1:0];
assign z = p;

always @(posedge clk)
begin: shift_register
   integer        i;

  begin
    if (load == 1'b1)
      int_x <= x;
    else if (ce_p == 1'b1)
    begin
```

```verilog
            for (i = 0; i <= K; i = i + 1)
                int_x[i] <= int_x[i + 1];
            int_x[K + 1] <= 1'b0;
        end
    end
end

assign xi = int_x[0];


always @(posedge clk)
begin: counter

  begin
    if (load == 1'b1)                //state2 时 load <= 1'b1;
        count <= K;                  //实际用了 K+2 次
    else if (ce_p == 1'b1)           //第二个 state3 开始
        count <= count - 1;
  end
end



always @(posedge clk)
begin: control_unit
  case (current_state)
    0, 1 :
       begin
          ce_p <= 1'b0;
          load <= 1'b0;
          load_timer <= 1'b1;
          done_reg <= !start;        //当 start 为 1 时为 0
       end
    2 :
       begin
          ce_p <= 1'b0;
          load <= 1'b1;
          load_timer <= 1'b1;
          done_reg <= 1'b0;
       end
    3 :
       begin
          ce_p <= 1'b1;              //运行信号 ce_p
          load <= 1'b0;
          load_timer <= 1'b1;
          done_reg <= 1'b0;
       end
    4 :
       begin
          ce_p <= 1'b0;
```

```verilog
                    load <= 1'b0;
                    load_timer <= 1'b0;
                    done_reg <= 1'b0;
                end
          endcase

        if (reset == 1'b1)
            current_state <= 0;
          else
            case (current_state)
              0 :
                if (start == 1'b0)
                  current_state <= 1;
              1 :
                if (start == 1'b1)          //输入有效了，输入没有缓存
                  current_state <= 2;
              2 :
                  current_state <= 3;
              3 :                           //下一个时钟开始工作
                if (equal_zero == 1'b1)
                  current_state <= 4;
              4 :
                if (time_out == 1'b1)
                  current_state <= 0;
            endcase
      end


    always @(posedge clk)
    begin: timer

      begin
        if (load_timer == 1'b1)
          timer_state <= DELAY;
        else
          timer_state <= timer_state - 1;
      end
    end

    assign time_out = (timer_state == ZERO) ? 1'b1 :
                        1'b0;

endmodule
/*
模块名称：基于Montgomery算法的RSA大数分解测试验证主文件
*/
`timescale 10ns/1ns

module test_rsa_montgomery;
```

```verilog
`include "montgomery_exponentiator_parameters.v"

integer i;
reg clk;
reg reset;

reg         in_valid;


reg [K-1:0] indata;
reg [K-1:0] inexp;
reg [K-1:0] inmod;
wire        out_valid;
wire [K-1:0] text_out;

Montgomery_exponentiator_lsb   Montgomery_exponentiator_lsb_ins(.x(inexp),  .y
(indata), .clk(clk), .reset(reset), .start(in_valid), .z(text_out), .done(out_valid));


//n:b0 49 fe 7f e2 e1 a4 d4 ff 62 dc 7e 81 0f 6c 91 b7 40 09 72 fa 6f 4f 9e 27 31
7c 66 b8 eb c1 ef ff 84 34 70 e2 67 e1 a6 b8 87 41 2d 50 c5 49 27 a0 82 d5 ad 9e 10 0c
4c 70 64 97 e8 28 4b e7 b3
//n:0xb049fe7fe2e1a4d4ff62dc7e810f6c91b7400972fa6f4f9e27317c66b8ebc1efff843470
e267e1a6b887412d50c54927a082d5ad9e100c4c706497e8284be7b3
//e:0x10001
//d:0x98172e7f6ce15990fdc8700ca0120b8be7ba52aca4cd35b66f7fed20be777fb7bc3ccf6e
f74c9a20901d3c9f3deca1a5c639c918572d41308b47861e3402b491

    initial

    begin
     clk = 'b0;
      reset = 'b0;
              in_valid = 'b0;
      indata = 'b0;
      inexp = 'b0;
      inmod = 'b0;
      @(posedge clk);
             in_valid = 'b0;
             indata                                                                    =
512'h33dfaee7be3fb0aa3567ec603b86b9d807ddb78ae4e50e9e1c7cce
90646acfe26a1cbd6d15a95d1efa4d1487393a2383ab0c2e28d5b16a26b3cabe51f96b87cb;
   //    indata =  512'h1f1652ba79a07bd98854cca11ad9321d3ca9c91725b705c58e62dbf6c
60ee5793526f9096a22cf76ded7240dd895e53d47d05283e649a3f3279b308e0cf5654e;
   //    inexp = 512'h91;

       inexp   = 512'h97c9eb8b5d7f62176918fe5e546c083ea0d763841d186a6318a66540c6
83f259e8a83e3be37687c35d9dc44d7fb75e8fc00f1b3b0714ff39ac597a1dbdf74f61;
```

```verilog
        //          inexp    = 512'h98172e7f6ce15990fdc8700ca0120b8be7ba52aca4cd35b66f7fed
20be777fb7bc3ccf6ef74c9a20901d3c9f3deca1a5c639c918572d41308b47861e3402b491;
        //          inmod    = 512'hb049fe7fe2e1a4d4ff62dc7e810f6c91b7400972fa6f4f9e27317c
66b8ebc1efff843470e267e1a6b887412d50c54927a082d5ad9e100c4c706497e8284be7b3;
        //   indata = 23;
        //   inexp  = 5;
        @(posedge clk);
        reset = #1 'b1;
        @(posedge clk);
        reset = #1 'b0;
        @(posedge clk);
        in_valid = #1 'b1;
        @(posedge clk);
        in_valid = #1 'b0;
        @(posedge clk);
        @(posedge clk);
        in_valid = #1 'b0;
        @(posedge clk);
        in_valid = #1 'b0;


            $display("Running test:");
            wait(out_valid);
                    @(posedge clk);
                    $display("%H",text_out);
                    @(posedge clk);

                    @(posedge clk);


        $finish;

    end

    always #5 clk = !clk;

endmodule
/*
模块名称：基于 Montgomery 算法的 RSA 大数分解参数配置模块
*/

parameter     K = 512;
parameter     logK = 9;
//parameter    K = 8;
//parameter    logK = 3;
parameter     DELAY = K/2;
//parameter              int_m  = 512'hb049fe7fe2e1a4d4ff62dc7e810f6c91b7400972fa
6f4f9e27317c66b8ebc1efff843470e267e1a6b887412d50c54927a082d5ad9e100c4c706497e8284be7b3;
parameter              int_m  = 512'ha8753cbb86bb3e1775588e6ee3617df4e81d033042
```

fd3d9784b6e66b403b9ef572233d8ea19faf1166f91767d5d441745343904ee8ddd605d20f3ace02fff793;

```verilog
    //parameter              int_m = 217;

  parameter [K-1:0]    M = int_m;
  //parameter [K:0]      minus_M = 2 ** K - int_m;
  //parameter [K:0]        minus_M = 512'h4fb601801d1e5b2b009d23817ef0936e48bff68d0590b061d8ce839947143e10007bcb8f1d981e594778bed2af3ab6d85f7d2a5261eff3b38f9b6817d7b4184d;
  //parameter [K:0]          minus_M = 512'h578ac3447944c1e88aa771911c9e820b17e2fccfbd02c2687b491994bfc4610a8ddcc2715e6050ee9906e8982a2bbe8bacbc6fb1172229fa2df0c531fd00086d;

  parameter [logK-1:0] ZERO = {logK{1'b0}};
  parameter [K-1:0]    one = 1;
  //parameter  [K-1:0]     exp_k  =   512'h8e8e07809197c7d70311b1877ab2e1276bbfd0c11bd371e93c0891fe63653650026af9cb93f897be655bba1d6c259239dd71d39be9afc281ce09087736847981;
  //parameter  [K-1:0]    exp_2k =   512'h5778a88c6686b2d42cfd1613dc3e1ca02bedb3d6919d77c32936444acee831eadd8fdeb1a8f2c26dfe2ec56a9adafd462e13f6696f6c42e093c62c3b6b14d50a;

  parameter [K-1:0]     exp_k  =     512'hd40939ad79c8b733feca966abb70c428f51ecde6e108e72e3b6997c7e9a463f532c8ea83641e59796297390fd0677460c6a9e268accfbdd13a49f2bee00328e;
  parameter [K-1:0]    exp_2k =   512'h58e7204b16323458a5a551d194b3cd913726b9801d6c3701594443c93b7954dd34d9911d6413fb08cdc22b53e32f8a767bb377bb04e3fe4b083d5e4bbed3a687;

  //parameter [K-1:0]    exp_k  = (2 ** (K+2)) % int_m;
  //parameter [K-1:0]    exp_2k = (2 ** (2 * (K+2))) % int_m;

  //constant DELAY: std_logic_vector(logK-1 downto 0) := conv_std_logic_vector(K/2, logK);
  /*
```
模块名称：基于Montgomery算法的RSA大数分解乘法控制模块
```verilog
  */

  `timescale 1 ns / 1 ns

  module Montgomery_exponentiator_lsb(x, y, clk, reset, start, z, done);
    include "montgomery_exponentiator_parameters.v"
    input [K-1:0]  x;
    input [K-1:0]  y;
    input          clk;
    input          reset;
    input          start;
    output [K-1:0] z;
    output         done;
    wire           done;
```

```verilog
    wire [K-1:0]    second;
    wire [K-1:0]    operand1;
    wire [K-1:0]    operand2;
    wire [K-1:0]    next_e;
    wire [K-1:0]    next_y;
    reg  [K-1:0]    e;
    reg  [K-1:0]    ty;
    reg  [K-1:0]    int_x;
    reg             start_mp1;
    reg             start_mp2;
    wire            mp1_done;
    wire            mp2_done;
     wire           mp_done;
    reg             ce_e;
    reg             ce_ty;
    reg             load;
    wire            update;
    wire            xi;
    wire            equal_zero;
    reg             first;
    reg             last;

    reg [0:15]      current_state;
    reg [logK:0]    count;

    reg done_r;

    assign done = done_r && !start;

    //最后一次去montgomery化
    assign second = (last == 1'b0) ? ty :
                    one;

    //第一次计算是计算ty的
    assign operand1 = (first == 1'b1) ? y :
                    ty;

    assign operand2 = (first == 1'b1) ? exp_2k :
                    ty;

    //主乘法
    Montgomery_multiplier_modif     main_component1(.x({2'b0,e}),    .y({2'b0,second}), .clk(clk), .reset(reset), .start(start_mp1), .z(next_e), .done(mp1_done));

    //平方
    Montgomery_multiplier_modif     main_component2(.x({2'b0,operand1}),   .y({2'b0,operand2}), .clk(clk), .reset(reset), .start(start_mp2), .z(next_y), .done(mp2_done));

    assign mp_done = mp1_done & mp2_done;
```

```verilog
assign z = next_e;

assign update = (current_state == 12);  //在下轮前

always @(posedge clk)
begin: register_e

  begin
    if (load == 1'b1)
       e <= exp_k;
    else if (ce_e == 1'b1)
       e <= next_e;
  end
end


always @(posedge clk)
begin: register_ty

  begin
    if (ce_ty == 1'b1)
       ty <= next_y;
  end
end


always @(posedge clk)
begin: shift_register

   begin
     if (load == 1'b1)
        int_x <= x;
     else if (update == 1'b1)
        int_x <= {1'b0, int_x[K - 1:1]};
   end
 end

assign xi = int_x[0];


always @(posedge clk)
begin: counter

   begin
    if(reset)
      count <= 0;
    else
     begin
       if (load == 1'b1)
```

```verilog
              count <= 512;              //循环次数等于加密钥长度
          else if (update == 1'b1)    //assign update = (current_state == 12); 13
时已经有效了
              count <= count - 1;
        end
     end
end
assign equal_zero = (count == 0) ? 1'b1 :
                    1'b0;

always @(posedge clk)
begin: control_unit
  if (reset == 1'b1)
     done_r <= 1'b0;
  else
  begin
   if(start)
      done_r <= 0;
   else if(mp_done == 1'b1&&current_state == 15)
      done_r <= 1;  //启动信号已经启动过，等待数据
  end

          case (current_state)
      0,1 :
         begin
            ce_e <= 1'b0;
            ce_ty <= 1'b0;
            load <= 1'b0;
            //update <= 1'b0;
            start_mp1 <= 1'b0;
            start_mp2 <= 1'b0;
            first <= 1'b0;
            last <= 1'b0;

        end
      2 :
         begin
            ce_e <= 1'b0;
            ce_ty <= 1'b0;
            load <= 1'b1;      //load有效
            //update <= 1'b0;
            start_mp1 <= 1'b0;
            start_mp2 <= 1'b0;
            first <= 1'b0;
            last <= 1'b0;
            // done <= 1'b0;    //????????
         end
      3 :
         begin
```

```verilog
              ce_e <= 1'b0;
              ce_ty <= 1'b0;
              load <= 1'b0;
              //update <= 1'b0;
              start_mp1 <= 1'b0;
              start_mp2 <= 1'b1;          //计算ty
              first <= 1'b1;              //第一次
              last <= 1'b0;
           //  done <= 1'b0;
              end
     4 :                                   //会开始判断mp2_done是否有效
           begin
              ce_e <= 1'b0;
              ce_ty <= 1'b0;
              load <= 1'b0;
              //update <= 1'b0;
              start_mp1 <= 1'b0;
              start_mp2 <= 1'b0;          // start_mp2 <= 1'b1; 才变有效
              first <= 1'b1;              //第一次
              last <= 1'b0;
           //  done <= 1'b0;
              end
     5 :
           begin
              ce_e <= 1'b0;
              ce_ty <= 1'b1;              //前提中的ty已经做好,下面进行平方了
              load <= 1'b0;
              //update <= 1'b0;
              start_mp1 <= 1'b0;
              start_mp2 <= 1'b0;
              first <= 1'b1;
              last <= 1'b0;
           //  done <= 1'b0;
              end
     6 :                                   //(xi == 1'b0) 6->7->8
           begin
              ce_e <= 1'b0;
              ce_ty <= 1'b0;
              load <= 1'b0;
              //update <= 1'b0;
              start_mp1 <= 1'b0;           //平方,不需要乘法
              start_mp2 <= 1'b1;           //被一起赋值
              first <= 1'b0;
              last <= 1'b0;
           //  done <= 1'b0;
              end
     7 :                                   //赋值有效了
           begin
              ce_e <= 1'b0;
```

```verilog
            ce_ty <= 1'b0;
            load <= 1'b0;
            //update <= 1'b0;
            start_mp1 <= 1'b0;
            start_mp2 <= 1'b0;
            first <= 1'b0;
            last <= 1'b0;
//          done <= 1'b0;
        end
8 :
        begin
            ce_e <= 1'b0;      //不更乘法结果
            ce_ty <= 1'b1;     //更新平方结果
            load <= 1'b0;
            //update <= 1'b0;
            start_mp1 <= 1'b0;
            start_mp2 <= 1'b0;
            first <= 1'b0;
            last <= 1'b0;
//          done <= 1'b0;
        end
9 :                            //(xi == 1'b1) 9->10->11
        begin
            ce_e <= 1'b0;
            ce_ty <= 1'b0;
            load <= 1'b0;
            //update <= 1'b0;
            start_mp1 <= 1'b1;
            start_mp2 <= 1'b1;
            first <= 1'b0;
            last <= 1'b0;
//          done <= 1'b0;
        end
10 :
        begin
            ce_e <= 1'b0;
            ce_ty <= 1'b0;
            load <= 1'b0;
            //update <= 1'b0;
            start_mp1 <= 1'b0;
            start_mp2 <= 1'b0;
            first <= 1'b0;
            last <= 1'b0;
//          done <= 1'b0;
        end
11 :
        begin
            ce_e <= 1'b1;      //更新乘法结果
            ce_ty <= 1'b1;     //更新平方结果
```

```
               load <= 1'b0;
               //update <= 1'b0;
               start_mp1 <= 1'b0;
               start_mp2 <= 1'b0;
               first <= 1'b0;
               last <= 1'b0;
            //   done <= 1'b0;
            end
        12:                       //完成一次
            begin
               ce_e <= 1'b0;
               ce_ty <= 1'b0;
               load <= 1'b0;
               //update <= 1'b1;
               start_mp1 <= 1'b0;
               start_mp2 <= 1'b0;
               first <= 1'b0;
               last <= 1'b0;
            //   done <= 1'b0;
            end
        13:                       //判断下一次的乘法
            begin
               ce_e <= 1'b0;
               ce_ty <= 1'b0;
               load <= 1'b0;
               //update <= 1'b0;
               start_mp1 <= 1'b0;
               start_mp2 <= 1'b0;
               first <= 1'b0;
               last <= 1'b0;
            //   done <= 1'b0;
            end
        14:                       //最后一次
            begin
               ce_e <= 1'b0;
               ce_ty <= 1'b0;
               load <= 1'b0;
               //update <= 1'b0;
               start_mp1 <= 1'b1;
               start_mp2 <= 1'b0;
               first <= 1'b0;
               last <= 1'b1;
            //   done <= 1'b0;
            end
        15:                       //最后一次是否结束?
            begin
               ce_e <= 1'b0;
               ce_ty <= 1'b0;
               load <= 1'b0;
```

```verilog
      //update <= 1'b0;
        start_mp1 <= 1'b0;
        start_mp2 <= 1'b0;
        first <= 1'b0;
        last <= 1'b1;
    //   done <= 1'b0;
      end
endcase

if (reset == 1'b1)
   current_state <= 0;
else
  case (current_state)
    0 :
      if (start == 1'b0)
         current_state <= 1;
    1 :
      if (start == 1'b1)
         current_state <= 2;
    2 :
      current_state <= 3;
    3 :
      current_state <= 4;
    4 :
      if (mp2_done == 1'b1)        //计算 ty 的
         current_state <= 5;
    5 :                            //开始计算了
      if (xi == 1'b0)
         current_state <= 6;
      else
         current_state <= 9;
    6 :                            //指数位为 0 的情况
      current_state <= 7;
    7 :
      if (mp2_done == 1'b1)
         current_state <= 8;
    8 :
      current_state <= 12;
    9 :                            //指数位为 1 的情况
      current_state <= 10;
    10 :
      if (mp_done == 1'b1)
         current_state <= 11;
    11 :
      current_state <= 12;
    12 :                           //完成一次乘法
      current_state <= 13;
    13 :
      if (equal_zero == 1'b1)      //是否结束？
```

```
                    current_state <= 14;
                else if (xi == 1'b0)
                    current_state <= 6;
                else
                    current_state <= 9;
            14 :
                current_state <= 15;
            15 :
                if (mp_done == 1'b1)
                    current_state <= 0;
        endcase
    end

endmodule
```

2.5.4 ECC

ECC 的 VHDL 代码示例如下。

```
------ECCP.vhd
library ieee;
use ieee.std_logic_1164.all;
use ieee.std_logic_arith.all;
use ieee.std_logic_unsigned.all;

entity ECCP is
    port
    (
        x1: in std_logic_vector(162 downto 0);
        clk,pmulstart: in std_logic;
        mdone,mbusy: out std_logic;
        y2: out std_logic_vector(162 downto 0));
end ECCP;

architecture marc of ECCP is
    type ctl_state is (idle,init,init1,highbit,highbitzero,highbitjmp,sm1,sm1w,
sm2,sm2w,sm3,sm3a,sm3b,sm3w,sm4,sm4w,sm4wa,sm4put,sm5,sm5a,sm5b,sm5w,sm5put,sm6,sm6
w,sm7,sm7w,sm8,sm8w,sm9,sm9w,--33
                                sm10,sm10w,sm11,sm11w,sm12,sm12w,sm13,sm13w,sm14,sm14w,
sm15,sm15w,sm16,sm16w,sm17,sm17w,sm18,sm18w,sm19,sm19w,sm20,sm20w,sm21,sm21w,sm21wa,--24
                                sm22,sm22w,sm23,sm23w,sm23wa,sm24,sm24w,smend,smout
);--14
    signal mstate:ctl_state;
    signal mulina,mulinb,mulresult,r0,r1,r2,r3,r4,r5,r6,r7,r8,r9,r10:std_logic_
vector(162 downto 0);
    signal mulstart,muldone,mulbusy:std_logic;
```

```vhdl
component multiply IS
PORT(
        clk   : IN std_logic;
        reset : IN std_logic;
        load  : IN std_logic;

        a_in  : IN std_logic_vector(162 downto 0);
        b_in  : IN std_logic_vector(162 downto 0);
        result: OUT std_logic_vector(162 downto 0);

        done  : out std_logic;
        busy  : out std_logic
);
END component ;


begin

 multiplier: multiply
 port map(
                  clk=>clk,
                  reset=>'0',
                  load=>mulstart,
        a_in=>mulina,
        b_in=>mulinb,
        result=>mulresult,
              done=>muldone,
              busy=>mulbusy);

eccpmachine : process ( clk )
      variable m:integer range 0 to 162;
      begin
            if(rising_edge(CLK))then
               case mstate is
      when idle =>
      //r0:k  r1:x   r2:y  r3:X1  r4:Z1  r5:X2  r6:Z2  r7:T1  r8:T2  r9:T3  r10:T4
                    if (pmulstart = '1') then
                         r1<=x1;
                         r2<=y1;
                         r0<=k;
                             mbusy<='1';
                             mdone<='0';
                             mstate<=init;
                      else
                             mstate <= idle;
                             mdone<='0';
                             mbusy<='0';
                      end if;
                when init =>
```

```
//3.r3:X1=x    r4:Z1=1    r5:X2=x^4+b    r6:Z2=x^2
            r0<=x1;
            r3<=r1;
            r4<=x"ffffffffffffffffffffffffffffffffffffffff"&"111";
            r5<=not(r1(160 downto 0)&r1(162 downto 161));
            r6<=r1(161 downto 0) & r1(162);
            m:=162;
            mstate<=init1;
    when init1=>
            mstate<=highbit;
    when highbit=>
                            r2<=x1;
            if r0(m)='1' then
                    mstate<=highbitjmp;
            else
                    m:=m-1;
                    mstate<=highbitzero;
            end if;
    when highbitzero=>
            if m=0 then
                    r5<=x"0000000000000000000000000000000000000000"&"000";
                    mstate<=smout;
            else
                    m:=m-1;
                    mstate<=highbit;
            end if;
    when highbitjmp=>
            if m=0 then
                    mstate<=sm6;
            else
                    m:=m-1;
                    mstate<=sm1;
            end if;
when sm1 =>--//1.T1=X1*Z2
     mulina<=r3;
     mulinb<=r6;
     mulstart<='1';
     mstate<=sm1w;
when sm1w =>
     mulstart<='0';
     if muldone='1' then
         mstate<=sm2;
     else
             mstate<=sm1w;
     end if;--//1.T1=X1*Z2
when sm2 =>
     r7<=mulresult;--上次乘法结果放入目的寄存器中
     mulina<=r5;
     mulinb<=r4;
```

```
            mulstart<='1';
            mstate<=sm2w;
      when sm2w =>
            mulstart<='0';
            if muldone='1' then
                  mstate<=sm3;
            else
                  mstate<=sm2w;
            end if; --//2.T2=X2*Z1
      when sm3 =>
            r8<=mulresult;
            mulina<=r7;
            mulinb<=mulresult;
            mulstart<='1';
            mstate<=sm3a;--//3.T3=T1*T2
      when sm3a =>
            mulstart<='0';
            r10<=r7 xor r8;
            mstate<=sm3b;--//3a.T4=T1+T2
      when sm3b =>
            r10<=r10(161 downto 0) & r10(162);
            mstate<=sm3w;--//3b.T4=T4^2
      when sm3w =>
            if muldone='1' then
                  mstate<=sm4;
            else
                  mstate<=sm3w;
            end if;--//3.T3=T1*T2
      when sm4 =>
            r9<=mulresult;
            mulina<=r1;
            mulinb<=r10;
            mulstart<='1';
            mstate<=sm4w;
      when sm4w =>
            mulstart<='0';
            if muldone='1' then
                  mstate<=sm4wa;
            else
                  mstate<=sm4w;
            end if;--//4.T1=x*T4
      when sm4wa =>
            r7<=mulresult;
            r8<=mulresult xor r9;
            mstate<=sm4put;--//4wa.T2=T1+T3
      when sm4put =>
            if r0(m)='1' then
                  r3<=r8;
                  r4<=r10;
```

```
                    r7<=r5(161 downto 0) & r5(162);
                    r9<=r6(161 downto 0) & r6(162);
            else
                    r5<=r8;
                    r6<=r10;
                    r7<=r3(161 downto 0) & r3(162);
                    r9<=r4(161 downto 0) & r4(162);
            end if;
            mstate<=sm5;
    when sm5 =>
            mulina<=r7;
            mulinb<=r9;
            mulstart<='1';
            mstate<=sm5a;--//5.T2=T1*T3
    when sm5a =>
            mulstart<='0';
            r7<=r7(161 downto 0) & r7(162);
            r9<=r9(161 downto 0) & r9(162);
            mstate<=sm5b;--//5a.T1=T1^2,T3=T3^2
    when sm5b =>
            r10<=r7 xor r9;
            mstate<=sm5w;--//5b.T4=T1+T3
    when sm5w =>
             if muldone='1' then
                    mstate<=sm5put;
             else
                    mstate<=sm5w;
             end if;
    when sm5put=>
            r8<=mulresult;
            if r0(m)='1' then
                    r6<=mulresult;
                    r5<=r10;
            else
                    r4<=mulresult;
                    r3<=r10;
            end if;
            if m=0 then
                    mstate<=sm6;
            else
                    m:=m-1;
                    mstate<=sm1;
            end if;

    when sm6=>--3.T1=x,4.T2=y,5.T3=Z1*Z2
            mulina<=r4;
            mulinb<=r6;
            r7<=r1;
            r8<=r2;
```

```
              mulstart<='1';
              mstate<=sm6w;
       when sm6w=>
              mulstart<='0';
              if muldone='1' then
                     mstate<=sm7;
              else
                     mstate<=sm6w;
              end if;
       when sm7=>--6.Z1=Z1*T1
              r9<=mulresult;
              mulina<=r4;
              mulinb<=r7;
              mulstart<='1';
              mstate<=sm7w;
       when sm7w=>
              mulstart<='0';
              if muldone='1' then
                     mstate<=sm8;
              else
                     mstate<=sm7w;
              end if;
       when sm8=>
              r4<=mulresult xor r3;--7.Z1=Z1+X1
              mulina<=r6;
              mulinb<=r7;
              mulstart<='1';
              mstate<=sm8w;--8.Z2=Z2*T1
       when sm8w=>
              mulstart<='0';
              if muldone='1' then
                     mstate<=sm9;
              else
                     mstate<=sm8w;
              end if;
       when sm9=>
              r6<=mulresult;
              mulina<=mulresult;
              mulinb<=r3;
              mulstart<='1';
              mstate<=sm9w;--9.X1=Z2*X1
       when sm9w=>
              mulstart<='0';
              if muldone='1' then
                     r6<=r6 xor r5;--10.Z2=Z2+X2
                     mstate<=sm10;
              else
                     mstate<=sm9w;
              end if;
```

```
                when sm10=>--11.Z2=Z2*Z1
                        r3<=mulresult;
                        mulina<=r6;
                        mulinb<=r4;
                        mulstart<='1';
                        r10<=r7(161 downto 0) & r7(162);--12.T4=T1^2
                        mstate<=sm10w;
                when sm10w=>
                        mulstart<='0';
                        if muldone='1' then
                                r10<=r10 xor r8;--13.T4=T4+T2
                                mstate<=sm11;
                        else
                                mstate<=sm10w;
                        end if;
                when sm11=>
                        r6<=mulresult;
                        mulina<=r10;
                        mulinb<=r9;
                        mulstart<='1';
                        mstate<=sm11w;--14.T4=T4*T3
                when sm11w=>
                        mulstart<='0';
                        if muldone='1' then
                                mstate<=sm12;
                        else
                                mstate<=sm11w;
                        end if;
                when sm12=>
                        --r10<=mulresult;
                        r10<=mulresult xor r6;--15.T4=T4+Z2
                        mulina<=r9;
                        mulinb<=r7;
                        mulstart<='1';--16T3=T3*T1
                        mstate<=sm12w;
                when sm12w=>
                        mulstart<='0';
                        if muldone='1' then
                                mstate<=sm13;
                        else
                                mstate<=sm12w;
                        end if;
--for 17.inverse(T3) inv(a)
                when sm13=>
                        r9<=mulresult;
                        mulina<=mulresult;
                        mulinb<=mulresult(161 downto 0)&mulresult(162);
                        mulstart<='1';--1.u1=a*a^2
                        mstate<=sm13w;
```

```
when sm13w=>
        mulstart<='0';
        if muldone='1' then
                mstate<=sm14;
        else
                mstate<=sm13w;
        end if;
when sm14=>
        r1<=mulresult;
        mulina<=mulresult;
        mulinb<=mulresult(160 downto 0)&mulresult(162 downto 161);
        mulstart<='1';
        mstate<=sm14w;--2.u2=(u1*u1^2)^2
when sm14w=>
        mulstart<='0';
        if muldone='1' then
                mstate<=sm15;
        else
                mstate<=sm14w;
        end if;
when sm15=>--3.u1=a*u2^2
        r2<=mulresult;
        mulina<=r9;
        mulinb<=mulresult(161 downto 0)&mulresult(162);
        mulstart<='1';
        mstate<=sm15w;
when sm15w=>
        mulstart<='0';
        if muldone='1' then
                mstate<=sm16;
        else
                mstate<=sm15w;
        end if;
when sm16=>--4.u2=u1*(u1^2)^5
        r1<=mulresult;
        mulina<=mulresult;
        mulinb<=mulresult(157 downto 0)&mulresult(162 downto 158);
        mulstart<='1';
        mstate<=sm16w;
when sm16w=>
        mulstart<='0';
        if muldone='1' then
                mstate<=sm17;
        else
                mstate<=sm16w;
        end if;
when sm17=>--5.u1=u2*(u2^2)^10
        r2<=mulresult;
        mulina<=mulresult;
```

```
                    mulinb<=mulresult(152 downto 0)&mulresult(162 downto 153);
                    mulstart<='1';
                    mstate<=sm17w;
            when sm17w=>
                    mulstart<='0';
                    if muldone='1' then
                            mstate<=sm18;
                    else
                            mstate<=sm17w;
                    end if;
            when sm18=>--6.u2=u1*(u1^2)^20
                    r1<=mulresult;
                    mulina<=mulresult;
                    mulinb<=mulresult(142 downto 0)&mulresult(162 downto 143);
                    mulstart<='1';
                    mstate<=sm18w;
            when sm18w=>
                    mulstart<='0';
                    if muldone='1' then
                            mstate<=sm19;
                    else
                            mstate<=sm18w;
                    end if;
            when sm19=>--7.u1=u2*(u2^2)^40
                    r2<=mulresult;
                    mulina<=mulresult;
                    mulinb<=mulresult(122 downto 0)&mulresult(162 downto 123);
                    mulstart<='1';
                    mstate<=sm19w;
            when sm19w=>
                    mulstart<='0';
                    if muldone='1' then
                            mstate<=sm20;
                    else
                            mstate<=sm19w;
                    end if;
            when sm20=>--8.u2=a*u1^2
                    r1<=mulresult;
                    mulina<=r9;
                    mulinb<=mulresult(161 downto 0)&mulresult(162);
                    mulstart<='1';
                    mstate<=sm20w;
            when sm20w=>
                    mulstart<='0';
                    if muldone='1' then
                            mstate<=sm21;
                    else
                            mstate<=sm20w;
                    end if;
```

```
                when sm21=>--9.u1=u2*(u2^2)^81
                        r2<=mulresult;
                        mulina<=mulresult;
                        mulinb<=mulresult(81 downto 0)&mulresult(162 downto 82);
                        mulstart<='1';
                        mstate<=sm21w;
                when sm21w=>
                        mulstart<='0';
                        if muldone='1' then
                                mstate<=sm21wa;
                        else
                                mstate<=sm21w;
                        end if;
                when sm21wa=>--10.a^(-1)=u1^2
                        r1<=mulresult;
                        r9<=mulresult(161 downto 0) & mulresult(162);
                        mstate<=sm22;
--end for 17.inverse(T3)
                when sm22=>--18.T4=T3*T4
                        mulina<=r9;
                        mulinb<=r10;
                        mulstart<='1';
                        mstate<=sm22w;
                when sm22w=>
                        mulstart<='0';
                        if muldone='1' then
                                mstate<=sm23;
                        else
                                mstate<=sm22w;
                        end if;
                when sm23=>--19.X2=X1*T3
                        r10<=mulresult;
                        mulina<=r3;
                        mulinb<=r9;
                        mulstart<='1';
                        mstate<=sm23w;
                when sm23w=>
                        mulstart<='0';
                        if muldone='1' then
                                mstate<=sm23wa;
                        else
                                mstate<=sm23w;
                        end if;
--     //r0:k   r1:x   r2:y   r3:X1   r4:Z1   r5:X2   r6:Z2   r7:T1   r8:T2   r9:T3   r10:T4
                when sm23wa=>--20.Z2=X2+T1
                        r5<=mulresult;
                        r6<=mulresult xor r7;
```

```
                    mstate<=sm24;
        when sm24=>--21.Z2=Z2*T4
            mulina<=r6;
            mulinb<=r10;
            mulstart<='1';
            mstate<=sm24w;
        when sm24w=>
            mulstart<='0';
            if muldone='1' then
                    mstate<=smend;
            else
                    mstate<=sm24w;
            end if;
        when smend=>--22.Z2=Z2+T2
            r6<=mulresult;
            r6<=mulresult xor r8;
            y2<=r5;
            mdone<='1';
            mstate<=smout;

        when smout=>
            y2<=r6;
            mdone<='0';
            mbusy<='0';
            mstate<=idle;

        when others =>
            mstate <= idle;

        end case;
    end if;
end process;

end marc;


---f_function.vhd
LIBRARY IEEE;
USE IEEE.std_logic_1164.all;



ENTITY f_function IS
    PORT(
            clk :       in std_logic;
        a_in,b_in :     IN std_logic_vector(162 downto 0);
          d_out:        OUT std_logic_vector(162 downto 0)
    );
END f_function;

ARCHITECTURE arch OF f_function IS
```

```vhdl
        signal d_out_s : std_logic_vector(162 downto 0);


BEGIN
process(clk)
      variable f: bit_vector(81 downto 0);
      variable a: bit_vector(162 downto 0);
      variable b: bit_vector(162 downto 0);
      variable d: bit_vector(162 downto 0);

      begin
       if(rising_edge(clk))then

                a := to_bitvector(a_in);
                b := to_bitvector(b_in);

                f(0):=a(161) and b(161);
                f(1):=(a(162) and b(0)) xor (a(0) and b(162));
                f(2):=(a(162) and b(1)) xor (a(1) and b(162));
                f(3):=(a(162) and b(2)) xor (a(2) and b(162));
                f(4):=(a(162) and b(3)) xor (a(3) and b(162));
                f(5):=(a(162) and b(4)) xor (a(4) and b(162));
                f(6):=(a(162) and b(5)) xor (a(5) and b(162));
                f(7):=(a(162) and b(6)) xor (a(6) and b(162));
                f(8):=(a(162) and b(7)) xor (a(7) and b(162));
                f(9):=(a(162) and b(8)) xor (a(8) and b(162));
                f(10):=(a(162) and b(9)) xor (a(9) and b(162));
                f(11):=(a(162) and b(10)) xor (a(10) and b(162));
                f(12):=(a(162) and b(11)) xor (a(11) and b(162));
                f(13):=(a(162) and b(12)) xor (a(12) and b(162));
                f(14):=(a(162) and b(13)) xor (a(13) and b(162));
                f(15):=(a(162) and b(14)) xor (a(14) and b(162));
                f(16):=(a(162) and b(15)) xor (a(15) and b(162));
                f(17):=(a(162) and b(16)) xor (a(16) and b(162));
                f(18):=(a(162) and b(17)) xor (a(17) and b(162));
                f(19):=(a(162) and b(18)) xor (a(18) and b(162));
                f(20):=(a(162) and b(19)) xor (a(19) and b(162));
                f(21):=(a(162) and b(20)) xor (a(20) and b(162));
                f(22):=(a(162) and b(21)) xor (a(21) and b(162));
                f(23):=(a(162) and b(22)) xor (a(22) and b(162));
                f(24):=(a(162) and b(23)) xor (a(23) and b(162));
                f(25):=(a(162) and b(24)) xor (a(24) and b(162));
                f(26):=(a(162) and b(25)) xor (a(25) and b(162));
                f(27):=(a(162) and b(26)) xor (a(26) and b(162));
                f(28):=(a(162) and b(27)) xor (a(27) and b(162));
                f(29):=(a(162) and b(28)) xor (a(28) and b(162));
                f(30):=(a(162) and b(29)) xor (a(29) and b(162));
                f(31):=(a(162) and b(30)) xor (a(30) and b(162));
```

```
f(32):=(a(162) and b(31)) xor (a(31) and b(162));
f(33):=(a(162) and b(32)) xor (a(32) and b(162));
f(34):=(a(162) and b(33)) xor (a(33) and b(162));
f(35):=(a(162) and b(34)) xor (a(34) and b(162));
f(36):=(a(162) and b(35)) xor (a(35) and b(162));
f(37):=(a(162) and b(36)) xor (a(36) and b(162));
f(38):=(a(162) and b(37)) xor (a(37) and b(162));
f(39):=(a(162) and b(38)) xor (a(38) and b(162));
f(40):=(a(162) and b(39)) xor (a(39) and b(162));
f(41):=(a(162) and b(40)) xor (a(40) and b(162));
f(42):=(a(162) and b(41)) xor (a(41) and b(162));
f(43):=(a(162) and b(42)) xor (a(42) and b(162));
f(44):=(a(162) and b(43)) xor (a(43) and b(162));
f(45):=(a(162) and b(44)) xor (a(44) and b(162));
f(46):=(a(162) and b(45)) xor (a(45) and b(162));
f(47):=(a(162) and b(46)) xor (a(46) and b(162));
f(48):=(a(162) and b(47)) xor (a(47) and b(162));
f(49):=(a(162) and b(48)) xor (a(48) and b(162));
f(50):=(a(162) and b(49)) xor (a(49) and b(162));
f(51):=(a(162) and b(50)) xor (a(50) and b(162));
f(52):=(a(162) and b(51)) xor (a(51) and b(162));
f(53):=(a(162) and b(52)) xor (a(52) and b(162));
f(54):=(a(162) and b(53)) xor (a(53) and b(162));
f(55):=(a(162) and b(54)) xor (a(54) and b(162));
f(56):=(a(162) and b(55)) xor (a(55) and b(162));
f(57):=(a(162) and b(56)) xor (a(56) and b(162));
f(58):=(a(162) and b(57)) xor (a(57) and b(162));
f(59):=(a(162) and b(58)) xor (a(58) and b(162));
f(60):=(a(162) and b(59)) xor (a(59) and b(162));
f(61):=(a(162) and b(60)) xor (a(60) and b(162));
f(62):=(a(162) and b(61)) xor (a(61) and b(162));
f(63):=(a(162) and b(62)) xor (a(62) and b(162));
f(64):=(a(162) and b(63)) xor (a(63) and b(162));
f(65):=(a(162) and b(64)) xor (a(64) and b(162));
f(66):=(a(162) and b(65)) xor (a(65) and b(162));
f(67):=(a(162) and b(66)) xor (a(66) and b(162));
f(68):=(a(162) and b(67)) xor (a(67) and b(162));
f(69):=(a(162) and b(68)) xor (a(68) and b(162));
f(70):=(a(162) and b(69)) xor (a(69) and b(162));
f(71):=(a(162) and b(70)) xor (a(70) and b(162));
f(72):=(a(162) and b(71)) xor (a(71) and b(162));
f(73):=(a(162) and b(72)) xor (a(72) and b(162));
f(74):=(a(162) and b(73)) xor (a(73) and b(162));
f(75):=(a(162) and b(74)) xor (a(74) and b(162));
f(76):=(a(162) and b(75)) xor (a(75) and b(162));
f(77):=(a(162) and b(76)) xor (a(76) and b(162));
f(78):=(a(162) and b(77)) xor (a(77) and b(162));
f(79):=(a(162) and b(78)) xor (a(78) and b(162));
f(80):=(a(162) and b(79)) xor (a(79) and b(162));
```

```
f(81):=(a(162) and b(80)) xor (a(80) and b(162));

d(0) := f(0) xor f(1);
d(1) := f(13);
d(2) := '0';
d(3) := f(9) xor f(71);
d(4) := f(40);
d(5) := f(17) xor f(60);
d(6) := f(22);
d(7) := f(21) xor f(43) xor f(58);
d(8) := f(33) xor f(61);
d(9) := f(3) xor f(20) xor f(73);
d(10) := f(35) xor f(63) xor f(77);
d(11) := f(54);
d(12) := '0';
d(13) := f(1) xor f(51);
d(14) := f(54);
d(15) := f(60);
d(16) := f(37) xor f(59) xor f(64);
d(17) := f(5) xor f(55) xor f(79);
d(18) := f(20);
d(19) := f(72) xor f(74);
d(20) := f(9) xor f(18);
d(21) := f(7) xor f(63);
d(22) := f(6) xor f(45) xor f(61) xor f(68);
d(23) := f(43);
d(24) := f(32);
d(25) := '0';
d(26) := f(30) xor f(36);
d(27) := f(33) xor f(45) xor f(61);
d(28) := f(47) xor f(49) xor f(57) xor f(76);
d(29) := f(35) xor f(63) xor f(74);
d(30) := f(26) xor f(56);
d(31) := f(32);
d(32) := f(24) xor f(31) xor f(44);
d(33) := f(8) xor f(27) xor f(44);
d(34) := f(49);
d(35) := f(10) xor f(29) xor f(50);
d(36) := f(26);
d(37) := f(16) xor f(61) xor f(68);
d(38) := f(41) xor f(79);
d(39) := f(47) xor f(76);
d(40) := f(4) xor f(72);
d(41) := f(38) xor f(71);
d(42) := f(47) xor f(57);
d(43) := f(7) xor f(23);
d(44) := f(32) xor f(33) xor f(67);
d(45) := f(22) xor f(27);
d(46) := f(47) xor f(48);
```

```
d(47):= f(28) xor f(39) xor f(42) xor f(46);
d(48):= f(46) xor f(72);
d(49):= f(28) xor f(34) xor f(80);
d(50):= f(35) xor f(77) xor f(81);
d(51):= f(13);
d(52):= f(54) xor f(66);
d(53):= f(65);
d(54):= f(11) xor f(14) xor f(52) xor f(78);
d(55):= f(17);
d(56):= f(30) xor f(75);
d(57):= f(28) xor f(42);
d(58):= f(7) xor f(63);
d(59):= f(16);
d(60):= f(5) xor f(15);
d(61):= f(8) xor f(22) xor f(27) xor f(37);
d(62):= f(73) xor f(76);
d(63):= f(10) xor f(21) xor f(29) xor f(58);
d(64):= f(16) xor f(68);
d(65):= f(53);
d(66):= f(52);
d(67):= f(44);
d(68):= f(22) xor f(37) xor f(64);
d(69):= '0';
d(70):= f(79);
d(71):= f(3) xor f(41) xor f(73);
d(72):= f(19) xor f(40) xor f(48);
d(73):= f(9) xor f(62) xor f(71) xor f(80);
d(74):= f(19) xor f(29) xor f(77);
d(75):= f(56);
d(76):= f(28) xor f(39) xor f(62) xor f(80);
d(77):= f(10) xor f(50) xor f(74);
d(78):= f(54);
d(79):= f(17) xor f(38) xor f(70);
d(80):= f(49) xor f(73) xor f(76);
d(81):= f(50);
d(82):= f(13);
d(83):= f(13);
d(84):= f(26);
d(85):= f(25) xor f(33) xor f(67);
d(86):= f(30);
d(87):= f(4);
d(88):= f(17) xor f(65) xor f(70);
d(89):= f(3) xor f(20);
d(90):= f(7);
d(91):= f(25);
d(92):= f(2) xor f(75);
d(93):= f(9) xor f(18) xor f(34) xor f(80);
d(94):= f(35) xor f(50) xor f(74) xor f(81);
d(95):= '0';
```

```
d(96):= f(18) xor f(56);
d(97):= f(42) xor f(46) xor f(72);
d(98):= f(23) xor f(39) xor f(62);
d(99):= f(4) xor f(58);
d(100):= '0';
d(101):= f(11) xor f(14) xor f(36);
d(102):= '0';
d(103):= f(52) xor f(78);
d(104):= f(65) xor f(69) xor f(70);
d(105):= f(5) xor f(41) xor f(79);
d(106):= f(49) xor f(57);
d(107):= f(19) xor f(40) xor f(77);
d(108):= f(66);
d(109):= f(24);
d(110):= f(12);
d(111):= f(2) xor f(14) xor f(51) xor f(78);
d(112):= f(60) xor f(66);
d(113):= f(27) xor f(31) xor f(44);
d(114):= f(31) xor f(57);
d(115):= f(24);
d(116):= '0';
d(117):= f(1) xor f(2) xor f(51);
d(118):= '0';
d(119):= f(23) xor f(69);
d(120):= '0';
d(121):= f(5) xor f(15) xor f(55);
d(122):= f(30) xor f(64);
d(123):= f(32) xor f(67);
d(124):= f(8) xor f(37) xor f(59);
d(125):= f(3) xor f(41);
d(126):= f(24) xor f(31);
d(127):= f(65);
d(128):= f(15) xor f(59);
d(129):= f(15) xor f(59);
d(130):= f(11) xor f(52);
d(131):= f(12);
d(132):= f(1);
d(133):= f(26) xor f(56);
d(134):= f(6) xor f(34) xor f(45);
d(135):= f(19) xor f(21) xor f(29) xor f(48);
d(136):= f(6) xor f(18) xor f(34);
d(137):= f(4) xor f(10) xor f(58);
d(138):= f(60) xor f(66);
d(139):= f(8);
d(140):= f(20) xor f(75);
d(141):= f(23) xor f(39) xor f(46);
d(142):= f(42);
d(143):= f(69);
d(144):= f(53) xor f(55);
```

```vhdl
                    d(145):= f(2) xor f(75) xor f(78);
                    d(146):= f(38) xor f(62) xor f(71);
                    d(147):= f(21) xor f(43) xor f(48);
                    d(148):= f(45);
                    d(149):= f(40);
                    d(150):= f(38) xor f(69) xor f(70);
                    d(151):= '0';
                    d(152):= f(43);
                    d(153):= f(25) xor f(53) xor f(67);
                    d(154):= f(11) xor f(64);
                    d(155):= f(25) xor f(53);
                    d(156):= f(14) xor f(36) xor f(51);
                    d(157):= f(16);
                    d(158):= f(12) xor f(55);
                    d(159):= f(36);
                    d(160):= f(6) xor f(68);
                    d(161):= '0';
                    d(162):= f(12);

                    d_out_s<=To_StdLogicVector(d);

            end if;
    end process;
        d_out <= d_out_s;
end arch;
--multiply.vhd
LIBRARY IEEE;
USE IEEE.std_logic_1164.all;


ENTITY f_function IS
        PORT(
                clk :       in std_logic;
          a_in,b_in :  IN std_logic_vector(162 downto 0);
                d_out:     OUT std_logic_vector(162 downto 0)
        );
END f_function;

ARCHITECTURE arch OF f_function IS
        signal d_out_s : std_logic_vector(162 downto 0);




BEGIN
process(clk)
        variable f: bit_vector(81 downto 0);
        variable a: bit_vector(162 downto 0);
        variable b: bit_vector(162 downto 0);
        variable d: bit_vector(162 downto 0);
```

```vhdl
begin
  if(rising_edge(clk))then

            a := to_bitvector(a_in);
            b := to_bitvector(b_in);

            f(0):=a(161) and b(161);
            f(1):=(a(162) and b(0)) xor (a(0) and b(162));
            f(2):=(a(162) and b(1)) xor (a(1) and b(162));
            f(3):=(a(162) and b(2)) xor (a(2) and b(162));
            f(4):=(a(162) and b(3)) xor (a(3) and b(162));
            f(5):=(a(162) and b(4)) xor (a(4) and b(162));
            f(6):=(a(162) and b(5)) xor (a(5) and b(162));
            f(7):=(a(162) and b(6)) xor (a(6) and b(162));
            f(8):=(a(162) and b(7)) xor (a(7) and b(162));
            f(9):=(a(162) and b(8)) xor (a(8) and b(162));
            f(10):=(a(162) and b(9)) xor (a(9) and b(162));
            f(11):=(a(162) and b(10)) xor (a(10) and b(162));
            f(12):=(a(162) and b(11)) xor (a(11) and b(162));
            f(13):=(a(162) and b(12)) xor (a(12) and b(162));
            f(14):=(a(162) and b(13)) xor (a(13) and b(162));
            f(15):=(a(162) and b(14)) xor (a(14) and b(162));
            f(16):=(a(162) and b(15)) xor (a(15) and b(162));
            f(17):=(a(162) and b(16)) xor (a(16) and b(162));
            f(18):=(a(162) and b(17)) xor (a(17) and b(162));
            f(19):=(a(162) and b(18)) xor (a(18) and b(162));
            f(20):=(a(162) and b(19)) xor (a(19) and b(162));
            f(21):=(a(162) and b(20)) xor (a(20) and b(162));
            f(22):=(a(162) and b(21)) xor (a(21) and b(162));
            f(23):=(a(162) and b(22)) xor (a(22) and b(162));
            f(24):=(a(162) and b(23)) xor (a(23) and b(162));
            f(25):=(a(162) and b(24)) xor (a(24) and b(162));
            f(26):=(a(162) and b(25)) xor (a(25) and b(162));
            f(27):=(a(162) and b(26)) xor (a(26) and b(162));
            f(28):=(a(162) and b(27)) xor (a(27) and b(162));
            f(29):=(a(162) and b(28)) xor (a(28) and b(162));
            f(30):=(a(162) and b(29)) xor (a(29) and b(162));
            f(31):=(a(162) and b(30)) xor (a(30) and b(162));
            f(32):=(a(162) and b(31)) xor (a(31) and b(162));
            f(33):=(a(162) and b(32)) xor (a(32) and b(162));
            f(34):=(a(162) and b(33)) xor (a(33) and b(162));
            f(35):=(a(162) and b(34)) xor (a(34) and b(162));
            f(36):=(a(162) and b(35)) xor (a(35) and b(162));
            f(37):=(a(162) and b(36)) xor (a(36) and b(162));
            f(38):=(a(162) and b(37)) xor (a(37) and b(162));
            f(39):=(a(162) and b(38)) xor (a(38) and b(162));
            f(40):=(a(162) and b(39)) xor (a(39) and b(162));
            f(41):=(a(162) and b(40)) xor (a(40) and b(162));
```

```
f(42):=(a(162) and b(41)) xor (a(41) and b(162));
f(43):=(a(162) and b(42)) xor (a(42) and b(162));
f(44):=(a(162) and b(43)) xor (a(43) and b(162));
f(45):=(a(162) and b(44)) xor (a(44) and b(162));
f(46):=(a(162) and b(45)) xor (a(45) and b(162));
f(47):=(a(162) and b(46)) xor (a(46) and b(162));
f(48):=(a(162) and b(47)) xor (a(47) and b(162));
f(49):=(a(162) and b(48)) xor (a(48) and b(162));
f(50):=(a(162) and b(49)) xor (a(49) and b(162));
f(51):=(a(162) and b(50)) xor (a(50) and b(162));
f(52):=(a(162) and b(51)) xor (a(51) and b(162));
f(53):=(a(162) and b(52)) xor (a(52) and b(162));
f(54):=(a(162) and b(53)) xor (a(53) and b(162));
f(55):=(a(162) and b(54)) xor (a(54) and b(162));
f(56):=(a(162) and b(55)) xor (a(55) and b(162));
f(57):=(a(162) and b(56)) xor (a(56) and b(162));
f(58):=(a(162) and b(57)) xor (a(57) and b(162));
f(59):=(a(162) and b(58)) xor (a(58) and b(162));
f(60):=(a(162) and b(59)) xor (a(59) and b(162));
f(61):=(a(162) and b(60)) xor (a(60) and b(162));
f(62):=(a(162) and b(61)) xor (a(61) and b(162));
f(63):=(a(162) and b(62)) xor (a(62) and b(162));
f(64):=(a(162) and b(63)) xor (a(63) and b(162));
f(65):=(a(162) and b(64)) xor (a(64) and b(162));
f(66):=(a(162) and b(65)) xor (a(65) and b(162));
f(67):=(a(162) and b(66)) xor (a(66) and b(162));
f(68):=(a(162) and b(67)) xor (a(67) and b(162));
f(69):=(a(162) and b(68)) xor (a(68) and b(162));
f(70):=(a(162) and b(69)) xor (a(69) and b(162));
f(71):=(a(162) and b(70)) xor (a(70) and b(162));
f(72):=(a(162) and b(71)) xor (a(71) and b(162));
f(73):=(a(162) and b(72)) xor (a(72) and b(162));
f(74):=(a(162) and b(73)) xor (a(73) and b(162));
f(75):=(a(162) and b(74)) xor (a(74) and b(162));
f(76):=(a(162) and b(75)) xor (a(75) and b(162));
f(77):=(a(162) and b(76)) xor (a(76) and b(162));
f(78):=(a(162) and b(77)) xor (a(77) and b(162));
f(79):=(a(162) and b(78)) xor (a(78) and b(162));
f(80):=(a(162) and b(79)) xor (a(79) and b(162));
f(81):=(a(162) and b(80)) xor (a(80) and b(162));

d(0):= f(0) xor f(1);
d(1):= f(13);
d(2):= '0';
d(3):= f(9) xor f(71);
d(4):= f(40);
d(5):= f(17) xor f(60);
d(6):= f(22);
d(7):= f(21) xor f(43) xor f(58);
```

2.5 本章参考代码

```
d(8) := f(33) xor f(61);
d(9) := f(3) xor f(20) xor f(73);
d(10):= f(35) xor f(63) xor f(77);
d(11):= f(54);
d(12):= '0';
d(13):= f(1) xor f(51);
d(14):= f(54);
d(15):= f(60);
d(16):= f(37) xor f(59) xor f(64);
d(17):= f(5) xor f(55) xor f(79);
d(18):= f(20);
d(19):= f(72) xor f(74);
d(20):= f(9) xor f(18);
d(21):= f(7) xor f(63);
d(22):= f(6) xor f(45) xor f(61) xor f(68);
d(23):= f(43);
d(24):= f(32);
d(25):= '0';
d(26):= f(30) xor f(36);
d(27):= f(33) xor f(45) xor f(61);
d(28):= f(47) xor f(49) xor f(57) xor f(76);
d(29):= f(35) xor f(63) xor f(74);
d(30):= f(26) xor f(56);
d(31):= f(32);
d(32):= f(24) xor f(31) xor f(44);
d(33):= f(8) xor f(27) xor f(44);
d(34):= f(49);
d(35):= f(10) xor f(29) xor f(50);
d(36):= f(26);
d(37):= f(16) xor f(61) xor f(68);
d(38):= f(41) xor f(79);
d(39):= f(47) xor f(76);
d(40):= f(4) xor f(72);
d(41):= f(38) xor f(71);
d(42):= f(47) xor f(57);
d(43):= f(7) xor f(23);
d(44):= f(32) xor f(33) xor f(67);
d(45):= f(22) xor f(27);
d(46):= f(47) xor f(48);
d(47):= f(28) xor f(39) xor f(42) xor f(46);
d(48):= f(46) xor f(72);
d(49):= f(28) xor f(34) xor f(80);
d(50):= f(35) xor f(77) xor f(81);
d(51):= f(13);
d(52):= f(54) xor f(66);
d(53):= f(65);
d(54):= f(11) xor f(14) xor f(52) xor f(78);
d(55):= f(17);
d(56):= f(30) xor f(75);
```

```
d(57) := f(28) xor f(42);
d(58) := f(7) xor f(63);
d(59) := f(16);
d(60) := f(5) xor f(15);
d(61) := f(8) xor f(22) xor f(27) xor f(37);
d(62) := f(73) xor f(76);
d(63) := f(10) xor f(21) xor f(29) xor f(58);
d(64) := f(16) xor f(68);
d(65) := f(53);
d(66) := f(52);
d(67) := f(44);
d(68) := f(22) xor f(37) xor f(64);
d(69) := '0';
d(70) := f(79);
d(71) := f(3) xor f(41) xor f(73);
d(72) := f(19) xor f(40) xor f(48);
d(73) := f(9) xor f(62) xor f(71) xor f(80);
d(74) := f(19) xor f(29) xor f(77);
d(75) := f(56);
d(76) := f(28) xor f(39) xor f(62) xor f(80);
d(77) := f(10) xor f(50) xor f(74);
d(78) := f(54);
d(79) := f(17) xor f(38) xor f(70);
d(80) := f(49) xor f(73) xor f(76);
d(81) := f(50);
d(82) := f(13);
d(83) := f(13);
d(84) := f(26);
d(85) := f(25) xor f(33) xor f(67);
d(86) := f(30);
d(87) := f(4);
d(88) := f(17) xor f(65) xor f(70);
d(89) := f(3) xor f(20);
d(90) := f(7);
d(91) := f(25);
d(92) := f(2) xor f(75);
d(93) := f(9) xor f(18) xor f(34) xor f(80);
d(94) := f(35) xor f(50) xor f(74) xor f(81);
d(95) := '0';
d(96) := f(18) xor f(56);
d(97) := f(42) xor f(46) xor f(72);
d(98) := f(23) xor f(39) xor f(62);
d(99) := f(4) xor f(58);
d(100) := '0';
d(101) := f(11) xor f(14) xor f(36);
d(102) := '0';
d(103) := f(52) xor f(78);
d(104) := f(65) xor f(69) xor f(70);
d(105) := f(5) xor f(41) xor f(79);
```

```
d(106) := f(49) xor f(57);
d(107) := f(19) xor f(40) xor f(77);
d(108) := f(66);
d(109) := f(24);
d(110) := f(12);
d(111) := f(2) xor f(14) xor f(51) xor f(78);
d(112) := f(60) xor f(66);
d(113) := f(27) xor f(31) xor f(44);
d(114) := f(31) xor f(57);
d(115) := f(24);
d(116) := '0';
d(117) := f(1) xor f(2) xor f(51);
d(118) := '0';
d(119) := f(23) xor f(69);
d(120) := '0';
d(121) := f(5) xor f(15) xor f(55);
d(122) := f(30) xor f(64);
d(123) := f(32) xor f(67);
d(124) := f(8) xor f(37) xor f(59);
d(125) := f(3) xor f(41);
d(126) := f(24) xor f(31);
d(127) := f(65);
d(128) := f(15) xor f(59);
d(129) := f(15) xor f(59);
d(130) := f(11) xor f(52);
d(131) := f(12);
d(132) := f(1);
d(133) := f(26) xor f(56);
d(134) := f(6) xor f(34) xor f(45);
d(135) := f(19) xor f(21) xor f(29) xor f(48);
d(136) := f(6) xor f(18) xor f(34);
d(137) := f(4) xor f(10) xor f(58);
d(138) := f(60) xor f(66);
d(139) := f(8);
d(140) := f(20) xor f(75);
d(141) := f(23) xor f(39) xor f(46);
d(142) := f(42);
d(143) := f(69);
d(144) := f(53) xor f(55);
d(145) := f(2) xor f(75) xor f(78);
d(146) := f(38) xor f(62) xor f(71);
d(147) := f(21) xor f(43) xor f(48);
d(148) := f(45);
d(149) := f(40);
d(150) := f(38) xor f(69) xor f(70);
d(151) := '0';
d(152) := f(43);
d(153) := f(25) xor f(53) xor f(67);
d(154) := f(11) xor f(64);
```

```vhdl
                    d(155):= f(25) xor f(53);
                    d(156):= f(14) xor f(36) xor f(51);
                    d(157):= f(16);
                    d(158):= f(12) xor f(55);
                    d(159):= f(36);
                    d(160):= f(6) xor f(68);
                    d(161):= '0';
                    d(162):= f(12);

                    d_out_s<=To_StdLogicVector(d);

          end if;
    end process;
         d_out <= d_out_s;
end arch;
---shirt_16.vhd
LIBRARY IEEE;
USE IEEE.std_logic_1164.all;


ENTITY shift_16 IS
       PORT(
              clk     : in std_logic;
           input    : IN std_logic_vector(162 downto 0);
              output  : OUT std_logic_vector(162 downto 0)
              );
END shift_16;

ARCHITECTURE arch OF shift_16 IS

BEGIN
process(clk)
       begin
       if(rising_edge(clk))then
       output<=input(146 downto 0)& input(162 downto 147);
end if;
end process;

end arch;
--sum.vhd
LIBRARY IEEE;
USE IEEE.std_logic_1164.all;

ENTITY sum IS
       PORT(
              Clk            : in std_logic;
           crl            : IN std_logic;--最后一次等于0，其余等于1。
           input15        : IN std_logic_vector(162 downto 0);
           input14        : IN std_logic_vector(162 downto 0);
```

```vhdl
            input13          : IN std_logic_vector(162 downto 0);
            input12          : IN std_logic_vector(162 downto 0);
            input11          : IN std_logic_vector(162 downto 0);
            input10          : IN std_logic_vector(162 downto 0);
            input9           : IN std_logic_vector(162 downto 0);
            input8           : IN std_logic_vector(162 downto 0);
            input7           : IN std_logic_vector(162 downto 0);
            input6           : IN std_logic_vector(162 downto 0);
            input5           : IN std_logic_vector(162 downto 0);
            input4           : IN std_logic_vector(162 downto 0);
            input3           : IN std_logic_vector(162 downto 0);
            input2           : IN std_logic_vector(162 downto 0);
            input1           : IN std_logic_vector(162 downto 0);
            input0           : IN std_logic_vector(162 downto 0);
            input4_0         : IN std_logic_vector(162 downto 0);
            input_sum        : IN std_logic_vector(162 downto 0);
            output           : OUT std_logic_vector(162 downto 0)
            );
    END sum;

    ARCHITECTURE arch OF sum IS
            signal out_s : std_logic_vector(162 downto 0);
    BEGIN
            process(clk,crl,input0,input1,input2,input3,input4,input5,input6,input7
,input8,input9,input10,input11,input12,input13,input14,input15,input_sum)
                  variable  in0,in1,in2,in3,in4,in5,in6,in7,in8,in9,in10,in11,in12,
in13,in14,in15,in_sum,out_v: bit_vector(162 downto 0);
                  begin
                     if(rising_edge(clk))then

                         in0:= to_bitvector(input0);
                         in1:= to_bitvector(input1);
                         in2:= to_bitvector(input2);
                         in3:= to_bitvector(input3);
                         in4:= to_bitvector(input4);
                         in5:= to_bitvector(input5);
                         in6:= to_bitvector(input6);
                         in7:= to_bitvector(input7);
                         in8:= to_bitvector(input8);
                         in9:= to_bitvector(input9);
                         in10:= to_bitvector(input10);
                         in11:= to_bitvector(input11);
                         in12:= to_bitvector(input12);
                         in13:= to_bitvector(input13);
                         in14:= to_bitvector(input14);
                         in15:= to_bitvector(input15);
                         in_sum:= to_bitvector(input_sum);

                         in1:=in1 rol 1;
                         in2:=in2 rol 2;
                         in3:=in3 rol 3;
```

```
                in4:=in4 rol 4;
                in5:=in5 rol 5;
                in6:=in6 rol 6;
                in7:=in7 rol 7;
                in8:=in8 rol 8;
                in9:=in9 rol 9;
                in10:=in10 rol 10;
                in11:=in11 rol 11;
                in12:=in12 rol 12;
                in13:=in13 rol 13;
                in14:=in14 rol 14;
                in15:=in15 rol 15;
                in_sum:=in_sum rol 16;

                if(crl='1')then
                    out_v:=in0 xor in1 xor in2 xor in3 xor in4 xor in5 xor in6 xor
in7 xor in8 xor in9 xor in10 xor in11 xor in12 xor in13 xor in14 xor in15 xor in_sum;
                else
                    out_v:=in13 xor in14 xor in15 xor in_sum;
                end if;

                    out_s<=To_StdLogicVector(out_v);

        end if;
    end process;
        output<=out_s;
end arch;
```

2.6 本章小结

本章介绍了密码、密码学、密码体制的相关概念和知识，并描述了主要的密码算法以及信息安全、信息安全产业、云计算安全和相关网络安全设施，最后给出了 DES、AES、RSA、ECC 的软硬件实现的参考代码。

2.7 本章参考文献

[1] William Stallings. 密码编码学与网络教程——原理与实践（第四版）. 北京：电子工业出版社，2006.
[2] Shor P W. Polynomial-time algorithms for prime factorization and discrete logarithms on a quantum computer[J]. SIAM journal on computing, 1997, 26(5): 1484–1509.

[3] IEEE P1363/D13, Standard Specifications for Public Key Cryptography. November 1999,pp:93-94.

[4] Coppersmith D.. The Data Encryption Standard (DES) and its strength against attacks IBM journal of research and development, IBM, 1994, 38, 243-250.

[5] Daemen J., Rijmen V.. The design of Rijndael: AES--the advanced encryption standard Springer-Verlag New York Inc, 2002.

[6] Rivest, R.; Shamir A., Adleman L.. A method for obtaining digital signatures and public-key cryptosystems. Communications of the ACM, ACM, 1978, 21, 120-126.

[7] Diffie W., Hellman M.. New directions in cryptography[J]. IEEE Transactions on Information Theory, 1976, 22(6): 644-654.

[8] Koblitz N.. Elliptic curve cryptosystems[J]. Mathematics of Computation, 1987, 48(177): 203-209.

[9] Miller V.. Uses of elliptic curves in cryptography[A]. Advances in Cryptology—CRYPTO'85 [C]. UK: Springer-Verlag, 1986: 417-426.

[10] ElGamal T.. A public key cryptosystem and a signature scheme based on discrete logarithms. Advances in Cryptology, 1985, 10-18.

[11] Merkle R. Secrecy, authentication, and public key systems[D]. Ph.D. dissertation, Dept. of Electrical Engineering, Stanford Univ., 1979.

[12] Hofftein J., Pipher J., and Silverman J.H.. NTRU: a ring based public key cryptosystem[J]. Algorithmic number theory, Springer, 1998(1423): 267-288.

[13] McEliece R.J.. A public-key cryptosystem based on algebraic coding theory[J]. DSN Progress Report, 1978, 42(44): 114-116.

[14] Bleichenbacher D.. Chosen ciphertext attacks against protocols based on the RSA encryption standard PKCS 1. Advances in Cryptology—CRYPTO'98, 1998, 1-12.

第 3 章 芯片设计基础

3.1 数字电路基础

3.1.1 现场可编程逻辑门电路

现场可编辑逻辑门阵列的英文简称是 FPGA（Fielol-Programmable Gate Array），它是目前最流行的研究和工业领域的硬件工具之一。

可编程阵列逻辑 PAL 是 70 年代末由 MMI 公司率先推出的一种低密度，一次性可编程逻辑器件，第一个具有典型实际意义的可编程逻辑器件（PLD-Programmable Logic Device）。它采用双极型工艺制作，熔丝编程方式。

通用阵列逻辑 GAL 是 Lattice 在 PAL 的基础上设计出来的器件。GAL 首次在 PLD 上采用了 EEPROM 工艺，使得其具有电可擦除重复编程的特点，彻底解决了熔丝型可编程器件的一次可编程问题。GAL 在"与-或"阵列结构上沿用了 PAL 的与阵列可编程、或阵列固定的结构，但对 PAL 的 I/O 结构进行了较大的改进，在 GAL 的输出部分增加了输出逻辑宏单元 OLMC。

复杂可编辑逻辑器件 CPLD 是从 PAL 和 GAL 器件发展出来的器件，相对而言规模大，结构复杂，属于大规模集成电路范围。它是一种用户根据各自需要而自行构造逻辑功能的数字集成电路，其基本设计方法是借助集成开发软件平台，用原理图、硬件描述语言等方法，生成相应的目标文件，通过下载电缆将代码传送到目标芯片中，实现设计的数字系统 FPGA，它作为专用集成电路（Application Specific Integrated Circuit, ASIC）领域的一种半定制电路的器件，既解决了定制电路先天不足的问题，又克服了原有可编辑器件中的门电路数有限的缺点。FPGA 能够实现现场可编辑的能力是因为它由多个可编辑元件组成，可以被用来实现一些基本的逻辑门电路或者更复杂一些的组合功能。我们不需要介入芯片的布局布线和工艺问题就可以根据需要通过可编辑的连接把 FPGA 内部的逻辑块连接起来，随时改变其逻辑功能，使用起来很灵活。使用 FPGA 设计电路的流程由以下 3 部分组成：

1）使用原理图、硬件描述语言 Verilog HDL 或 VHDL 对要设计的数字系统建模；
2）运用 EDA 工具对建立的模型进行仿真和综合，生成基于一些标准库的网络表；
3）将综合后的结果配置到 FPGA 的芯片即可使用。

这也是现代集成电路设计验证的技术主流。主流的 FPGA 芯片分为两大类：

1）第一类侧重低成本应用，容量中等，性能可以满足一般的逻辑设计要求；
2）第二类侧重于高性能应用，容量大，性能可以满足各类高端应用。

主流的 FPGA 生产商是 Xilinx 和 Altera 公司，Xilinx 公司的 FPGA 的市场占有率更高。

1）Xilinx 公司的主要产品是 Virtex 系列，其中最经典的产品是 2002 年推出的 Virtex II，它被认为是高速低耗的最理想产品；最新的产品是 Virtex 7，是业界密度最高的 FPGA，比现在的 FPGA 都高出 2.5 倍。Xilinx 公司针对它的 FPGA 芯片开发的 EDA 工具是 Xilinx ISE Design Suite 系列工具。

2）Altera 公司的主要产品是 FLEX10K 系列、Cyclone 系列和 Stratix 系列，其中 Cyclone 系列是现今成本最低和经成品认证的 FPGA，Stratix 系列是 Altera 公司的高端产品。Altera 的 EP2S180F1020C5(Stratix II)则提供约 180K 个逻辑单元，大概相当于 500 万个可编程晶体管；EP2C35F484C8(Cyclone II)则有 35K 个逻辑单元，约相当 100 万门。Stratix IV 采用 40nm 工艺，高达 820K 个逻辑单元；Stratix V 则采用 28nm 工艺，高达 1000K 个逻辑单元。Altera 公司针对它的 FPGA 芯片开发的 EDA 工具是 Altera Quartus II 系列工具。

用于 FPGA 电路设计的仿真工具是 Mentor 公司的 ModelSim 系列工具。用户可对 FPGA 内部的逻辑模块和 I/O 模块重新配置，以实现用户的逻辑。因而也被用于对 CPU（或其他 ASIC）的模拟。用户对 FPGA 的编程数据放在 Flash 芯片中，通过上电加载到 FPGA 中，对其进行初始化；也可在线对其编程，实现系统在线重构，这一特性可以构建一个根据计算任务不同而实时定制的 CPU（或 ASIC）。

3.1.2 专用集成电路

专用集成电路的英文简称是 ASIC，它是根据特定的电路使用者的要求和特定电子系统的需要而设计和制造的电路。

ASIC 的主要特点是体积小、功耗低、可靠性高、性能高、保密性强和成本低等。在批量生产芯片时，ASIC 比通用集成电路（例如可编辑器件等）更加有优势。但是由于单个 ASIC 芯片的生产成本很高，如果出货量较小，采用 ASIC 在经济上不合算，这种情况下可以使用可编辑逻辑器件（如 FPGA）来作为目标硬件，以实现集成电路设计。

ASIC 分为全定制和半定制设计。

1）全定制 ASIC 设计需要设计者完成所有的电路设计，需要大量的人力和物力，虽然灵活性比较好但开发效率较为低下。如果设计的电路较为理想，全定制的设计能够比半定制的设计的运行速度更快。

2）半定制 ASIC 设计使用库里的标准逻辑单元，在进行电路设计时可以从标准逻辑单元库中选择使用多种逻辑单元（例如逻辑门电路、加法器、比较器、总线、存储器、乘法器、微控制器和 IP 核等）。这些逻辑单元已经布局完毕，而且它们的设计较为可靠，设计者可以方便地利用它们完成需要的系统设计。

用于 ASIC 电路设计的 EDA 工具是 Synopsys Design Vision DC，用于电路仿真的工具是 Mentor 公司的 ModelSim 系列工具。

3.1.3 硬件编程语言

用于设计硬件的编程语言一般是 Verilog HDL（Verilog Hardware Description Language）或 VHDL（Very-High-Speed Hardware Description Language）。Verilog HDL 和 VHDL 作为业界广泛认可和同为电气和电子工程师协会（Institute of Electrical and Electronics Engineers，IEEE）标准的硬件描述语言，有着各自的特点。

VHDL 又被称作超高速集成电路硬件描述语言，除了应用于 FPGA，它还在基于复杂可编辑逻辑器件和 ASIC 的数字系统设计中有着广泛的应用。它诞生于 1983 年，在 1987 年被美国国防部和 IEEE 确定为标准的硬件描述语言。然后，IEEE 发布了 VHDL 的第一个标准版本——IEEE 1076-1987（1987 年）。在 1987 年以后，它在电子设计行业得到了广泛的认同，各大电子设计自动化（Electronic Design Automation，EDA）公司先后推出了自己支持 VHDL 的 EDA 工具。在这之后，IEEE 又发布了 IEEE 1076-1993（1993 年）和 IEEE 1076-2000（2000 年）版本。

VHDL 主要用于描述数字系统的结构、行为、功能和接口。除了含有许多具有硬件特征的语句外，VHDL 的语言形式、描述风格以及语法十分类似于一般的计算机高级语言。VHDL 的程序结构特点是将一项工程设计或设计实体（可以是一个元件、一个电路模块或一个系统）分成外部（或称可视部分及端口）和内部（或称不可视部分），即涉及实体的内部功能和算法完成部分。在对一个设计实体定义了外部界面后，一旦其内部开发完成后，其他的设计就可以直接调用这个实体。这种将设计实体分成内外部分的概念是 VHDL 系统设计的基本点。

Verilog HDL 是一种用于描述和设计电子系统的硬件描述语言，它具有简明和高效的代码风格。由于 Verilog HDL 与 C 语言在语法上有相似之处，因此具有 C 语言基础的设计人员更容易掌握它。它是由民间商业公司的私有产品发展为 IEEE 标准的，因此在商用领域的市场占有量更大。

Verilog 是由 Gateway 设计自动化公司的工程师于 1983 年末创立的。当时，Gateway 设计自动化公司叫作自动集成设计系统（Automated Integrated Design Systems），1985 年公司将名字改成了前者。该公司的菲尔·莫比（Phil Moorby）完成了 Verilog 的主要设计工作。1990 年，Gateway 设计自动化被 Cadence 公司收购。

1990 年代初，开放 Verilog 国际（Open Verilog International, OVI）组织（即现在的 Accellera）成立，Verilog 面向公有领域开放。1992 年，该组织寻求将 Verilog 纳入 IEEE 标准。最终，Verilog 成为了 IEEE 1364-1995 标准，即通常所说的 Verilog-95。

设计人员在使用这个版本的 Verilog 的过程中发现了一些可改进之处。为了解决用户在使用此版本 Verilog 过程中反映的问题，Verilog 进行了修正和扩展，这部分内容后来再次被提交给 IEEE。这个扩展后的版本成为了 IEEE 1364-2001 标准，即通常所说的 Verilog-2001。Verilog-2001 是对 Verilog-95 的一个重大改进版本，它具备一些新的实用功能，例如敏感列表、多维数组、生成语句块、命名端口连接等。目前，Verilog-2001 是 Verilog 的最主流版本，被大多数商业电子设计自动化软件包支持。

2005 年，Verilog 再次进行了更新，即 IEEE 1364-2005 标准。该版本只是对上一版本的细微修正。这个版本还包括了一个相对独立的新部分，即 Verilog-AMS。这个扩展使得传统的 Verilog 可以对集成的模拟和混合信号系统进行建模，其中容易与 IEEE 1364-2005 标准混淆的是加强硬件验证语言特性的 SystemVerilog（IEEE 1800-2005 标准）。它是 Verilog-2005 的一个超集，是硬件描述语言、硬件验证语言（针对验证的需求，特别加强了面向对象特性）的一个集成。

2009 年，IEEE 1364-2005 和 IEEE 1800-2005 两个部分合并为 IEEE 1800-2009，成为了一个新的、统一的 SystemVerilog 硬件描述验证语言（Hardware Description and Verification Language, HDVL）。

3.1.4 有限状态机技术

有限状态机技术是一种专为时序逻辑电路设计而创建的建模技术，是设计时序逻辑电路的通用模型，在设计数字控制模块电路系统时非常有效。

如图 3-1 所示，有限状态机分为两部分：组合逻辑电路和时序逻辑电路。其中，组合逻辑电路的输入为当前状态 Pr_state 和外部输入信号，组合逻辑电路输出为下一个状态 Nx_state 和输出信号。时序逻辑电路的输入信号有 3 个：

1) 时钟信号 Clock；
2) 重置信号 Reset 和下一个状态 Nx_state；
3) 输出信号为当前状态 Pr_state。

有限状态机可以分为米勒型状态机和摩尔型状态机两种，可以由输入和输出的关系进行区分。

1) 米勒型状态机的输出信号是由当前状态和输入信号决定，它的输出信号会在其输入信号发生变化后立即改变。

2) 摩尔型状态机的输出信号只由当前状态来决定，其输出信号在时钟跳变后才变化，并且会在一个时钟周期内保持不变。

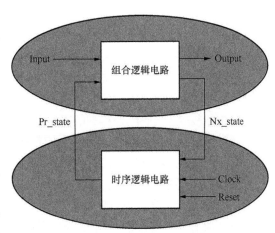

图 3-1 有限状态机结构图

基于状态机的电路从结构上可以分为两个部分，在代码结构上也可以分为两部分。时序逻辑电路部分的代码可以在 PROCESS 中进行顺序描述，对于组合逻辑部分可以并行地实现逻辑。Clock 和 Reset 通常应该出现在用来实现时序逻辑电路功能的 PROCESS 的信号列表中。当 Reset 有效时，状态将会强制回到系统初始状态；当 Reset 无效时，每当出现时钟的上升沿，寄存器将存储下一个状态，并通过前一个状态反馈给组合逻辑电路。

状态机在数字电路系统中得到的广泛应用与其良好的性能有必然关系。状态机主要有 6 个优点：

1）状态机的逐级工作模式易于 VHDL 综合器的优化；
2）状态机完好的代码风格易构成性能良好的时序逻辑模块；
3）状态机的结构模式简单，层次分明，易于排错；
4）利用 FPGA 中的同步时序和全局时钟可实现高速状态机的设计；
5）状态机的运行模式类似于 CPU，易于顺序控制；
6）状态机有很高的可靠性，易于非法状态的控制。

状态机结构清晰，易于实现。时序逻辑电路和组合逻辑可以独立进行设计。此时需要定义一个枚举数据类型，包含所有状态机需要的状态。

时序逻辑部分包含寄存器，所以 Clock 与 Reset 都与之相连，输入信号是 Nx_state，输出信号为 Pr_state，时序电路的逻辑部分都在 PROCESS 中实现。在 PROCESS 中任何一种顺序描述语句都可以使用，算法 3-1 描述了 PROCESS 的过程。

算法 3-1　时序逻辑设计

```
PROCESS (reset, clock)
BEGIN
    IF (reset = '1') THEN
            pr_state <= state0;
    ELSIF (clock' EVENT AND clock = '1') THEN
            pr_state <= nx_state;
    END IF;
END PROCESS;
```

算法 3-1 中的代码非常简单，包含了决定系统初始状态的 state0 和异步复位信号 reset，由时钟上升沿触发 nx_state 的同步存储。这种时序逻辑的设计风格是标准的，而且占用的寄存器数量较少。代码综合得到的寄存器数量等于对状态机所有的状态进行编码所需的位数。如果使用二进制编码方式，需要的寄存器数量是 $\lceil \log n \rceil$，其中 n 为状态的总数。

对于组合逻辑电路来说，通常可以使用并发代码来实现功能，也可以在 PROCESS 中使用顺序代码来实现。组合逻辑只做两件事情：对输出端口赋值和确定状态机的下一个状态，它遵守组合逻辑电路的基本要求。输入信号必须出现在 PROCESS 的信号列表中，并且所有输入输出信号组合必须被完整地列出。在整个代码中，由于没有任何信号的赋值是通过其他某个信号的跳变来触发的，所以不会生成寄存器。

米勒型状态机和摩尔型状态机设计的主要不同之处在组合逻辑的设计部分。米勒型状态机不需要中间信号，输出只取决了当前的输入，输出随着输入的变化而变化。摩尔型状态机则不同，它适用于有很多个中间状态寄存器的情况，即需要同步的寄存器输出，输出信号只有在时钟边沿出现时才能够更新。在组合逻辑中，只需要加入辅助信号，通过辅助信号来计算电路的输出值即可。

3.2 硬件编程语言 VHDL

3.2.1 VHDL 概述

VHDL 语言作为一种标准的硬件描述语言，具有结构严谨、描述能力强的特点，由于 VHDL 语言来源于 C、Fortran 等计算机高级语言，在 VHDL 语言中保留了部分高级语言的原语句，如 if 语句、子程序和函数等，便于阅读和应用。具体特点如下：

1）它支持从系统级到门级电路的描述，既支持自底向上（bottom-up）的设计也支持从顶向下（top-down）的设计，同时也支持结构、行为和数据流三种形式的混合描述。

2）VHDL 的设计单元的基本组成部分是实体（entity）和结构体（architecture），实体包含设计系统单元的输入和输出端口信息，结构体描述设计单元的组成和行为，便于各模块之间数据传送。利用单元（componet）、块（block）、过程（procure）和函数（function）等语句，用结构化层次化的描述方法，使复杂电路的设计更加简便。采用包的概念，便于标准设计文档资料的保存和广泛使用。

3）VHDL 语言有常数、信号和变量三种数据对象，每一个数据对象都要指定数据类型，VHDL 的数据类型丰富，有数值数据类型和逻辑数据类型，有位型和位向量型。既支持预定义的数据类型，又支持自定义的数据类型，其定义的数据类型具有明确的物理意义，VHDL 是强类型语言。

4）数字系统有组合电路和时序电路，时序电路又分为同步和异步，电路的动作行为有并行和串行动作，VHDL 语言常用语句分为并行语句和顺序语句，完全能够描述复杂的电路结构和行为状态。

3.2.2 标识符

VHDL 中的标识符可以是常数、变量、信号、端口、子程序或参数的名字。使用标识符要遵守如下法则：

1）标识符由字母（A...Z；a...z）、数字和下划线字符组成。

2）任何标识符必须以英文字母开头。
3）末字符不能为下划线。
4）不允许出现两个连续下划线。
5）标识符中不区分大小写字母。
6）VHDL 定义的保留子或称关键字，不能用作标识符。

以下是非法标识符：
1）-Decoder　　　—起始不能为非英文字母
2）3DOP　　　　 —起始不能为数字
3）Large#number　—"#"不能成为标识符的构成符号
4）Data__bus　　 —不能有双下划线
5）Copper_　　　 —最后字符不能为下划线
6）On　　　　　 —关键字不能用作标识符。

在 VHDL 语言中不区分大小写，所以写程序时应该养成习惯，应用关键字时用大写，自己定义的标识符用小写。

3.2.3　数据类型

VHDL 语言提供了许多标准的数据类型，用户也可自定义数据类型，这样使 VHDL 语言的描述能力和自由度进一步提高。但 VHDL 语言的数据类型的定义相当严格，不同类型之间的数据不能直接代入。

1）整数（Integer）

整数类型的数代表正整数、负整数和零，它与算术整数相似，可进行"+"、"-"、"*"、"/"等算术运算，不能用于逻辑运算。

2）实数（Real）

实数类型也类似于数学上的实数，或称浮点数。

3）位（Bit）

在数字系统中信号通常采用一个位来表示，取值是 1 或 0。

4）位矢量（Bit_Vector）

位矢量是用双引号括起来的一组位数据，使用位矢量必须注明位宽。

5）布尔量（Boolean）

一个布尔量具有两个状态："真"或"假"。布尔量不属于数值，因此不能用于运算，它只能通过关系运算符获得。一般这一类型的数据初始值总为 FALSE。

6）字符（Character）

字符也是一种数据类型，字符类型通常用单引号引起来，如 'A'。字符类型区分大小写，如 'B' 不同于 'b'。

7）字符串（String）

字符串是由双引号括起来的一个字符序列，也称字符矢量或字符串数组。常用于程序的提示和说明，如"STRING"等。

8）时间（Time）

时间是一个物理数据。完整的时间类型包括整数和单位两部分。在系统仿真时，时间数据很有用，可用它表示信号延时，从而使模型系统能更逼近实际系统的运行环境。

9）错误等级（Severity Level）

在 VHDL 仿真器中，错误等级用来指示设计系统的工作状态，它有四种：NOTE（注意）、WARNING（警告）、ERROR（出错）、FAILURE（失败）。在仿真过程中，可输出这四种状态以提示系统当前的工作状态。

10）自然数（Natural）和正整数（Positive）

自然数是整数的一个子类型，非负的整数，即为零和正整数。而 Positive 只能为正整数。

3.2.4 数据对象

1）常数（Constant）

常数是一个固定的值，主要是为了使设计实体中的常数更容易阅读和修改。常数一被赋值就不能在改变。一般格式：

CONSTANT 常数名：数据类型：=表达式；

例：CONSTANT Vcc: REAL: =5.0;

—设计实体的电源电压指定

常数所赋得值应与定义的数据类型一致。

常量的使用范围取决于它被定义的位置。程序包中定义的常量具有最大的全局化特性，可以用在调用此程序包的所有设计实体中；设计实体中某一结构体中定义的常量只能用于此结构体；结构体中某一单元定义的常量，如一个进程中，这个常量只能用在这一进程中。

2）变量（Variable）

变量是一个局部变量，它只能在进程语句、函数语句和进程语句结构中使用。用作局部数据存储。在仿真过程中。它不像信号那样，到了规定的仿真时间才进行赋值，变量的赋值是立即生效的。变量常用在实现某种算法的赋值语句中。

一般格式：

VARIABLE 变量名 数据类型 约束条件：=表达式；

例：VARIABLE x,y:INTEGER; —定义 x,y 为整数变量

VARIABLE count: INTEGER RANGE 0 TO 255:=10; —定义计数变量范围

变量的适用范围仅限于定义了变量的进程或子程序中。若将变量用于进程之外，必须该值赋给一个相同的类型的信号，即进程之间传递数据靠的信号。

变量赋值语句的语法格式如下：

目标变量：=表达式；

变量赋值符号是"：="。赋值语句右方的表达式必须是一个与目标变量有相同数据类型的数值。变量不能用于硬件连线和存储元件。

3）信号（Signal）

信号是描述硬件系统的基本数据对象，它类似于连接线，它除了没有数据流动方向说明以外，其他性质与实体的端口（Port）概念一致。变量的值可以传递给信号，而信号的值不能传递给变量。信号通常在构造体、包集合和实体中说明。信号说明格式为：

SIGNAL 信号名：数据类型；

信号初始值的设置不是必需的，而且初始值仅在VHDL的行为仿真中有效。

4）信号与变量的区别

信号赋值可以有延迟时间，变量赋值无时间延迟；信号除当前值外还有许多相关值，如历史信息等，变量只有当前值；进程对信号敏感，对变量不敏感；信号可以是多个进程的全局信号，但变量只在定义它之后的顺序域可见；信号可以看作硬件的一根连线，但变量无此对应关系。

3.2.5 运算符

VHDL语言共四类操作符，可以分别进行逻辑运算（Logic）、关系运算（Relational）、算术运算（Arithmetic）和并置运算（Concatenation）。被操作符所操作的对象是操作数，且操作数的类型应该和操作符所要求的类型相一致。

1）逻辑运算符

运算符在VHDL语言中逻辑运算符有6种，分别为：

NOT（非）

OR（或）

AND（与）

NOR（或非）

NAND（与非）

XOR（异或）

2）关系运算符

=（等于）

/=（不等于）

<（小于）

<=（小于等于）

>（大于）

\>=（大于等于）

3）算术运算符

+（加）

/（除）

SLL（逻辑左移）

ROR（逻辑循环右移）

-（减）

MOD（求模）

SRL（逻辑右移）

ABS（取绝对值）

*（乘）

REM（取余）

SLA（算术左移）

4）其他运算符

<=（信号赋值）

：=（信号赋值）

-（负）

+（正）

&（并置运算符，用于位的连接）

=>（并联运算符）

3.2.6 VHDL 的结构

一个 VHDL 语言的设计程序描述的是一个电路单元，这个电路单元可以是一个门电路，或者是一个计数器，也可以是一个 CPU。一般情况下，一个完整的 VHDL 语言程序至少要包含程序包、实体和结构体三个部分。实体给出电路单元的外部输入输出接口信号和引脚信息，结构体给出了电路单元的内部结构和信号的行为特点，程序包定义在设计结构体和实体中将用到的常数、数据类型、子程序和设计好的电路单元等。

一位全加器的逻辑表达式是：

$$S = A \oplus B \oplus C_i$$
$$C_o = AB + AC_i + BC_i$$

全加器的 VHDL 程序的文件名称是 fulladder.VHD，其中 VHD 是 VHDL 程序的文件扩展名，程序如下：

```
LIBRARY IEEE;  --IEEE 标准库
USE IEEE.STD_LOGIC_1164.ALL;
```

```
USE IEEE.STD_LOGIC_ARITH.ALL;
USE IEEE.STD_LOGIC_UNSIGNED.ALL;
ENTITY fulladder IS    -- fulladder 是实体名称
  PORT(
    A, B, Ci : IN STD_LOGIC; --定义输入/输出信号
    Co, S : OUT STD_LOGIC
  );
END fulladder;
ARCHITECTURE addstr OF fulladder IS    --addstr 是结构体名
  BEGIN
    S <= A XOR B XOR Ci;
    Co <= (A AND B) OR (A AND Ci) OR (B AND Ci);
END addstr;
```

从这个例子中可以看出，一段完整的 VHDL 代码主要由以下几部分组成：

第一部分是程序包，程序包是用 VHDL 语言编写的共享文件，定义在设计结构体和实体中将用到的常数、数据类型、子程序和设计好的电路单元等，放在文件目录名称为 IEEE 的程序包库中。

第二部分是程序的实体，定义电路单元的输入/输出引脚信号。程序的实体名称 fulladder 是任意取的，但是必须与 VHDL 程序的文件名称相同。实体的标识符是 ENTITY，实体以 ENTITY 开头，以 END 结束。其中，定义 A、B、Ci 是输入信号引脚，定义 Co 和 S 是输出信号引脚。

第三部分是程序的结构体，具体描述电路的内部结构和逻辑功能。结构体有三种描述方式，分别是行为（BEHAVIOR）描述、数据流（DATAFLOW）描述方式和结构（STRUCTURE）描述方式，其中数据流（DATAFLOW）描述方式又称为寄存器（RTL）描述方式，例中结构体的描述方式属于数据流描述方式。结构体以标识符 ARCHITECTURE 开头，以 END 结尾。结构体的名称 addstr 是任意取的。

VHDL 每条语句是以分号";"作为结束符的，并且 VHDL 对空格是不敏感的，所以符合之间空格的数目是可以自己设定的。可以按自己的习惯任意添加，增强代码可读性。

VHDL 中的注释由两个连续的短线（--）开始，直到行尾。

3.3　Altera FPGA 开发环境 Quartus II

3.3.1　Quartus II 介绍

Quartus II 可以在 Windows、Linux 以及 UNIX 上使用，除了可以使用 TCL（Tool Command Language）脚本完成设计流程外，还提供了完善的用户图形界面设计方式。具

3.3 Altera FPGA 开发环境 Quartus II

有运行速度快、界面统一、功能集中、操作简单等特点。

Quartus II 支持 Altera 的 IP 核，包含了 LPM/MegaFunction 宏功能模块库，使用户可以充分利用成熟的模块，简化了设计的复杂性、加快了设计的速度。它对第三方 EDA 工具的良好支持也使用户可以在设计流程的各个阶段使用熟悉的第三方 EDA 工具。Altera 在 Quartus II 中包含了许多诸如 SignalTap II、Chip Editor 和 RTL Viewer 的设计辅助工具，集成了 SOPC 和 HardCopy 设计流程，并且继承了 Maxplus II 友好的图形界面及简便的使用方法。Altera Quartus II 作为一种可编程逻辑的设计环境，有着强大的设计能力和直观、易用的接口，越来越受到数字系统设计者的欢迎。

此外，Quartus II 通过和 DSP Builder 工具与 Matlab/Simulink 相结合，可以方便地实现各种 DSP 应用系统。它支持 Altera 的片上可编程系统（SOPC）开发，集系统级设计、嵌入式软件开发、可编程逻辑设计于一体，是一种综合性的开发平台。

3.3.2 Quartus II 使用例解

1）如图 3-2 所示，打开 Quartus II 软件。

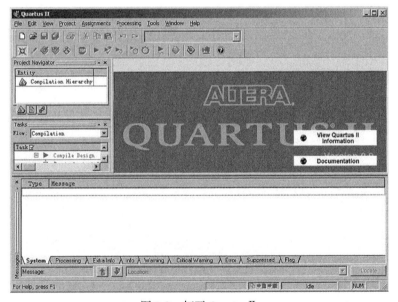

图 3-2　打开 Quartus II

2）接下来，如图 3-3 和图 3-4 所示，建立一个新工程。

第3章 芯片设计基础

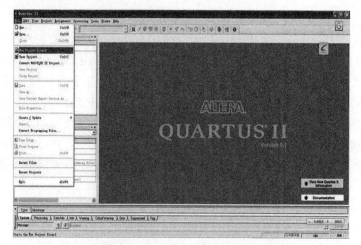

图 3-3 单击菜单的选项卡 File

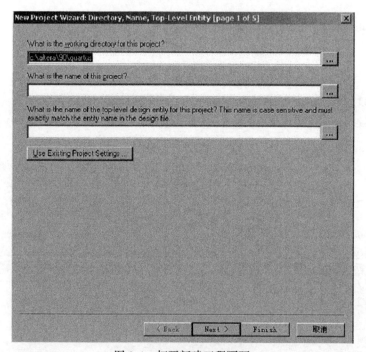

图 3-4 打开新建工程页面

3）如图 3-5 所示，第一行是所建工程的路径，第二行是工程项目名称，第三行是实体名称。

4）下边一直单击 Next 按钮，直到出现图 3-6 所示的界面。

3.3 Altera FPGA 开发环境 Quartus II

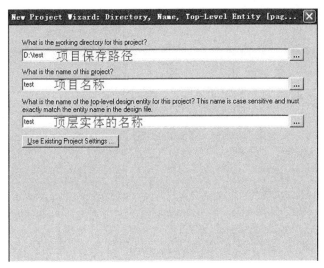

图 3-5 填写项目信息

5）在 Family 里边选择 Sratix II，Available devices 里边选择 EP2S60F672C5（具体内容根据你所使用的芯片所决定）。

6）接着单击 Next 按钮，不需要做任何修改，一直到单击 Finish 按钮。到此为止，工程已经建立完成。

7）接着，需要建立一个 Block Diagram/Schematic File，单击 File→New 出现图 3-7 所示的界面。

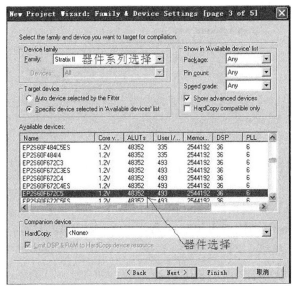

图 3-6 选择器件

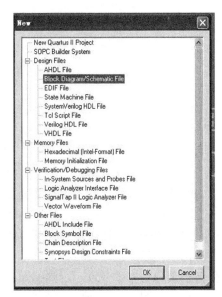

图 3-7 新建一个 Block Diagram/Schematic File

8）单击 OK 按钮，完成创建，工程中出现一个 Block1.bdf 文件，如图 3-8 所示。

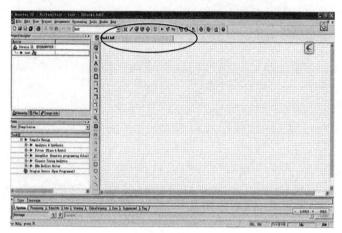

图 3-8　新建 Block1.bdf

9）如图 3-9 所示，在这个 bdf 文件空白处双击鼠标，或者右键单击鼠标，接着单击 Insert→Symbol。

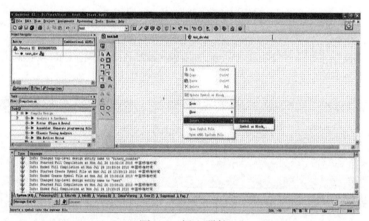

图 3-9　插入器件

10）如图 3-10 所示，这里边的器件很多，可以在里边输入所需的器件，也可以直接单击分类，根据分类查找你需要的器件。

11）单击 File→New，选择 VHDL File（根据使用的编程语言），如图 3-11 所示。

12）单击 OK 按钮后，在下边的界面就可以编写 VHDL 程序了，如图 3-12 所示。

13）当然，也可以根据自己掌握的语言种类进行编程。注意，VHDL 语言保存的文件名必须与实体名一致，否则编译会出错，如图 3-13 所示。

14）如图 3-14 所示，设置当前为最高实体。

3.3 Altera FPGA 开发环境 Quartus II

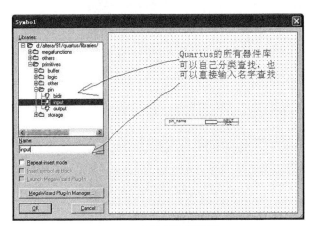

图 3-10 选择所需的器件

图 3-11 新建 VHDL 文件

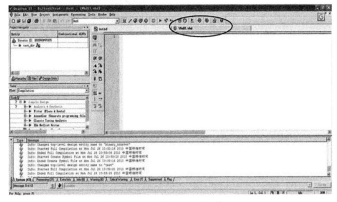

图 3-12 编写 VHDL 代码

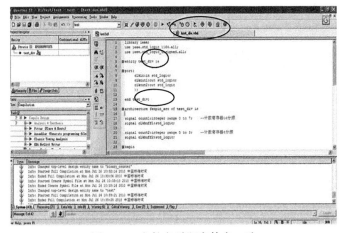

图 3-13 文件名需跟实体名一致

第3章 芯片设计基础

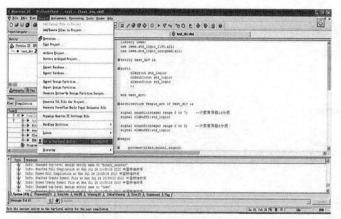

图 3-14　设置当前为最高实体

15）如图 3-15 所示，单击紫色的三角，开始编译。

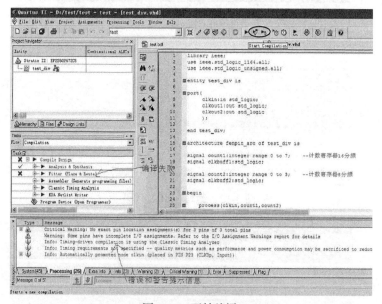

图 3-15　开始编译

16）如图 3-16 所示，产生模块，单击就可以进行。

17）模块生成后，回到 bdf 主界面。双击该界面，在 Project 下拉框中就会出现刚才编译文件生成的模块，单击就可以将其放入主原理图实体中。今后如果重新改变 VHDL 程序，必须先设置最高实体，然后编译、产生模块，最后添加。放置模块的时候，通过自己的程序编译产生的模块，会在 Project 目录下显示，如图 3-17 所示。

18）如果需要改动已经做好的并且放入到原理图的模块程序，那么改动后也必须先编译，后产生模块，最后按照图 3-18 所示进行模块更新。

3.3 Altera FPGA 开发环境 Quartus II

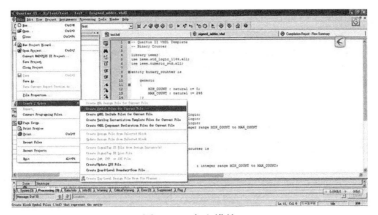

图 3-16 产生模块

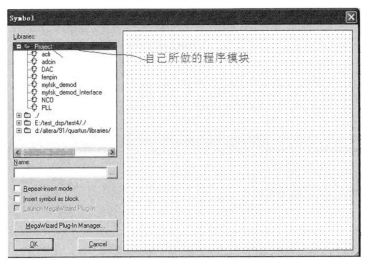

图 3-17 查看模块

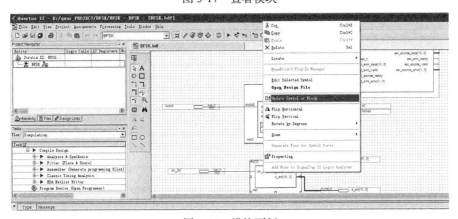

图 3-18 模块更新

19）根据需求选择更新模块的方式，如图 3-19 所示。

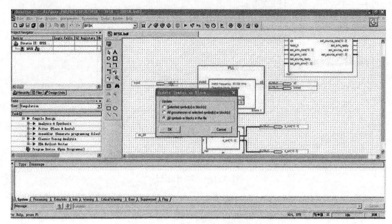

图 3-19　选择更新模块的方式

20）如图 3-20 所示，以后每次要用的时候，都可以双击鼠标，进入 Project 进行选择并使用。

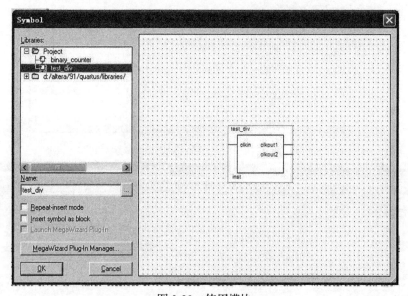

图 3-20　使用模块

21）如图 3-21 所示，右键单击模块，接着单击 Generate Pins of Symbol Ports。
22）如图 3-22 所示，也可以自己命名输入输出引脚。

3.3 Altera FPGA 开发环境 Quartus II

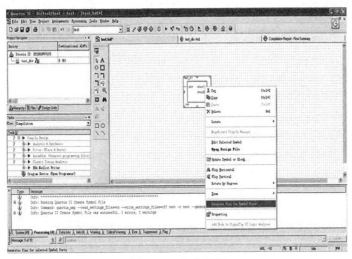

图 3-21 生成引脚

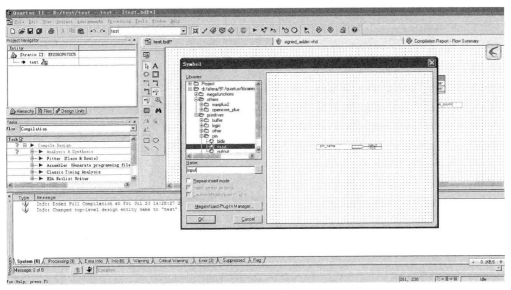

图 3-22 命名引脚

23）如图 3-23 所示，设置当前实体为最高实体，再次进行编译。

24）编译完成后，需要分配引脚。分配引脚通常直接在工程中分配，这种方式在引脚较少时比较方便，如图 3-24 所示。

25）如图 3-25 所示，选择 Pin。

第3章 芯片设计基础

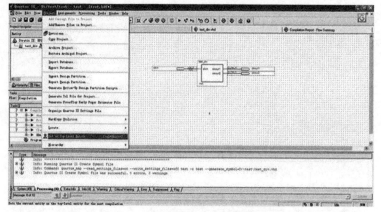

图 3-23　设置当前实体为最高实体

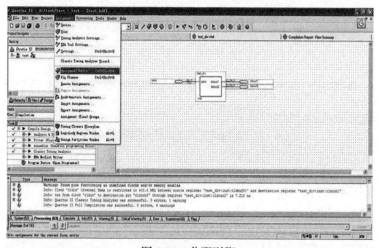

图 3-24　分配引脚

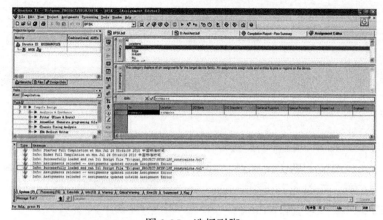

图 3-25　选择引脚

26）如图 3-26 所示，双击引脚分配处的 To 和 Location，就可以确定应用的 FPGA 引脚分配情况。

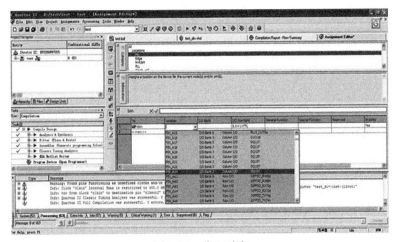

图 3-26　分配引脚

27）如图 3-27 所示，分配好引脚后，单击保存按钮。

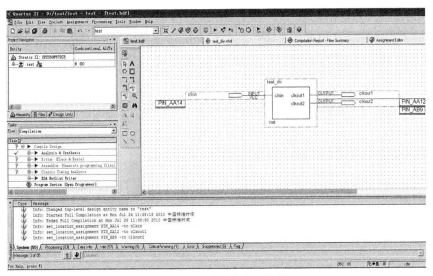

图 3-27　查看引脚分配情况

28）然后再编译。最后在单击 Tool→Programmer，或者直接在工具栏中的单击下载图标，如图 3-28 所示。

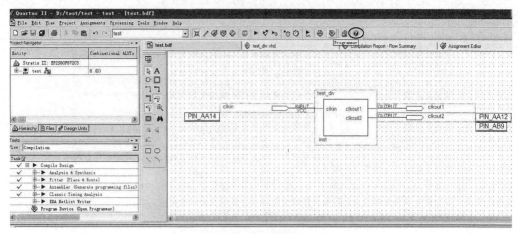

图 3-28　编译和下载

29）如图 3-29 所示，出现下载界面。

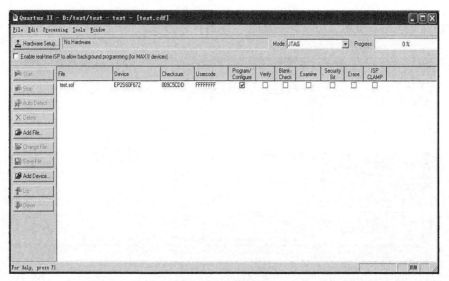

图 3-29　下载界面

30）如图 3-30 所示，单击 Hardware Setup 按钮。如果你没插 USB-Blaster，打开后不会有显示；如果插了，这里就会显示有一个硬件可以选择。右上位置选择下载方式。注意，JTAG 模式和 AS 模式接口是不同的。

31）如图 3-31 所示，选择好 USB-Blaster 后，单击 Close 按钮，然后单击 Start 按钮。

3.3 Altera FPGA 开发环境 Quartus II

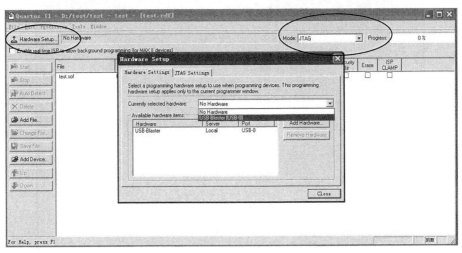

图 3-30 单击 Hardware Setup 按钮

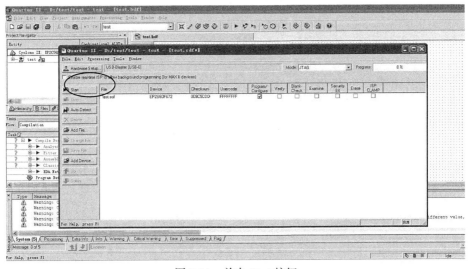

图 3-31 单击 Start 按钮

32）当前选择的是 JTAG 模式，因此程序被下载到 RAM，可以查看调试结果。至此，FPGA 的原理图制作完成，代码编写流程及下载流程已经全部完毕。

33）接下来，新建一个简单的分频器。如同单片机的最小系统一样，FPGA 的系统需要一个时钟源作为支撑。FPGA 内部有个锁相环 PLL（Phase Locked Loop）资源，这个 PLL 可以对输入频率进行倍频。因此，几乎在每个系统设计的时候，都需要对这个 PLL 进行设置，如图 3-32 所示。

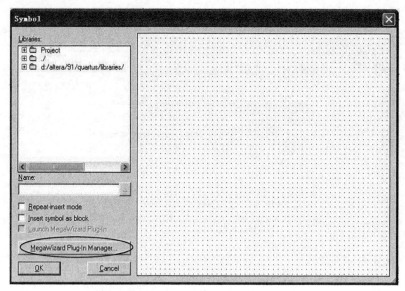

图 3-32　新建一个分频器

34）如图 3-33 所示，选择第一个选项。

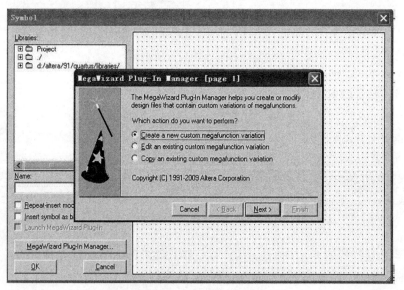

图 3-33　选择第一个选项

35）如图 3-34 所示，单击 Next 按钮，接着选择 I/O 栏目下的 ALTPLL，给这个模块起一个名字叫 PLL，然后单击 Next 按钮。

3.3 Altera FPGA 开发环境 Quartus II

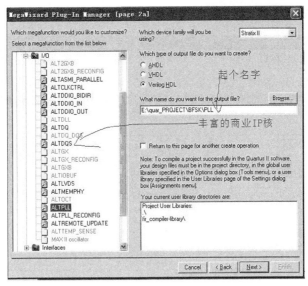

图 3-34 选择 ALTPLL

36）如图 3-35 所示，根据提示进行选择，然后单击进入下一个设置。

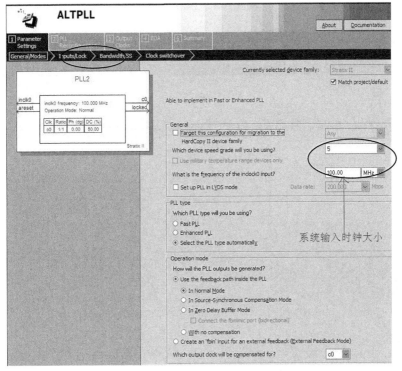

图 3-35 根据需求输入系统时钟频率

37）如图 3-36 所示，在这里，把所有的勾选去掉，然后一直单击 Next 按钮。

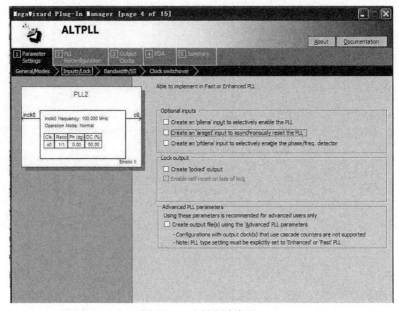

图 3-36　去掉所有勾选

38）一直单击 Next 按钮，至图 3-37，在其中设置时钟参数。

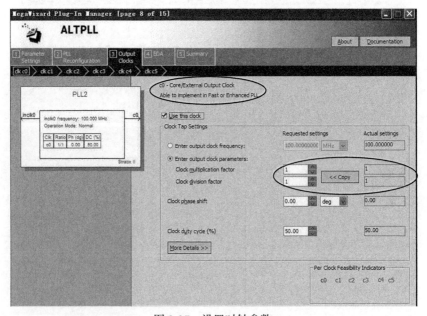

图 3-37　设置时钟参数

39）每个芯片可以设置输出的频率个数不同，当前总共有 2 个 PLL，每个 PLL 可以设置 6 个不同的频率输出。现在就可以一直单击 Next 按钮，直到单击 Finish 按钮就可以了。然后在 Project 里边将 PLL 放置到原理图上，如图 3-38 所示。

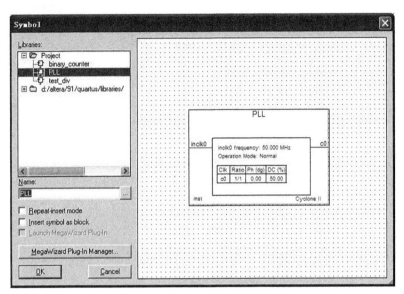

图 3-38　将 PLL 放置到原理图上

40）新建一个 test_div 的程序，程序代码如下。

```
library ieee;
use ieee.std_logic_1164.all;
use ieee.std_logic_unsigned.all;

entity test_div is

port(
    clkin:in std_logic;
    clkout1:out std_logic;
    clkout2:out std_logic
    );

end test_div;

architecture fenpin_arc of test_div is

signal count1:integer range 0 to 7;     --计数寄存器16分频
signal clkbuff1:std_logic;

signal count2:integer range 0 to 3;     --计数寄存器8分频
```

```vhdl
    signal clkbuff2:std_logic;

begin

    process(clkin,count1,count2)

    begin
        if rising_edge(clkin) then      --计数、分频1

            if (count1 >= 7) then
                count1 <= 0;
                clkbuff1 <= not clkbuff1;
            else
                count1 <= count1 + 1;
                clkout1 <= clkbuff1;
            end if;

        end if;

        if rising_edge(clkin) then      --计数、分频2

            if (count2 >= 3) then
                count2 <= 0;
                clkbuff2 <= not clkbuff2;
            else
                count2 <= count2 + 1;
                clkout2 <= clkbuff2;
            end if;

        end if;


    end process;

end fenpin_arc;
```

41）保存程序，设置当前为最高实体，并进行编译，编译后产生模块，最终也可以将其放在原理图上。用鼠标将所需要连接的线连接起来，然后设置当前为最高实体，并进行编译、分配引脚、编译，最后下载就可以完成了。

42）同时，除了下载进 FPGA 中进行调试外，我们还可以提前利用 Quartus 进行时序仿真。首先生成一个 Block，并将其放置在原理图上，然后再加上一个 PLL，连接起来后，如图 3-39 所示。

43）如图 3-40 所示，单击 Processing→Simulation Debug→Current Vector Inputs。

44）如图 3-41 所示，在 Name 处单击鼠标右键，在弹出的菜单中选择 Inset→Inset Node or Bus。

3.3 Altera FPGA 开发环境 Quartus II

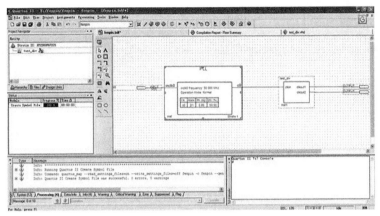

图 3-39 原理图

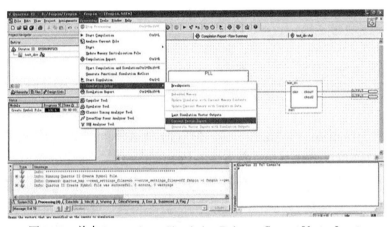

图 3-40 单击 Processing→Simulation Debug→Current Vector Inputs

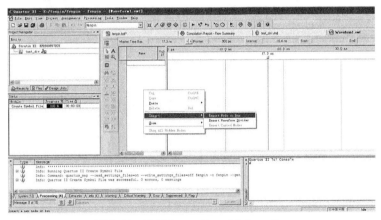

图 3-41 选择 Inset→Inset Node or Bus

45）如图 3-42 所示，单击 Node Finder。

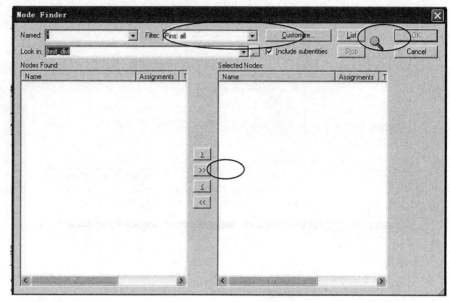

图 3-42　单击 Node Finder

46）如图 3-43 所示，可以选择引脚，在 Filter 处选择 Pins:all，当然也可以选择一些需要的引脚，其他的引脚不显示，然后单击 List 按钮，最后单击加入符号。

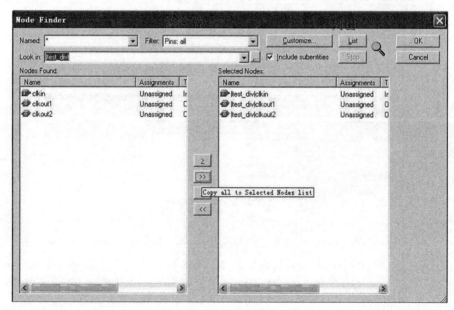

图 3-43　显示引脚

47）在上图中单击 OK 按钮，回到仿真页面，如图 3-44 所示，给输入信号加所需信号，如 clkin 是时钟信号，直接单击时钟符号，就可以进行设置。

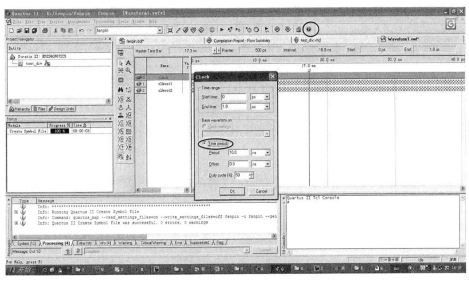

图 3-44　设置引脚

48）如图 3-45 所示，设置好输入后，可以选择 Edit 菜单下的 End Time 设置仿真时间。

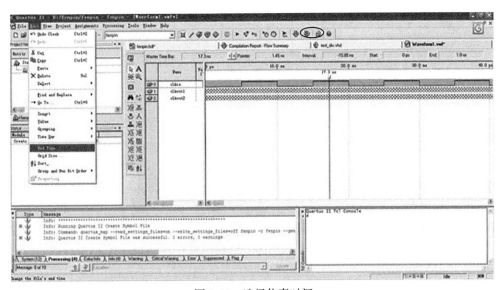

图 3-45　选择仿真时间

49）如图 3-46 所示，全部设置好后，单击保存按钮，起好名字，然后单击 Start Simulation 按钮，开始进行时序仿真。仿真结束后，可以观察信号时序。

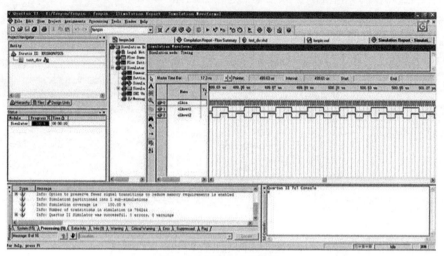

图 3-46　进行时序仿真

3.4　集成电路产业

3.4.1　集成电路

集成电路（Integrated Circuit）是一种微型电子器件或部件。采用一定的工艺，把一个电路中所需的晶体管、二极管、电阻、电容和电感等元件及布线互连，设置在一小块或几小块半导体晶片或介质基片上，然后封装在一个管壳内，成为具有所需电路功能的微型结构；其中所有元件在结构上已组成一个整体，这样，整个电路的体积大大缩小，且引出线和焊接点的数目也大为减少，从而使电子元件向着微小型化、低功耗和高可靠性方面迈进了一大步。集成电路具有体积小、重量轻、引出线和焊接点少、寿命长、可靠性高、性能好等优点，同时成本低，便于大规模生产。它不仅在工用、民用电子设备，如收录机、电视机、计算机等方面得到广泛的应用，同时在军事、通信、遥控等方面也得到了广泛的应用。用集成电路来装配电子设备，其装配密度可比晶体管提高几十倍至几千倍，设备的稳定工作时间也可大大提高。它在电路中用字母"IC"表示。

集成电路发明者为杰克·基尔比（基于硅的集成电路）和罗伯特·诺伊思（基于锗的集成电路）。当今半导体工业大多数应用的是基于硅的集成电路。集成电路按其功能、结构的不同，可以分为模拟集成电路和数字集成电路两大类。模拟集成电路用来产生、放大和处理各种模拟信号，而数字集成电路用来产生、放大和处理各种数字信号。

3.4.2 产业发展现状

集成电路作为信息产业的核心，其重要性不言而喻。但是长久以来，国内的集成电路市场被国外企业垄断，即使在政务、金融、民政、公安这样的关键领域，我们也广泛使用外国芯片，信息安全隐患巨大。

在近十几年来，国家对本土集成电路产业的发展不断给予政策支持，更把集成电路发展上升到了国家安全战略的高度。2000 年，国务院出台的 18 号文《鼓励软件产业和集成电路产业发展的若干政策》(国发[2000]18 号)，在审批程序、税收支持、进出口等方面给予了集成电路产业重点扶持。2008 年，国家在 863 计划、973 计划和《国家中长期科学和技术发展规划纲要（2006—2020 年）》中，通过重大科技专项的方式对集成电路产业研究和产业发展给予重点支持，并重点提出了到 2020 年我国在高端通用芯片、基础软件和核心电子器件领域基本形成具有国际竞争力的高新技术研发与创新体系，进一步缩小与世界先进水平差距，并开拓国际市场。2011 年，国务院出台 4 号文《进一步鼓励软件产业和集成电路产业发展的若干政策》(国发[2011]4 号)，在原有 18 号文的基础上再次强调了对集成电路产业的重点支持，提出了从财税政策、投融资、研究开发、进出口、人才政策、知识产权等 8 个方面给予集成电路研发系统性扶持。

2001 年～2010 年这 10 年间，我国集成电路产量的年均增长率超过 25%，集成电路销售额的年均增长率则达到 23%。2007 年全年国内生产总值 246 619 亿元，比 2006 年增长 11.4%，加快 0.3%，连续五年增速达到或超过 10%。2008 年全球半导体产业经受了近 20 年来最严峻的挑战。根据 SIA 发布的数据，2008 全年全球半导体市场规模为 2 486.03 亿美元，同比下跌了 2.8%。2010 年国内集成电路产量达到 640 亿块，销售额超过 1 430 亿元，分别是 2001 年的 10 倍和 8 倍。中国集成电路产业规模已经由 2001 年不足世界集成电路产业总规模的 2%提高到 2010 年的近 9%。中国成为过去 10 年世界集成电路产业发展最快的地区之一。

国内集成电路市场规模也由 2001 年的 1 140 亿元扩大到 2010 年的 7 350 亿元，增长了 6.5 倍。国内集成电路产业规模与市场规模之比始终未超过 20%。如扣除集成电路产业中接受境外委托代工的销售额，则中国集成电路市场的实际国内自给率还不足 10%，国内市场所需的集成电路产品主要依靠进口。近几年，国内集成电路进口规模迅速扩大，2010 年已经达到创纪录的 1 570 亿美元，集成电路已连续两年超过原油成为国内最大宗的进口商品。与巨大且快速增长的国内市场相比，中国集成电路产业虽发展迅速但仍难以满足内需要求。

在美国的棱镜门事件爆发之后，国家更是高度担心国家安全问题，提出要快速提高国产芯片市场占比。在国家政策的指引和近几年在智能机渗透率快速提升的刺激下，国内芯片设计领域市场规模一路高奏凯歌。据中国半导体协会统计，2013年国内芯片设计市场规模已经达到809亿，较2004年增长了近10倍，年复合增长率为29%。据IC Insights统计，2013年全球前25大Fabless IC设计公司中，中国厂商华为海思和展讯占据两席。其中，华为海思以13.55亿美元的销售额排在第12位，较2012年增长了15%，近5年年复合增长率为24%。随着华为智能终端出货量的高速增长以及华为海思芯片在内部占比的快速提高，华为海思的销售收入迎来爆发式增长。展讯以10.7亿美元紧跟其后，并以48%的同比增速成为增长最快的公司。展讯在2012年上半年首次成功推出智能机SoC芯片SC8810，在超低端智能手机芯片领域优势明显。

随着市场规模的进一步扩大以及国内集成电路企业自身实力的提升，2015年，中国集成电路市场规模创纪录地达到11 024亿元，同比增长6.1%，成为全球为数不多的仍能保持增长的区域市场。2015年，中国集成电路产业全年销售额达到3 690.8亿元，设计、制造、封测3个环节销售额分别为1 325亿元、900.8亿元及1 384亿元，其中设计和制造环节增速明显快于封测，占比进一步上升，产业结构更趋平衡。华为海思、展讯作为国内集成电路设计的行业龙头，均有望进入全球Fabless企业前十名，而国内制造业龙头中芯国际也在2015年顺利突破28nm制程工艺，开始进入量产阶段。

2015年，除中国集成电路产业传统的环渤海、长三角和珠三角三大产业集聚区域之外，若干区域外城市纷纷将集成电路产业作为当地"十三五"期间重点发展产业，有望成为中国集成电路产业发展的"第4极"，将为国内集成电路产业发展注入新的动力。

除了手机等移动通信终端之外，芯片在其他领域的应用也非常广泛，例如，公交卡（深圳通、羊城通等）、IC银行卡、智能锁（车锁、门锁）、监控摄像头、无人机、网络设备、传感器、二代身份证等。

3.4.3 产业发展前景

行业进入壁垒不断提高：集成电路行业是一个典型的资本/技术双密集型行业，且随着技术进步，行业进入壁垒一直在不断提高。集成电路技术进步主要遵循摩尔定律，即集成电路芯片上的晶体管数目，约每18个月增加1倍，性能也提升1倍，而价格降低到原来的一半。集成电路行业是需要不断投入巨额资金的行业，设备费用和研发费用都非常大。特别是集成电路制造业对工艺和环境要求很高，堪称"吞金业"。随着集成电路技术的深化，以及电路结构的越来越复杂，加工工艺也将越来越复杂。新一代生产线所需的投资额成倍甚至数十倍地增加。一条12英寸生产线所需投资额就高达20亿美元。

上游行业垄断程度高：集成电路行业的上游行业主要是硅材料行业和集成电路设备制造业。制造设备市场呈高度垄断的局面，高端设备往往只有几家制造商有能力生产，如全球浸润式光刻机仅有3家企业能供应：荷兰ASML、日本佳能与美国尼康。中国本土企业还无法提供类似产品,而一台浸润式光刻机价格高达4～5亿元。在这种高度垄断的市场格局下,我国集成电路行业处于明显的弱势地位。

行业内竞争将加剧:集成电路行业规模经济特征明显,企业规模越大,越能节约单位生产成本,越有助于增强竞争力。集成电路国际巨头的规模一直在不断扩大。而我国集成电路企业普遍产能偏低、规模偏小,因此,国内众多集成电路生产企业在努力提高技术水平的同时,纷纷制定产能扩张计划,规划新建生产线,以提高产能、扩大生产规模。虽然在经济危机的冲击下,集成电路陷入了低谷,迫于资金压力,集成电路不得不暂停在建项目,取消扩张计划,降低产能。但行业的低迷也带来了行业重组的机会,我国集成电路企业正在酝酿着大规模的兼并重组。在经济复苏后,将形成新的产业格局,企业规模将得到提升。同时,随着国际集成电路制造商纷纷在中国投资设厂,国际集成电路制造业向中国转移,我国集成电路行业内竞争将进一步加剧。

设计业区域发展势头不一,产业集中度进一步提高:2015年,珠三角和环渤海区域的设计业规模增速超过30%,一直以来占据设计业最大份额的长三角地区仅取得7%左右的增长,3个区域的差距进一步缩小。中西部地区在合肥、武汉、长沙等地的带动下,也取得了18.2%的增长。2016年,各区域的发展势头保持平稳,中西部地区随着引入企业的落地运营,有较大幅度的增长。2015年,中国大陆集成电路设计业前10大设计企业的总销售规模占全行业总收入的比例进一步提高到43.79%,相比美国的超80%和中国台湾地区的超70%仍有一定差距。2016年,国内的设计业在资本的催动下继续整合并购的浪潮,大型企业将进一步快速成长,设计业的集中程度进一步提高。

制造业发展提速,助力配套环节实现突破:受惠于政府的政策支撑、广阔的市场需求以及技术水平的提升,2015年中国集成电路产业,特别是制造业得到了快速发展。中芯国际、联电、力晶、台积电等国际集成电路制造企业纷纷在北京、厦门、合肥、南京等地建设或筹划建设12英寸晶圆厂,以加快扩大国内集成电路制造业规模,优化产业结构。到2016年,各地新建的晶圆厂将处于生产配套或试运行阶段,这必将对集成电路装备、材料等相关配套产品产生巨大需求。国内集成电路配套企业将在政府及客户的共同推动下,通过自主技术研发、合作并购等方式,突破技术壁垒,生产出具有一定替代能力的国产集成电路装备、材料等配套产品,从而加速国内集成电路的全产业链建设。

工业控制、汽车电子、网络通信引领市场增长,新兴应用领域爆发尚需时日:随着国内整体工业水平的升级,以及"中国制造2025"等战略的实施,工业控制领域的集成电路产品需求将持续走高。随着工业物联网的推广,在传统的工业控制领域,连接类的集成电路产品将成为市场关注的焦点。根据国际电信联盟预测,5G通信将在2020年正式进入商用,在这之前将会迎来5G通信基础设施建设的高峰,5G通信相关的集成电路产品市场前景广阔。国内通信设备企业如华为、中兴等均具有一定的终端市场地位,本身又有强大的芯片自研能力,将为中国集成电路产业在通信领域应用争得一席之地。汽

第 3 章　芯片设计基础

车领域，高级驾驶辅助系统的普及，以及新能源汽车、车联网、自动驾驶等技术的不断发展，对集成电路产品提出了新的要求。随着我国整车销售量的持续走高，汽车消费水平的不断提升以及电动汽车的大力普及，汽车电子领域的集成电路产品将保持高速增长态势。除上述集成电路传统应用领域外，智能手机市场增速将进一步放缓，而被市场寄予厚望的平板电脑市场在大屏手机的冲击下已经进入下降通道，可穿戴设备市场一直未达到市场预期。至于无人机、VR（虚拟现实）等新兴热点应用，受限于其技术成熟度不足、应用领域较窄、产业生态尚未建立等因素，暂时难以成为引领中国集成电路产业发展的主要动力。

海外并购继续升温，我国加快融入全球产业重构浪潮：智能手机市场增速放缓、个人电脑持续萎缩，云计算、大数据、物联网、智能制造、新能源汽车等新兴增长点正逐步形成，全球集成电路市场进入换挡期，为我国集成电路产业发展创造了良好机遇。为实现产业升级、快速发展，我国对海外并购将持续增强。一方面我国集成电路产业在高性能处理器、存储器、传感器、功率器件等领域或处于重点端地位或存在产品空白。为实现制造强国战略、信息安全战略，我国对关键核心产品的自主可控有着迫切需求，有的放矢的海外并购是实现产业快速升级的必然。另一方面，全球集成电路产业重构正在进行，在激烈的竞争和平台化解决方案战略的影响下，将出现大量优质投资标的。一是国际大厂不断剥离非核心业务，例如瑞萨剥离移动芯片业务，意法半导体剥离电视机顶盒业务；二是巨头并购规避反垄断检查，出售优质产品线，例如恩智浦出售原RFPower部门；三是只提供单一产品线的企业难以与平台化战略竞争，成为被收购对象，例如功率器件大厂仙童半导体。优质标的出现将为我国产业实施并购提供良好机遇。产业快速发展的迫切需求和投资标的必然出现将促使我国集成电路产业海外并购持续升温，同时成为我国产业融入全球产业格局、缩短差距、快速升级的关键。

垂直整合成为趋势，集成电路产业融入电子信息大生态：融入电子信息大生态是分担风险的关键，垂直整合将成为集成电路产业发展的途径之一。从近几年的发展态势来看，全行业的垂直整合成为高端产品取得成功的关键。以苹果手机为例，从工业设计、操作系统、到核心芯片，均由自身完成，构建了独特的垂直整合模式，确保了用户体验、销售量和利润，从而能够吸引最优秀的人才，实现高速的可持续发展。海思无疑是中国最成功的集成电路企业，它的快速发展，与华为构建了一条通信设备、终端、芯片、软件的垂直整合模式不无关系。一方面，通信领域的技术积累和客户关系以及快速增长的终端销量为海思芯片初期的技术积累和运营发展提供了基本的条件；另一方面，高性能的芯片又使华为能够减轻对高通、联发科的依赖，具有更多的话语权。当前，中国拥有全球第一大手机产业集群和全球第二大的互联网产业集群，垂直整合模式和大生态的构建已初显端倪。小米联手联芯，互联网公司频频从内容提供商和服务商向智能硬件方向整合。集成电路产业积极融入电子信息大生态，贴近需求、分担风险将成为未来几年发展的重要趋势。

3.5 本章参考代码

3.5.1 VHDL 参考例子

带进位的加法器：

```
ibrary ieee; -- 可以实现两个16位数相加，输出带进位的32位

use ieee.std_logic_1164.all;
use ieee.std_logic_arith.all;
use ieee.std_logic_unsigned.all;
entity vector_add_test is
     port(ivector1, ivector2 : in std_logic_vector(15 downto 0);
              ovector : out std_logic_vector(31 downto 0));
end vector_add_test;
architecture one of vector_add_test is
shared variable outv : integer;
begin
     process(ivector1, ivector2)
     begin
          outv := conv_integer(ivector1) + conv_integer(ivector2);
     end process;
     ovector <= conv_std_logic_vector(outv, 32);
end one;
```

分频器：

```
LIBRARY IEEE;
USE IEEE.STD_LOGIC_1164.ALL;
USE IEEE.STD_LOGIC_UNSIGNED.ALL;
ENTITY fp IS
PORT(clk: IN STD_LOGIC;
    fp1024,fp512,fp64,fp4,fp1: OUT STD_LOGIC);
END ENTITY fp;
ARCHITECTURE a OF fp IS
SIGNAL buf:STD_LOGIC_VECTOR(9 DOWNTO 0);
BEGIN
fp1<=buf(9);
fp4<=buf(7);
fp64<=buf(3);
fp512<=buf(0);
fp1024<=clk;
PROCESS(clk) IS
```

```
BEGIN
IF(clk'EVENT AND clk='1') THEN
    IF(buf="1111111111") THEN LIBRARY IEEE;
```

报时电路：

```
USE IEEE.STD_LOGIC_1164.ALL;
USE IEEE.STD_LOGIC_UNSIGNED.ALL;
ENTITY bs IS
PORT(min,sin: IN STD_LOGIC_VECTOR(7 DOWNTO 0);
    clk,bs512,bs1024: IN STD_LOGIC;
    bsout: OUT STD_LOGIC);
END ENTITY bs;
ARCHITECTURE a OF bs IS
BEGIN
PROCESS(clk) IS
BEGIN
IF(clk'EVENT AND clk='1') THEN
    IF(min="01011001") THEN
        CASE sin IS
            WHEN "01010000"=>bsout<=bs512;
            WHEN "01010010"=>bsout<=bs512;
            WHEN "01010100"=>bsout<=bs512;
            WHEN "01010110"=>bsout<=bs512;
            WHEN "01011000"=>bsout<=bs512;
            WHEN "01011001"=>bsout<=bs1024;
            WHEN OTHERS=>bsout<='0';
        END CASE;
    END IF;
END IF;
END PROCESS;
END ARCHITECTURE a;
        buf<="0000000000";
    ELSE
        buf<=buf+'1';
    END IF;
END IF;
END PROCESS;
END ARCHITECTURE a;
```

BCD 码的加法器：

```
ENTITY bcdadd IS
PORT (
op1, op2 :IN INTEGER RANGE 0 TO 9 ;
result :OUT INTEGER RANGE 0 TO 31
);
END bcdadder;
ARCHITECTURE a OF bcdadder IS
CONSTANT adj :INTEGER := 6 ; --定义常数adj=6
```

```
SIGNAL binadd :INTEGER RANGE 0 TO 18 ;
--定义信号 binadd 的取值范围是 0～18
BEGIN
binadd <=op1+op2; --求 op1+op2 和运算
PROCESS (binadd)
VARIABLE tmp : INTEGER:=0; --定义变量 tmp 是整数型，初值是 0
BEGIN
IF binadd > 9 THEN --如果 binadd 大于 9，结果要调整

tmp := adj ; --方法是和加 6，否则，结果加 0。
ELSE
tmp := 0 ;
END IF ;
result <=binadd+tmp ; --给外部信号赋值
END PROCESS;
END a;
```

计算 Y=AB+C⊕D：

```
ENTITY loga IS
PORT (
A, B, C, D : IN BIT;
Y : OUT BIT
);
END loga;  --定义 A, B, C, D 是输入端口信号，Y 是输出端口信号
ARCHITECTURE stra OF loga IS
SIGNALE : BIT; --定义 E 是内部信号
BEGIN
Y <=(A AND B) OR E; --以下两条并行语句与顺序无关

E <=C XOR D;
END stra;
```

四选一数据选择器：

```
LIBRARY IEEE;
USE IEEE.STD_LOGIC_1164.ALL;
USE IEEE.STD_LOGIC_UNSIGNED.ALL;
ENTITY mux4 IS
PORT(
a0, a1, a2, a3 :IN STD_LOGIC;
s :IN STD_LOGIC_VECTOR (1 DOWNTO 0);
y :OUT STD_LOGIC
);
END mux4;
ARCHITECTURE archmux OF mux4 IS
BEGIN

y <= a0 WHEN s = "00" else  --当 s=00 时，y=a0
```

```
            a1 WHEN s = "01" else   --当s=01时，y=a1

            a2 WHEN s = "10" else   --当s=10时，y=a2

            a3;    --当s取其他值时，y=a3
            END archmux;
```

D 触发器：

```
LIBRARY IEEE;
USE IEEE.STD_LOGIC_1164.ALL;
USE IEEE.STD_LOGIC_ARITH.ALL;
USE IEEE.STD_LOGIC_UNSIGNED.ALL;
ENTITY dff1 IS
PORT(
CLK, D : IN STD_LOGIC;
Q : OUT STD_LOGIC
);
END dff1;
ARCHITECTURE a OF dff1 IS
BEGIN
PROCESS (CLK)
BEGIN
IF CLK'EVENT AND CLK='1' THEN
Q <= D;
END IF;
END PROCESS;
END a;
```

奇偶校验器中的奇校验：

```
LIBRARY IEEE;
USE IEEE.STD_LOGIC_1164.ALL;
USE IEEE.STD_LOGIC_ARITH.ALL;
USE IEEE.STD_LOGIC_UNSIGNED.ALL;
ENTITY loop1 IS
PORT(
D : IN STD_LOGIC_VECTOR(0 TO 7);  --输入D是八位二进制数
Y : OUT STD_LOGIC
);
END loop1;
tmp := .0.;
BEGIN
FOR I IN 0 TO 7 LOOP
tmp := tmp XOR D(I);  --变量赋值语句是立即赋值，tmp=tmp⊕D(I)
END LOOP;
Y<= tmp ;
END PROCESS;
END a;
```

四位移位寄存器:

```
LIBRARY IEEE;
USE IEEE.STD_LOGIC_1164.ALL;
USE IEEE.STD_LOGIC_ARITH.ALL;
USE IEEE.STD_LOGIC_UNSIGNED.ALL;
ENTITY shift IS
PORT(
DIN, CLK : IN STD_LOGIC;
DOUT : OUT STD_LOGIC;
Q : BUFFER STD_LOGIC_VECTOR(3 DOWNTO 0)
);
END shift;
ARCHITECTURE B OF shift IS
COMPONENT dff1
PORT ( D, CLK : IN STD_LOGIC;
Q : OUT STD_LOGIC );
END COMPONENT;
SIGNAL D : STD_LOGIC_VECTOR(0 TO 4);
BEGIN
D(0)<= DIN;
gen2: FOR I IN 0 TO 3 GENERATE
fx: dff1 PORT MAP (D(I),CLK,D(I+1));
END GENERATE ;
Q(0) <= D(1);
Q(1) <= D(2);
Q(2) <= D(3);
Q(3) <= D(4);
DOUT<=D(4)
END B;
```

3.5.2 Verilog 参考例子

8 位全加器:

```
module adder8(cout,sum,ina,inb,cin);
output[7:0] sum;
output cout;
input[7:0] ina,inb;
input cin;

assign {cout,sum}=ina+inb+cin;     //全加

endmodule
```

8 位计数器:

```
module counter8(out,cout,data,load,cin,clk);
```

```
output[7:0] out;
output cout;
input[7:0] data;
input load,cin,clk;
reg[7:0] out;
always @(posedge clk)
    begin
        if(load)
            out=data;
        else
            out=out+cin;
    end
assign cout=&out&cin;
endmodule
```

4 选 1 数据选择器：

```
module mux4_1(out,in0,in1,in2,in3,sel);
output out;
input in0,in1,in2,in3;
input[1:0] sel;
reg out;
always @ (in0 or in1 or in2 or in3 or sel)
    case(sel)
        2'b00: out=in0;
        2'b01: out=in1;
        2'b10: out=in2;
        2'b11: out=in3;
        default: out=2'bx;
    endcase
endmodule
```

3.6 本章小结

本章介绍了 FPGA 和 ASIC 以及芯片设计重要方法状态机设计方法，然后介绍了 VHDL 和 Verilog 两种硬件编程语言，以及 FPGA 开发设计工具 Quartus，最后对国内芯片产业的发展进行了描述。

3.7 本章参考文献

[1] A. Hodjat, P. Schaumont, I. Verbauwhede. Architectural Design Features of a Programmable

High Throughput AES Coprocessor[C]. Proceedings of the International Conference on Information Technology: Coding and Computing (ITCC'04).

[2] Gerardo Orlando. Efficient Elliptic Curve Processor Architectures for Field Programmable Logic[O]. 2002-3.

[3] A. Hodjat, I. Verbauwhede, Minimum Area Cost for a 30 to 70 Gbits/s AES Processor, Proceedings of IEEE computer Society Annual Symposium on VLSI[C], Pages: 83-88, February 2004.

[4] A. Satoh, K. Takano, A Scalable Dual-Field Elliptic Curve Cryptographic Processor [J]. IEEE Transactions on Computers, vol. 52, no. 4, April 2003, pp. 449-460.

[5] H. Eberle, N. Gura, S. Shantz, V. Gupta, L. Rarick. A Public-key Cryptographic Processor for RSA and ECC[C]. Proceedings of the 15th IEEE International Conference on Application-Specific Systems, Architectures and Processors (ASAP'04).

[6] 曲英杰，刘卫东，战嘉瑾. 可重构密码协处理器指令系统的设计方法[J]. 计算机工程与应用，2004.2(1):10-12.

[7] 田泽，张怡浩，于敦山，盛世敏，仇玉林. 单片密码数据处理器系统级体系结构的研究[J]. 固体电子学研究与进展，2003,4(23):427-433.

[8] 林容益. CPU/SOC 及外围电路应用设计[M]. 北京航空航天大学出版社，2004.

[9] Brickell E.. A survey of hardware implementations of RSA. Advances in Cryptology—CRYPTO'89 Proceedings, 1990, 368-370.

[10] Rudra A, Dubey P, Jutla C, et al. Efficient Rijndael encryption implementation with composite field arithmetic[A]. In: Cryptographic Hardware and Embedded Systems - CHES 2001[C], 2001: 171-184.

[11] Zhang X, Parhi K. On the Optimum Constructions of Composite Field for the AES Algorithm[J]. IEEE Trans. Circuits and Systems II: Express Briefs, 2006, 53(10): 1153 -1157.

[12] Schroeppel R, Orman H, O'Malley S W, et al. Fast Key Exchange with Elliptic Curve Systems[A]. In: Proceedings of the 15th Annual International Cryptology Conference on Advances in Cryptology[C]. London, UK: Springer-Verlag. 1995. CRYPTO '95.

[13] Satoh A, Morioka S, Takano K, et al. A compact Rijndael hardware architecture with S-box optimization[A]. In: Advances in Cryptology—ASIACRYPT 2001[C]. Springer, 2001: 239-254.

[14] Jeong Y J, Burleson W P. VLSI array algorithms and architectures for RSA modular multiplication[J]. IEEE Transactions on Very Large Scale Integration (VLSI) Systems, 1997, 5(2): 211-217.

[15] Kim H W, Lee S. Design and implementation of a private and public key crypto processor and its application to a security system[J]. IEEE Transactions on Consumer Electronics, 2004,

50(1): 214-224.

[16] Giraud C. An RSA implementation resistant to fault attacks and to simple power analysis[J]. IEEE Transactions on Computers, 2006, 55(9): 1116-1120.

[17] Zhang C N. Integrated approach for fault tolerance and digital signature in RSA[J]. IEE ProceedingsComputers and Digital Techniques, 1999, 146(3): 151-159.

[18] Ma K, Liang H, Wu K. Homomorphic property-based concurrent error detection of RSA: a countermeasure to fault attack[J]. IEEE Transactions on Computers, 2012, 61(7): 1040-1049.

[19] Yen S M, Kim S, Lim S, et al. RSA speedup with Chinese remainder theorem immune against hardware fault cryptanalysis[J]. IEEE Transactions on Computers, 2003, 52(4): 461-472.

[20] Sewell R. Bulk encryption algorithm for use with RSA[J]. Electronics Letters, 1993, 29(25):2183-2185.

[21] Bajard J, Imbert L. A full RNS implementation of RSA[J]. IEEE Transactions on Computers, 2004, 53(6): 769-774.

[22] Chang C C, Hwang M S. Parallel computation of the generating keys for RSA cryptosystems[J]. Electronics Letters, 1996, 32(15): 1365-1366.

[23] [76] Hong J H, Wu C W. Cellular-array modular multiplier for fast RSA public-key cryptosystem based on modified Booth's algorithm[J]. IEEE Transactions on Very Large Scale Integration (VLSI) Systems, 2003, 11(3): 474-484.

[24] Selby A, Mitchell C. Algorithms for software implementations of RSA[J]. IEE Proceedings E Computers and Digital Techniques, 1989, 136(3): 166-170.

[25] Takagi N, Yajima S. Modular multiplication hardware algorithms with a redundant representation and their application to RSA cryptosystem[J]. IEEE Transactions on Computers, 1992, 41(7): 887-891.

[26] Aydos M, Yanık T, Koc C. High-speed implementation of an ECC-based wireless authentication protocol on an ARM microprocessor[J]. IEEE Proceedings-Communications, 2001, 148(5): 273-279.

[27] Suzuki T, Yamagami Y, Hatanaka I, et al. A sub-0.5-V operating embedded SRAM featuring a multibit-error-immune hidden-ECC scheme[J]. IEEE Journal of Solid-State Circuits, 2006, 41(1): 152-160.

[28] Cilardo A, Coppolino L, Mazzocca N, et al. Elliptic curve cryptography engineering [J]. Proceedings of the IEEE, 2006, 94(2): 395-406.

[29] Lai J Y, Huang C T. A highly efficient cipher processor for dual-field elliptic curve cryptography[J]. IEEE Transactions on Circuits and Systems II: Express Briefs, 2009, 56(5): 394-398.

[30] Hassan M N, Benaissa M. Embedded software design of scalable low-area elliptic-curve cryptography[J]. IEEE Embedded Systems Letters, 2009, 1(2): 42-45.

[31] Cheung R C, Telle N J b, Luk W, et al. Customizable elliptic curve cryptosystems[J]. IEEE Transactions on Very Large Scale Integration (VLSI) Systems, 2005, 13(9): 1048-1059.

[32] Lai J Y, Huang C T. Energy-adaptive dual-field processor for high-performance elliptic curve cryptographic applications[J]. IEEE Transactions on Very Large Scale Integration (VLSI) Systems, 2011, 19(8): 1512-1517.

[33] Ananyi K, Alrimeih H, Rakhmatov D. Flexible hardware processor for elliptic curve cryptography over NIST prime fields[J]. IEEE Transactions on Very Large Scale Integration (VLSI) Systems, 2009, 17(8): 1099-1112.

[34] 张红霞, 凌兰兰. 智能卡技术及其应用. 洛阳大学学报, 2006 年 2 月, 第 17 卷 (第 2 期): 42～43.

第 4 章　多变量公钥密码技术

4.1　多变量公钥密码概述

多变量公钥密码的研究最早起源于 20 世纪 80 年代。在 1988 年，日本学者 Tsutomu Matsumoto 和 Hideki Imai 共同提出了第一个多变量公钥密码方案，即著名的 MI 加密算法。在 1995 年，法国学者 Jacques Patarin 用一种线性化方程的攻击方法破解了 Tsutomu Matsumoto 和 Hideki Imai 共同提出的 MI 加密算法，证明了它不安全。随后，Jacques Patarin 通过改进 MI 加密算法，提出了 HFE 密码算法，事实上它也不安全，倍受攻击。在 1997 年，Jacques Patarin 在线性化方程攻击的基础上，提出了油醋签名算法，这是目前多变量油醋家族的起源。

美国 University of Cincinnati 学者 Jintai Ding 在 2009 年的著作《Multivariate Public Key Cryptosystems》中系统地总结了多变量公钥密码体制的发展概况，他根据陷门的构造方式将目前提出的方案分成 3 类：

1）第一类中心映射建立在扩域上方案，典型方案有 MI 方案和 HFE 方案等；

2）第二类中心映射建立在基域上的方案，如三角形体制 TTM 方案和非平衡油醋 UOV 方案；

3）第三类介于第一类和第二类之间，其中心映射是建立在中间的域上的方案，如 MFE 方案和可逆循环体制 ℓ-IC 方案。

目前，多变量公钥密码的研究重点在于通过多种变形的方法设计安全性较高的多变量方案。典型的多变量公钥密码的变形方法有加方法、减方法、醋变量方法和内部扰动方法等，其中减方法和醋变量方法主要用于设计多变量数字签名方案，内部扰动方法则是由美国学者 Jintai Ding 提出的一种系统化的增强多变量公钥密码的安全性的方法。通过变形后的最著名的多变量公钥密码算法是 SFlash 签名算法（从 MI 减加密算法发展而来），曾被 NESSIE 确定为 2004 年低功耗智能卡的安全标准，但 Dubois 在 2007 年结合差分攻击和线性化方程将 SFLASH 方案攻破。其他变形方案包括 Rainbow 签名算法（从非平衡油醋签名算法发展而来）、PMI 加密算法（对 MI 加密算法进行内部扰动的变体）和 PMI+加密算法（通过整合内部扰动和加变形对 PMI 进行改造而来）。

4.1.1　多变量加密

当前比较流行的多变量公钥密码体制的加密方案是 MI 体制的变型。Jintai Ding 在

2004 年通过对 MI 的中心映射增加内部扰动增强了 MI 的安全性，产生了针对 MI 密码体制的新变体 PMI 加密体制，而 PMI 加密体制很快被 Fouque 等人在 2005 年使用差分攻击的方法攻破。Jintai Ding 在 2006 年对 PMI 的中心映射引入新的外部扰动，产生 PMI+ 加密体制，安全性得到了很大的提升。到目前为止，PMI+ 加密体制还是安全的。Jintai Ding 在 2004 年也通过对 HFE 的中心映射增加内部扰动增强了安全性，产生了针对 HFE 密码体制的新变体 IPHFE 加密体制。Jintai Ding 等人在 2007 年对 IPHFE 方案进行了安全性分析，发现 Jintai Ding 给出的参数不能达到安全性要求，他们给出了新的参数。到目前为止，IPHFE 加密体制还是安全的。

4.1.2 多变量签名

当前比较流行的多变量公钥密码体制的签名方案是油醋签名体制，这类签名体制比较典型。Jacques Patarin 史无前例地找到了一种跟以往多变量公钥密码思想完全不同的方法去构造新的方案，即油醋签名方案。油醋签名体制分为平衡油醋和非平衡油醋两种方案。Patarin 提出的平衡油醋方案很快被攻破了，Kipnis、Patarin 和 Goubin 在其基础上提出的非平衡油醋签名方案也被攻破。Jintai Ding 等人在 2005 年提出了一种多层非平衡油醋方案，即安全高效的 Rainbow 签名。到目前为止，Rinbow 签名还是安全有效的。Petzoldt 等人在 2011 年提出对 Rainbow 的公钥进行缩短的方案可以在不降低安全性的前提下将 Rainbow 的公钥大小减少到原来的 86%。TTS 签名体制是 Rainbow 签名的一个特例，最早由 Chen 和 Yang 于 2002 年在 TTM 的基础上进行改进并提出，之后他们又在 2004 年指出早期版本的 TTS 不能满足安全性要求，并给出了新的方案。

4.1.3 多变量公钥密码芯片

多变量公钥密码的研究主要集中在算法设计和安全性分析领域，而相应的密码硬件的研究和设计并不是特别多，这导致多变量公钥密码的商业和工业应用远远不如 AES、RSA 和椭圆曲线密码。

事实上，因为多变量公钥密码一般是在小的有限域进行运算，所以它比在大的有限域的运算的密码更加高效，在很多方面可以媲美甚至超越 RSA 和椭圆曲线密码。因为多变量公钥密码的高效性，它的算法可以广泛应用于各类硬件和设备，如现场可编辑逻辑门阵列（Field Programmable Gate Array，FPGA）、专用集成电路（Application Specific Integrated Circuits，ASIC）、通用处理器、单片机、嵌入式设备、手持设备、无线传感器网络、射频识别标签和智能卡等。常用的 FPGA 芯片包括 Xilinx FPGA 和 Altera FPGA，它们提供了灵活的电路设计能力。台湾积体电路制造公司的 TSMC-0.18μm 标准库 CMOS

ASIC 是常用工艺库，它提供了全套可选择的电路设计的工艺标准。

4.2 多变量公钥密码系统

公钥密码体制的核心问题是陷门的构造。RSA 公钥密码体制是目前使用最为广泛的公钥密码体制，其陷门为大整数分解的困难问题。椭圆曲线公钥密码体制是当前被认为替代 RSA 公钥密码体制的最佳候选。与 RSA 公钥密码体制相比，椭圆曲线公钥密码体制的密钥长度更短，加密速度更快，并且随着加密强度的提高，密钥长度增长不大。椭圆曲线公钥密码体制的陷门为求解椭圆曲线点群上的离散对数问题是困难问题。Shor 在 1997 年提出的量子分解算法，可在多项式时间内解决大整数分解问题或离散对数问题，因此 RSA 和椭圆曲线密码的陷阱将不再成为陷门，从而不会再有安全性可言。

多变量公钥密码体制的陷门为 MQ 问题。Patarin 等人在 1997 年证明了 MQ 问题在任意有限域上都是 NP 完全问题。量子分解算法对解 MQ 问题并没有优势，所以在量子时代，多变量公钥密码体制的安全性仍然能得到保证。并且，多变量公钥密码体制主要运算为矩阵运算，具有计算速度快和适合硬件实现等优良特性，非常适合在低端智能卡等硬件设备上实现。

Jingtai Ding 在 2009 年的著作中对多变量公钥密码算法的发展概况做了详细的介绍和系统的总结。他根据陷门的构造方式提出将目前提出的方案分成 3 类：双极系统、混极系统和多项式同构系统，其中，双极系统是多变量公钥密码研究的热点，现存的多变量公钥密码方案中，双极系统占了大多数。

双极系统构造的基本思想是利用 L_1 和 L_2 对中心映射进行隐藏。双极系统中的中心映射可以很容易被求出原像，并且要满足一定的结构，所以中心映射的选择非常有限，攻击者可以很容易地猜测中心映射进行攻击。

双极系统的基本构造方法如下：定义 k 为有限基域，在双极系统中，通过定义一个从 k^n 扩域到 k^m 扩域的一个非线性映射 $\bar{F}$，将明文映射到密文空间。$\bar{F}$ 可以定义为 $\bar{F}(x_1, x_2, ..., x_m) = (\bar{f}_1, \bar{f}_2, ..., \bar{f}_n)$，其中任何一个 $\bar{f}_i$ 都是一个二次多项式，系数属于有限域 k。

双极系统的中心映射 F 构造起来比较简单，F 也是一个从 k^n 扩域到 k^m 扩域的一个映射，需要满足以下两个条件：

1）需要满足 $F(x_1, x_2, ..., x_m) = (f_1, f_2, ..., f_n)$，其中任何一个 f_i 都是一个二次多项式，系数属于有限基域 k；

2）利用 F 构造出方程组 $F(x_1, x_2, ..., x_m) = (y_1, y_2, ..., y_n)$，选取任意的 $(y_1, y_2, ..., y_n)$ 方程组都可以很容易地求解，也就是说，可以很容易找到 $(y_1, y_2, ..., y_n)$ 的原像或者找到方程组逆映射 $F^{-1}(y_1, y_2, ..., y_n) = (x_1, x_2, ..., x_m)$。

这样，我们就构造出了中心映射 F，然后随机选取两个可逆映射 L_1 和 L_2，最终我们

得到 $\bar{F}(x_1,x_2,...,x_m) = L_1 \circ F \circ L_2(x_1,x_2,...,x_m) = (y_1,y_2,...,y_n)$。$\bar{F}$ 作为系统的公钥，L_1 和 L_2 的逆映射 L_1^{-1} 和 L_2^{-1}，中心映射的逆映射 F^{-1} 作为系统的私钥被保存起来。图 4-1 展示了双极系统的多变量公钥密码构造过程。

双极系统的加密解密过程也比较简单和直接。加密只需将明文 $(x_1,x_2,...,x_n)$ 加入公钥多项式 $\bar{F}$ 中，就可以很容易计算出密文 $(y_1,y_2,...,y_n)$。解密相对复杂一些，需要求解方程组 $\bar{F}(x_1,x_2,...,x_m) = (y_1,y_2,...,y_n)$，首先将 $(y_1,y_2,...,y_n)$ 代入 L_1^{-1} 得到 $(y_1',y_2',...,y_n')$，然后再将 $(y_1',y_2',...,y_n')$ 代入 F^{-1} 得到 $(y_1'',y_2'',...,y_n'')$，最后将 $(y_1'',y_2'',...,y_n'')$ 代入 L_2^{-1} 之后得到明文 $(x_1,x_2,...,x_m)$。图 4-2 展示了双极系统加密解密的过程。

签名与验证签名的过程刚好和加密与解密的过程相反，双极系统的签名和验证的过程也非常简单。签名等同于解密，对消息 $(y_1,y_2,...,y_n)$ 进行签名，生成 $(x_1,x_2,...,x_m)$，需要求解 $\bar{F}(x_1,x_2,...,x_m) = (y_1,y_2,...,y_n)$，首先将 $(y_1,y_2,...,y_n)$ 代入 L_1^{-1} 得到 $(y_1',y_2',...,y_n')$，然后再将 $(y_1',y_2',...,y_n')$ 代入 F^{-1} 得到 $(y_1'',y_2'',...,y_n'')$，最后将 $(y_1'',y_2'',...,y_n'')$ 代入 L_2^{-1} 之后得到签名 $(x_1,x_2,...,x_m)$。验证签名等同于加密，只需要将签名 $(x_1,x_2,...,x_m)$ 代入公钥多项式 $\bar{F}$ 中，就可以很容易计算出消息 $(\tilde{y}_1,\tilde{y}_2,...,\tilde{y}_n)$，将其与原消息 $(y_1,y_2,...,y_n)$ 比较，如果相等，则签名验证成功，该签名是有效的，否则签名无效。

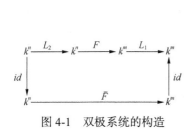

图 4-1 双极系统的构造　　　图 4-2 双极系统的加密解密过程

4.3 多变量公钥密码典型算法

本节我们将介绍几种具有代表性的方案，这些方案各自代表了一系列的陷门构造方式。这些陷门的构造方式分为三大类：

1）中心映射建立在扩域上方案，典型方案有 MI 方案和 HFE 方案等；

2）中心映射建立在基域上的方案，如三角形体制 TTM 方案和非平衡油醋 UOV 方案；

3）介于第一类和第二类之间，其中心映射是建立在中间的域上的方案，如 MFE 方案和可逆循环体制 ℓ-IC 方案。下面将介绍这几种具有代表性的方案的构造方式。

4.3.1 MI 密码算法

MI 公钥密码体制在 1988 年由 Matsumoto 和 Imai 提出，它是第一个具有实际意义的多变量公钥密码方案，在多变量公钥密码体制的发展历程中占据了重要的地位。MI 方案的基本思想是利用有限扩域 k 上的隐藏域结构和向量空间 k^n 来构造出陷门。MI 公钥密码体制在 1996 年被 Patarin 利用求解线性化方程的方法攻破了，但是其对多变量公钥密码的发展起了很大的作用。它将最新的数学思想引入公钥密码体系，引发了多变量公钥密码的研究热潮，后面产生了很多 MI 方案的变体，如 Akkar 等人在 2002 年提出的 SFLASH 签名方案；Ding 等人在 2004 在对 MI 方案加入内部扰动后提出 PMI 加密解密方案；在 2006 年提出 PMI 的改进版 PMI+加密方案，PMI+方案到目前为止还是安全的。下面我们将对 MI 方案的构造进行详细的介绍。

MI 的构造如下。

设有限域 k 的阶为 2，基为 q，$g(x) \in k[x]$ 是不可约多项式。有限域 K 为有限基域 k 的 n 次扩域，有限扩域 K 到 n 维向量空间 k^n 的映射可以表示为 $\phi: K \rightarrow k^n$，即 $\phi(a_0 + a_1 x + \ldots + a_{n-1} x^{n-1}) = (a_0, a_1, \ldots, a_{n-1})$。

首先，选择参数 θ，满足条件：$0 < \theta < n$ 和 $\gcd(q^\theta + 1, q^n - 1) = 1$；

然后，选择合适的有限扩域 K 上的映射函数 $\widetilde{F}(x) = x^{1+q^\theta}$，由 θ 的性质可得 $\widetilde{F}(x)$ 是可逆的，$\widetilde{F}^{-1}(X) = X^t$，中心映射 F 为 n 维向量空间 k^n 上的映射，可以表示为 $F(x) = \phi \circ \widetilde{F} \circ \phi^{-1}(x)$。

最后，在 n 维向量空间 k^n 上随机选择两个可逆仿射 L_1 和 L_2，则 $\overline{F}(x) = L_1 \circ F \circ L_2(x)$ 作为 MI 的公钥，私钥是中心映射 F，L_1 的逆映射 L_1^{-1} 和 L_2 的逆映射 L_2^{-1}。

MI 的加密解密过程如下。

加密：只需将明文 x 代入公钥多项式 $\overline{F}$ 中，就可以很容易计算出密文 y。

解密：解密的过程相对来说复杂一些，需要求解方程组 $\overline{F}(x) = (y)$，分为以下几步：首先将 y 代入 L_1^{-1} 得到 y'，然后再将 y' 代入 F^{-1} 得到 y''，最后将 y'' 代入 L_2^{-1} 之后得到明文 x。

4.3.2 PMI+密码算法

PMI+加密体制是 Ding 等人在 PMI 加密体制被 Fouque 等人在 2005 年使用差分攻击攻破之后，对其进行改进而产生的一种新的多变量公钥加密体制。到目前为止，该方案还是安全的。PMI+的基本思想是对 MI 方案的中心映射加上内部扰动和外部扰动来抵御线性化方程攻击和差分攻击。由于 PMI+加密体制加密产生的密文是唯一确定的，也可以用于构造多变量公钥签名体制。

PMI+的构造如下。

令 k 是一个有限域（通常 k 的阶数为 2），K 是有限域 k 的 n 次扩域。定义 $\phi: K \to k^n$ 是从有限域 K 到 n 维向量空间 k^n 的同构映射，即：

$$\phi(a_0 + a_1 x + \ldots + a_{n-1} x^{n-1}) = (a_0, a_1, \ldots, a_{n-1})$$

选择参数 θ，θ 满足 $0 < \theta < n$，且 $\gcd(q^\theta + 1, q^n - 1) = 1$。

选择有限域 K 上的映射函数 $\widetilde{F}(X) = X^{1+q^\theta}$，则 $\widetilde{F}$ 的逆映射可以很容易地求得：

$$t(1 + q^\theta) = 1 \bmod (q^n - 1)$$
$$\widetilde{F}^{-1}(X) = X^t$$

定义 n 维向量空间 k^n 上的映射：

$$F'(x_1, x_2, \ldots, x_n) = \phi \circ \widetilde{F} \circ \phi^{-1}(x_1, x_2, \ldots, x_n)$$

定义线性方程组：

$$z_1(x_1, x_2, \ldots, x_n) = \sum_{j=1}^{n} \alpha_{j1} x_j + \beta_1$$

$$\ldots$$

$$z_r(x_1, x_2, \ldots, x_n) = \sum_{j=1}^{n} \alpha_{jr} x_j + \beta_r$$

其中 r 为小的随机整数，$z_i - \beta_i$ 是线性无关的。

令映射 $Z: k^n \to k^r$ 定义为：

$$Z(x_1, x_2, \ldots, x_n) = (z_1(x_1, x_2, \ldots, x_n), z_2(x_1, x_2, \ldots, x_n), \ldots, z_r(x_1, x_2, \ldots, x_n))$$

映射 Z 为内部扰动源头。

随机选 n 个多项式 $\hat{f}_1, \hat{f}_2, \ldots, \hat{f}_n \in k[z_1, z_2, \ldots, z_r]$，其阶为 2，令映射 $\hat{F}: k^r \to k^n$ 定义为：

$$\hat{F}: (z_1, z_2, \ldots, z_r) = (\hat{f}_1(z_1(x_1, x_2, \ldots, x_n), z_2(x_1, x_2, \ldots, x_n), \ldots, (z_r(x_1, x_2, \ldots, x_n))))$$

定义扰动映射：

$$F^*(x_1, x_2, \ldots, x_n) = \hat{F} \circ Z(x_1, x_2, \ldots, x_n) = (f_1^*, f_2^*, \ldots, f_n^*)$$

定义映射：

$$F(x_1, x_2, \ldots, x_n) = (F' + F^*) = (x_1, x_2, \ldots, x_n)$$

给中心映射 F 随机选择 a 个关于变量 $x_1, x_2, \ldots, x_n$ 非线性方程作为外部扰动，相对于 PMI，攻击 PMI+ 的复杂度增加了 C_n^a 倍。在 $n+a$ 维向量空间 k^{n+a} 上随机选择可逆仿射 L_1，在 n 维向量空间 k^n 上随机选择可逆仿射 L_2，则 $\overline{F}(x_1, x_2, \ldots, x_n) = L_1 \circ F \circ L_2(x_1, x_2, \ldots, x_n)$ 为 PMI+ 的公钥，即为 $n+a$ 个二次多项式。系数和变量都属于基域 $GF(2)$，私钥是中心映射 F、内部扰动映射 $\hat{F}$ 以及 L_1 的逆映射 L_1^{-1} 和 L_2 的逆映射 L_2^{-1}。

PMI+加密的算法比较简单，给定明文分组 $(x_1, x_2, \ldots, x_n)$，要对该明文进行加密，只需要把明文分组代入公钥映射即可：

$$y_1 = \overline{f}_1(x_1, x_2, \ldots, x_n)$$
$$\ldots$$
$$y_{n+a} = \overline{f}_{n+a}(x_1, x_2, \ldots, x_n)$$

计算 $n+a$ 个多项式的值，获得密文为 $(y_1, y_2, \ldots, y_{n+a})$。

PMI+解密算法相对复杂一些，给定密文分组 $(y_1, \ldots, y_n, \ldots, y_{n+a})$，要对密文进行解密，等价于求解方程组：

$$\overline{F}(x_1, x_2, \ldots, x_n) = L_1 \circ F \circ L_2(x_1, x_2, \ldots, x_n) = (y_1, y_2, \ldots, y_{n+a})$$

可以将其转化为计算：

$$X = (x_1, x_2, \ldots, x_n) = L_2^{-1} \circ F^{-1} \circ L_1^{-1}(y_1, y_2, \ldots, y_{n+a})$$

具体的过程如下。

1）计算 L_1 的逆仿射函数：$Y' = L_1^{-1}(Y) = (y_1', y_2', \ldots, y_{n+a}')$。

2）将 a 个外部扰动多项式从 Y' 中去掉，得到 $\overline{Y} = (\overline{y}_1, \overline{y}_2, \ldots, \overline{y}_n)$，本步骤是 PMI+算法与 PMI 算法的重要区别。

3）内部扰动映射 $\hat{F}$ 是向量 k^r 到向量空间 k^n 的映射，即对每一个 $\mu \in k^r$ 都存在 $\lambda \in k^n$，满足 $\hat{F}(\mu) = \lambda$。对每一个 (λ, μ) 对，计算 $(y_{\lambda 1}, \ldots, y_{\lambda n}) = F^{-1}((\overline{y}_1, \overline{y}_2, \ldots, \overline{y}_n) + \lambda)$，得到 λ 之后计算 $\mu' = Z(\lambda)$，判断 μ' 是否等于 μ，如果不相等，则继续循环，否则进入下一步骤。

4）将 λ 代入公钥多项式中外部扰动的 a 个多项式中，验证是否与密文中的相应的元素相等，如果验证成功，则进入下一步骤，否则返回上一步计算。

5）计算 L_2 的逆仿射函数：$X = L_2^{-1}(\lambda) = (x_1, x_2, \ldots, x_n)$，则 X 为解密所得明文。

PMI+算法的复杂度如表 4-1 所示。

表 4-1　　PMI+算法的复杂度

PMI+算法	复杂度
生成公私钥	$O(n^4)$
加密	$O(n^3)$
解密	$O(kn^2)$

4.3.3　HFE 密码算法

HFE 公钥加密方案是由 Patarin 在 1996 年利用线性化方程攻破 MI 方案之后，在 MI 方案的基础上变种得来。HFE 方案相对 MI 方案的不同之处主要在于中心映射的构造：MI 方案的中心映射为单变量单项式结构，而 HFE 方案的中心映射则是单变量多项式结构。HFE 加密解密方案已经被 Kipins 等人提出的再线性化方法攻破，而 Faugère 等人则利用 Gröbner 基算法成功破解了"HFE 挑战一"算法，根据明文解出了对应的密文。Patarin 等人在 2001 年将 Oil-Vinegar 和 Minus 方法结合在一起，提出了 HFE 加密体制的变形方案，即 Quartz 签名方案，该方案曾在 2004 年被 NESSIE 推荐为首选短签名方案。Ding 等人在 2004 年对 HFE 加入内部扰动而提出 IPHFE 方案，该方案至今还是安全的，但是其效率比较低。下面我们将对最初的 HFE 方案的构造进行详细的介绍。

HFE 的构造如下。

设有限域 k 的阶为 2，基阶为 q，$g(x) \in k[x]$ 是不可约多项式。K 是有限域 k 的 n 次扩域，有限扩域 K 到 n 维向量空间 k^n 的映射可以表示为 $\phi: K \to k^n$，即 $\phi(a_0 + a_1 x + \ldots + a_{n-1} x^{n-1}) = (a_0, a_1, \ldots, a_{n-1})$。HFE 的设计思想和 MI 非常相似，主要区别是映射函数的设计。HFE 的映射函数可以表示如下：

$$\widetilde{F}(X) = \sum_{i=0}^{r_2-1} \sum_{j=0}^{i} a_{ij} X^{q^i + q^j} + \sum_{i=0}^{r_1-1} b_i X^{q^i} + c$$

其中 a_{ij}, b_i, c 是随机选择系数变量，r_1, r_2 的选择需要令 $\widetilde{F}(X)$ 的度小于参数 d。d 主要是用来控制映射函数求逆的复杂度，根据 Berlekamp 算法可以得出 $\widetilde{F}^{-1}(X)$ 的计算复杂度为 $o(nd^2 \log d + d^3)$，如果参数 d 过大，则 $\widetilde{F}^{-1}(X)$ 的计算复杂度会急剧升高。

中心映射 F 是 n 维向量空间 k^n 上的映射，可以表示为 $F(x_1, x_2, \ldots, x_n) = \phi \circ \widetilde{F} \circ \phi^{-1} (x_1, x_2, \ldots, x_n)$。

在 n 维向量空间 k^n 上随机选择两个可逆仿射 L_1 和 L_2，则 $\overline{F}(x_1, x_2, \ldots, x_n) = L_1 \circ \widetilde{F} \circ L_2 (x_1, x_2, \ldots, x_n)$ 为 HFE 的公钥，私钥为中心映射 F，L_1 的逆映射 L_1^{-1} 和 L_2 的逆映射 L_2^{-1}。

HFE 的加密解密过程如下。

加密同 MI 方案的加密方式类似，只需将明文 $(x_1, x_2, \ldots, x_n)$ 带入公钥多项式 $\overline{F}$ 中，就可以很容易计算出密文 $(y_1, y_2, \ldots, y_n)$。

解密相对来说复杂一些，需要求解方程组 $\overline{F}(x_1, x_2, \ldots, x_n) = (y_1, y_1, \ldots, y_n)$：

首先，将 $(y_1, y_2, \ldots, y_n)$ 代入 L_1^{-1} 得到 $(y_1', y_2', \ldots, y_n')$；

然后，令 $Y' = \phi^{-1}(y_1', y_2', \ldots, y_n')$，计算集合 $Z = \{Z \in K | \widetilde{F}(Z) = Y'\}$，在这里可以利用 Berlekamp 算法来计算 Z；

最后，对每一个 $Z_i \in Z$，计算得到得到的候选明文 $(x_{i1}, x_{i2}, \ldots, x_{in}) = L_2^{-1} \circ \phi(Z_i)$，这里可能存在多组解，可以使用哈希函数和 Plus 法来找到真正的明文。

4.3.4 ℓ-IC 密码算法

ℓ-IC 公钥密码方案是在 2007 年由 Ding 等人在 PKC2007 会议上首次提出的，方案的陷门的构造采用了中间域，其中心映射是利用 Cremona 变换映射的变形映射来构造的。Cremona 变换映射可以表示为 $(A_1, A_2, A_3) \rightarrow (A_2A_3, A_3A_1, A_1A_2)$。

当 $A_1A_2A_3 \neq 0$ 时，Cremona 变换映射一定可以求出逆映射，这是一个非常好的性质，非常适合构造双极系统的中心映射。由于 Cremona 变换映射具有轮换的特性，所以 ℓ-IC 公钥密码方案中有轮换的结构。ℓ-IC 公钥密码方案的速度快于 Quartz 和 SFLASH 方案，而且公钥长度较小，所以比较适合在资源受限的硬件中实现。下面我们将对最基本的 ℓ-IC 方案的构造进行详细的介绍。

ℓ-IC 公钥密码系统的构造如下。

1）设有限域 k 的阶为 q，有限域 $K = k^r$ 是 k 的 r 次扩域，r 是正整数。

2）设 $(\theta_1, \theta_2, \ldots, \theta_l)$ 属于 $(0, 1, \ldots, r-1)$，l 是循环的次数，ℓ-IC 的中心映射函数可以表示为 $\hat{F}: k^l \rightarrow k^l$，即为 $\hat{F}(x_1, x_2, \ldots, x_l) = (X_1^{q^{\theta_1}} X_2, \ldots, X_{l-1}^{q^{\theta_{l-1}}} X_l, X_l^{q^{\theta_l}} X_{l+1})$，这样中心映射就被构造出来了。

3）中心映射 F 为 n 维向量空间 k^n 上的映射，可以将中心映射表示为 $F(x_1, x_2, \ldots, x_n) = \phi \circ \widetilde{F} \circ \phi^{-1}(x_1, x_2, \ldots, x_n)$。

4）在 n 维向量空间 k^n 上随机选择两个可逆仿射 L_1 和 L_2，则 $\overline{F}(x_1, x_2, \ldots, x_n) = L_1 \circ \widetilde{F} \circ L_2(x_1, x_2, \ldots, x_n)$ 为 ℓ-IC 的公钥，私钥为中心映射 F，L_1 的逆映射 L_1^{-1} 和 L_2 的逆映射 L_2^{-1}。

ℓ-IC 的加密解密过程如下。

加密只需将明文 $(x_1, x_2, \ldots, x_n)$ 代入公钥多项式 $\overline{F}$ 中，就可以很容易计算出密文 $(y_1, y_2, \ldots, y_n)$。

解密相对来说复杂一些，需要求解方程组 $\overline{F}(x_1, x_2, \ldots, x_n) = (y_1, y_2, \ldots, y_n)$，分为以下几步：首先将 $(y_1, y_2, \ldots, y_n)$ 代入 L_1^{-1} 得到 $(y_1', y_2', \ldots, y_n')$，然后再将 $(y_1', y_2', \ldots, y_n')$ 代入 F^{-1} 得到 $(y_1'', y_2'', \ldots, y_n'')$，最后将 $(y_1'', y_2'', \ldots, y_n'')$ 代入 L_2^{-1} 之后得到明文 $(x_1, x_2, \ldots, x_n)$。

ℓ-IC 要保证安全性，其中参数 r 要保证足够大，而为了公钥小，参数 l 要尽量小，推荐的选择是 3 到 5。ℓ-IC 方案还有很多变种。

4.3.5　TTM 密码算法

TTM 公钥密码系统属于三角形体制。三角形体制的公钥密码系统的构造方式比较特殊，它的基本思想是基于代数几何学的。

TTM 密码系统在 1999 年由 Moh 首次提出，其构造的思想可以追溯到 Fell 和 Diffie 在 1985 年的工作，当时他们并没有找到高效率和安全的三角形方案。Tsujii 等人利用其思想构造了多变量公钥密码系统，但是被 Hasegawa 等人在 1987 年攻破。Shamir 在 1993

年也用三角形映射构造了签名方案,在 1997 年被 Coppersmith 等人攻破。

三角形体制方案的主要思想是基于求解一组可逆非线性多项式映射的困难性,类似著名的雅可比猜想。雅可比猜想是数学中关于多变量多项式的非常著名的一个猜想,由 Keller 在 1939 年首次提出,接着由 Shreeram 改进为一个基于代数几何的问题。

雅可比猜想的描述如下:

给定一个正整数 $n>1$ 和多项式映射 $G:C^n \to C^n$,可以表示为 $G(x_1,x_2,\ldots,x_n)=(g_1(x_1,x_2,\ldots,x_n),g_2(x_1,x_2,\ldots,x_n),\ldots,g_n(x_1,x_2,\ldots,x_n))$,$G$ 的雅可比行列式 $J(x_1,x_2,\ldots,x_n)$ 是一个 $n\times n$ 的矩阵 $[g_{ij}]$,其中 $g_{ij}=\dfrac{\partial g_i}{\partial x_j}$ 是 $g_i(x_1,x_2,\ldots,x_n)$ 对 x_j 求偏导。

TTM 公钥密码系统的构造如下。

1)设 $F:k^n \to k^n$ 是由一组 l 个可逆非线性多项式映射复合而成的,形如 $F(x_1,x_2,\ldots,x_n)=G_1 \circ G_2 \circ \ldots \circ G_l(x_1,x_2,\ldots,x_n)=(f_1,f_2,\ldots,f_n)$,其中 G_i 是从向量空间 k^n 到 k^n 的可逆非线性映射。

2)$F:k^n \to k^n$ 还需要满足性质:给定 $(x_1',x_2',\ldots,x_n')$ 属于 k^n,计算 $F(x_1',x_2',\ldots,x_n')$ 是高效的;给定 F 的多项式组 $(f_1,f_2,\ldots,f_n)$,找到一组多项式映射 $(G_1,G_2,\ldots,G_l)$ 的困难的。

3)设 $L:k^m \to k^n$ 为可逆仿射,可以表示为 $L(x_1,x_2,\ldots,x_m)=(x_1,\ldots,x_m,0,\ldots,0)$,则 $\overline{F}(x_1,x_2,\ldots,x_m)=F \circ L(x_1,x_2,\ldots,x_m)$ 为 TTM 的公钥,私钥为 G_i 的逆映射 G_i^{-1} 和可逆仿射 L 的逆仿射 L^{-1}。

TTM 的加密解密过程如下。

加密比较简单,只需将明文 $(x_1,x_2,\ldots,x_n)$ 代入公钥多项式 $\overline{F}$ 中,就可以很容易计算出密文 $(y_1,y_2,\ldots,y_n)$。

解密相对来说复杂一些,需要求解方程组 $\overline{F}(x_1,y_1,\ldots,x_n)=(y_1,y_2,\ldots,y_n)$,分为以下几步:

1)将 $(y_1,y_2,\ldots,y_n)$ 代入 L^{-1} 得到 $(y_1',y_2',\ldots,y_n')$。

2)需要依次进行 G_i^{-1} 的逆映射计算,在进行了 l 次之后,得到明文 $(x_1,x_2,\ldots,x_n)$,完成整个解密过程。

4.3.6 TTS 签名算法

TTS 签名体制在 2002 年由 Chen 等人首次提出,主要是结合了 TTM 公钥密码体制的基本思想(雅可比猜想)和减法的思想来构造方案的。其构造方式与 Rainbow 签名有一定的相似之处,但又属于三角形公钥密码体制。TTS 签名体制的效率较高,其运算速度快于 Rainbow 签名方案,但安全性没有 Rainbow 高。下面我们将对最基本的 TTS 签名方案的构造进行详细的介绍。

TTS 的中心映射的定义为 $F: k^m \to k^n$，可以表示成以下一组多项式：

$$y_1 = x_1$$
$$y_2 = x_2 + q_2(x_1)$$
$$y_3 = x_3 + q_3(x_1, x_2)$$
$$\ldots$$
$$y_n = x_n + q_n(x_1, x_2, \ldots, x_{n-1})$$
$$y_{n+1} = q_{n+1}(x_1, x_2, \ldots, x_n)$$
$$\ldots$$
$$y_m = q_m(x_1, x_2, \ldots, x_n)$$

其形如三角形，q_i 为二次多项式。在 n 维向量空间 k^n 上随机选择可逆仿射 L_1，在 m 维向量空间 k^m 上随机选择可逆仿射 L_2，则 $\overline{F}(x_1, x_2, \ldots, x_n) = L_1 \circ F \circ L_2(x_1, x_2, \ldots, x_n)$ 为 TTS 的公钥，私钥为中心映射 F，L_1 的逆映射 L_1^{-1} 和 L_2 的逆映射 L_2^{-1}。

TTS 签名方案的签名和验证签名过程如下。

签名过程相比验证签名计算量大很多。

1）令向量空间 k^m 中的一组向量 $(y_1, y_2, \ldots, y_m)$ 为待签名消息。

2）将 $(y_1, y_2, \ldots, y_m)$ 代入可逆仿射 L_1^{-1} 中，计算出 $(y_1', y_2', \ldots, y_m')$。

3）将 $(y_1', y_2', \ldots, y_n')$ 代入中心映射变换，计算出 $(x_1', x_2', \ldots, x_n')$。

4）将 $(\hat{x}_1', \hat{x}_2', \ldots, x_n')$ 代入可逆仿射 L_2^{-1} 中，计算出签名 $(x_1, x_2, \ldots, x_n)$。

5）将 $\hat{x}_1, \hat{x}_2, \ldots, \hat{x}_n$ 代入可逆仿射 L^{-1} 中，计算出签名 $(z_1, z_2, \ldots, z_n)$。

验证签名过程非常简单，只需将 $(x_1, x_2, \ldots, x_n)$ 代入公钥多项式，验证 $\overline{F}(x_1, x_2, \ldots, x_n) = (y_1, y_2, \ldots, y_m)$ 是否成立，如果成立则签名验证成功，否则签名无效。

4.3.7　en-TTS 签名算法

en-TTS 签名由中国台湾地区的两位学者 Bo-Yin Yang 和 Jiun-Ming Chen 在 2005 年共同提出，它属于三角签名家族，可以被看成多变量公钥加密算法（Tame Transformation Mehod，TTM）的变体。

假定消息的散列值是 $y(y_0, y_1, \ldots, y_{19})$，它的长度是 20B，其中 $y_0, y_1, \ldots, y_{19}$ 是 $GF((2^4)^2)$ 的元素。而且，我们假定签名是 $x(x_0, x_1, \ldots, x_{27})$，它的长度是 28B，其中 $x_0, x_1, \ldots, x_{27}$ 是 $GF((2^4)^2)$ 的元素。

为了对消息的散列值 $y(y_0, y_1, \ldots, y_{19})$ 签名，我们需要进行如下计算：

$$F \circ L_2(x_0, x_1, \ldots, x_{27}) = L_1^{-1}(y_0, y_1, \ldots, y_{19})$$

4.3 多变量公钥密码典型算法

这里，F 是一个中心映射变换，L_1 和 L_2 是两个可逆的仿射变换。
为了实现上述计算，我们首先要做可逆仿射 L_1 的逆变换，计算过程如下：

$$\overline{y} = L_1^{-1}(y_0, y_1, \ldots, y_{19})$$

这里，$\overline{y}(\overline{y}_0, \overline{y}_1, \ldots, \overline{y}_{19})$ 是变换的结果，$\overline{y}$ 的长度是 20B。
可逆仿射 L_1 的逆变换 L_1^{-1} 有如下形式：

$$\overline{y} = Ay + b$$

这里，A 是一个规模为 20×20 的矩阵，b 是一个维度为 20 的向量，A 和 b 都被当作私钥来运算。
通过求解中心映射 F 的逆变换，我们获得 $\overline{x}(\overline{x}_0, \overline{x}_1, \ldots, \overline{x}_{27})$，$\overline{x}$ 的长度是 28B。

$$\overline{x} = F^{-1}(\overline{y}_0, \overline{y}_1, \ldots, \overline{y}_{19})$$

中心映射变换 F 由 20 个多变量多次多项式 $(f_0, f_1, \ldots, f_{19})$ 组成，它有如下形式：

$$F(\overline{x}_0, \overline{x}_1, \ldots, \overline{x}_{27}) = (f_0, f_1, \ldots, f_{19})$$

那么，我们将 $\overline{y}(\overline{y}_0, \overline{y}_1, \ldots, \overline{y}_{19})$ 代入中心映射的变换中，它变成如下形式：

$$\overline{y}(\overline{y}_0, \overline{y}_1, \ldots, \overline{y}_{19}) = f(f_0, f_1, \ldots, f_{19})$$

中心映射变换 F 的 20 个多变量多次多项式 $(f_0, f_1, \ldots, f_{19})$ 的定义如下：

$$f_{i-8} = \overline{x}_i + \sum_{j=1}^{7} p_{ij} \overline{x}_j \overline{x}_{8+((i+j) \bmod 9)} \quad i = 8, 9, \ldots, 16$$

$$f_9 = \overline{x}_{17} + p_{17,1} \overline{x}_1 \overline{x}_6 + p_{17,2} \overline{x}_2 \overline{x}_5 + p_{17,3} \overline{x}_3 \overline{x}_4 + p_{17,4} \overline{x}_9 \overline{x}_{16} + p_{17,5} \overline{x}_{10} \overline{x}_{15} + p_{17,6} \overline{x}_{11} \overline{x}_{14} + p_{17,7} \overline{x}_{12} \overline{x}_{13}$$

$$f_{10} = \overline{x}_{18} + p_{18,1} \overline{x}_2 \overline{x}_7 + p_{18,2} \overline{x}_3 \overline{x}_6 + p_{18,3} \overline{x}_4 \overline{x}_5 + p_{18,4} \overline{x}_{10} \overline{x}_{17} + p_{18,5} \overline{x}_{11} \overline{x}_{16} + p_{18,6} \overline{x}_{12} \overline{x}_{15} + p_{18,7} \overline{x}_{13} \overline{x}_{14}$$

$$f_{i-8} = \overline{x}_i + p_{i,0} \overline{x}_{i-11} \overline{x}_{i-9} + \sum_{j=19}^{i} p_{i,j-18} \overline{x}_{2(i-j)} \overline{x}_j + \sum_{j=i+1}^{27} p_{i,j-18} \overline{x}_{i-j+19} \overline{x}_j, i = 19, 20, \ldots, 27$$

这里，p_{ij} 是多变量多次多项式的系数，被当作私钥使用。
中心映射变换 F 的 20 个多变量多次多项式 $(f_0, f_1, \ldots, f_{19})$ 可以被分成 3 组，分组情况如下：

1）$f_i | i = 0, 1, \ldots, 8$，共 9 个多项式；
2）$f_i | i = 9, 10$，共 2 个多项式；
3）$f_i | i = 11, 12, \ldots, 19$，共 9 个多项式。

同时，中心映射变换 F 中的 $\overline{x}(\overline{x}_0, \overline{x}_1, \ldots, \overline{x}_{27})$ 也可以分成 4 组，分组情况如下：

1）$\overline{x}_i | i = 0, 1, \ldots, 7$，共 8 个元素；
2）$\overline{x}_i | i = 8, 9, \ldots, 16$，共 9 个元素；
3）$\overline{x}_i | i = 17, 18$，共 2 个元素；
4）$\overline{x}_i | i = 19, 20, \ldots, 27$，共 9 个元素。

计算中心映射变换的步骤如下：

1）我们为 $\bar{x}_i$ 的第一组的 8 个元素 $\bar{x}_0, \bar{x}_1, \ldots, \bar{x}_7$ 选择随机的数值；

2）我们将 $\bar{x}_0, \bar{x}_1, \ldots, \bar{x}_7$ 的数值代入中心映射变换 F 的第一组 9 个多变量多次多项式 $f_0, f_1, \ldots, f_8$，并计算它们的系数；

3）$f_0, f_1, \ldots, f_8$ 被转化为只关于 $\bar{x}_i$ 的第二组的 9 个元素 $\bar{x}_8, \bar{x}_9, \ldots, \bar{x}_{16}$ 的线性方程组，它的系数矩阵的规模是 9×9，然后我们求解这个线性方程组；

4）我们将 $\bar{x}_i$ 的第一组和第二组共 17 个元素的值代入中心映射变换 F 的第二组 2 个多变量多次多项式 f_9, f_{10} 中，就可以计算 $\bar{x}_i$ 的第三组的 2 个元素 $\bar{x}_{17}, \bar{x}_{18}$ 的值；

5）我们将 $\bar{x}_i$ 的第一组、第二组和第三组共 19 个元素的值代入中心映射变换 F 的第三组 9 个多变量多次多项式 $f_{11}, f_{12}, \ldots, f_{19}$ 中，并计算它们的系数；

6）$f_{11}, f_{12}, \ldots, f_{19}$ 被转化为只关于 $\bar{x}_i$ 的第四组的 9 个元素 $\bar{x}_{19}, \bar{x}_{20}, \ldots, \bar{x}_{27}$ 的线性方程组，它的系数矩阵的规模是 9×9，然后我们求解这个线性方程组。

到此为止，$\bar{x}(\bar{x}_0, \bar{x}_1, \ldots, \bar{x}_{27})$ 的 28 个元素都计算完成，求得数值。

最后，我们将 $\bar{x}(\bar{x}_0, \bar{x}_1, \ldots, \bar{x}_{27})$ 的 28 个元素代入可逆仿射 L_2 的逆变换 L_2^{-1}，计算它的逆变换得 $x(x_0, x_1, \ldots, x_{27})$，即消息的散列值的 en-TTS 签名：

$$x = L_2^{-1}(\bar{x}_0, \bar{x}_1, \ldots, \bar{x}_{27})$$

可逆仿射的逆变换 L_2^{-1} 有如下形式：

$$x = C\bar{x} + d$$

这里，C 是一个规模为 28×28 的矩阵，d 是一个长度为 28 的向量，C 和 d 都被当作私钥来运算。

所以 $x(x_0, x_1, \ldots, x_{27})$ 是 $y(y_0, y_1, \ldots, y_{19})$ 的签名。

4.3.8 油醋签名算法

当 Patarin 在 1996 年利用线性化方法攻破 MI 体制之后，多变量公钥密码体制的发展并没有停滞不前。Patarin 史无前例地找到了一种跟以往多变量公钥密码思想完全不同的方法去构造新的方案，这就是油醋签名方案。油醋方案分为 3 种：平衡油醋方案，非平衡油醋方案和 Rainbow 方案。前两种方案已经被发现有安全性问题，Rainbow 签名是一种多层结构的方案，每一层都使用非平衡油醋的思想，具有高性能和高安全性。下面我们将对最基本的油醋方案的构造进行详细的介绍。

油醋签名方案的构造的重点在于油醋多项式的构造。油醋多项式是二次的，油变量在油醋多项式中是以一次的形式出现的，当醋变量的值被固定时，油醋多项式会被化简成变量为油变量的线性多项式。我们可以通过一组这样的线性多项式解出油变量，产生一个签名。

油醋方案的构造如下。

设有限域 k 的元素个数为 q，$x_1, x_2, \ldots, x_o$ 为油变量，$\hat{x}_1, \hat{x}_2, \ldots, \hat{x}_v$ 为醋变量，均属于有限域 k，其中 $n = o + v$。则油醋多项式定义如下：

$$f = \sum_{i=1}^{o} \sum_{j=1}^{v} a_{ij} x_i \hat{x}_j + \sum_{i=1}^{v} \sum_{j=1}^{v} b_{ij} \hat{x}_i \hat{x}_j + \sum_{i=1}^{o} c_i x_i + \sum_{j=1}^{v} d_j \hat{x}_j + e$$

其中 a_{ij}、b_{ij}、c_i、d_j 和 e 也均属于有限域 k。油醋方案的多项式映射可以定义为 $F: k^n \to k^o$，称为油醋映射，可以描述为 $F(x_1, x_2, \ldots, x_o, \hat{x}_1, \hat{x}_2, \ldots, \hat{x}_v) = (f_1, f_2, \ldots, f_o)$，$f_1, f_2, \ldots, f_o$ 是油醋多项式。油醋多项式和线性化方程有相似之处，它们在某种意义上是一致的：密文相当于油醋多项式中的醋变量，明文相当于油变量，因为油醋变量是正交的，它们不会有重合的部分。

油醋映射有一个特性，如果 $F: k^n \to k^o$ 的系数是随机选择的，给定一组向量空间 k^o 中的一组向量 $(y_1', y_2', \ldots, y_o')$，我们可以随机选择醋变量 $\hat{x}_1, \hat{x}_2, \ldots, \hat{x}_v$，代入到油醋映射：$F(x_1, x_2, \ldots, x_o, \hat{x}_1, \hat{x}_2, \ldots, \hat{x}_v) = (y_1', y_2', \ldots, y_o')$，解出油变量，但有一定概率是不能解出油变量的，所以需要重新选择醋变量 $\hat{x}_1, \hat{x}_2, \ldots, \hat{x}_v$ 进行一轮运算。

要构造油醋签名方案，还需要随机选择一个可逆仿射 $L: k^n \to k^n$ 来隐藏中心映射 $F: k^n \to k^o$，可以表示为 $(x_1, x_2, \ldots, x_o, \hat{x}_1, \hat{x}_2, \ldots, \hat{x}_v) = L(z_1, z_2, \ldots, z_n)$，最终二次映射 $\bar{F}$ 可以表示为 $\bar{F}(x_1, x_2, \ldots, x_o, \hat{x}_1, \hat{x}_2, \ldots, \hat{x}_v) = F \circ L(x_1, x_2, \ldots, x_o, \hat{x}_1, \hat{x}_2, \ldots, \hat{x}_v)$，因为油醋中心映射 $F: k^n \to k^o$ 是随机选择的，所以不需要左可逆仿射来隐藏。二次映射 $\bar{F}$ 为油醋签名方案的公钥，私钥为中心映射 F 和可逆仿射 L。

油醋方案的签名和验证签名过程如下。

签名过程相比验证签名计算量大很多。

1）令向量空间 k^o 中的一组向量 $(y_1, y_2, \ldots, y_o)$ 为待签名消息，签名者随机选择醋变量 $\hat{x}_1, \hat{x}_2, \ldots, \hat{x}_v$。

2）将 $(y_1, y_2, \ldots, y_o)$ 和 $\hat{x}_1, \hat{x}_2, \ldots, \hat{x}_v$ 代入 $F(x_1, x_2, \ldots, x_o, \hat{x}_1, \hat{x}_2, \ldots, \hat{x}_v) = (y_1', y_2', \ldots, y_o')$，求解油变量 $x_1, x_2, \ldots, x_v$，如果无解则需再次选择醋变量 $\hat{x}_1, \hat{x}_2, \ldots, \hat{x}_v$ 重新求解。

3）将 $\hat{x}_1, \hat{x}_2, \ldots, \hat{x}_v$ 和 $x_1, x_2, \ldots, x_v$ 代入可逆仿射 L^{-1} 中，计算出签名 $(z_1, z_2, \ldots, z_n)$。

验证签名过程非常简单，只需将 $(z_1, z_2, \ldots, z_n)$ 代入公钥多项式，验证 $\bar{F}(z_1, z_2, \ldots, z_n) = (y_1, y_2, \ldots, y_o)$ 是否成立，如果成立则签名验证成功，否则签名无效。

4.3.9 UOV 签名算法

非平衡油醋签名算法 UOV 是由 3 位学者 Aviad Kipnis、Jacques Patarin 和 Louis Goubin 在 1999 年共同提出的，它属于油醋签名家族。

接下来我们介绍非平衡油醋签名方案。我们假定消息的散列值是 $y(y_0, y_1, \ldots, y_{m-1})$，

它的长度是 m 字节，其中 $y_0, y_1, \ldots, y_{m-1}$ 是有限域的元素。而且，我们假定签名是 $x(x_0, x_1, \ldots, x_{n-1})$，它的长度是 n 字节，其中 $x_0, x_1, \ldots, x_{n-1}$ 是有限域的元素。

为了对消息的散列值 $y(y_0, y_1, \ldots, y_{m-1})$ 签名，我们需要进行如下计算：

$$L(x_0, x_1, \ldots, x_{n-1}) = F^{-1}(y_0, y_1, \ldots, y_{m-1})$$

这里，F 是一个中心映射变换，L 是一个可逆的仿射变换。

为了实现上述计算，我们首先要做中心映射的逆变换，计算过程如下：

$$\bar{x} = F^{-1}(y_0, y_1, \ldots, y_{m-1})$$

这里，$\bar{x}(\bar{x}_0, \bar{x}_1, \ldots, \bar{x}_{n-1})$ 是变换的结果，$\bar{x}$ 的长度是 n 节字。

中心映射变换 F 由 m 个多变量多次多项式 $(f_0, f_1, \ldots, f_{m-1})$ 组成，它有如下的形式：

$$F(\bar{x}_0, \bar{x}_1, \ldots, \bar{x}_{n-1}) = (f_0, f_1, \ldots, f_{m-1})$$

事实上 $\bar{x}(\bar{x}_0, \bar{x}_1, \ldots, \bar{x}_{n-1})$ 是醋变量和油变量的有限集合。

1）$\bar{x}_0, \bar{x}_1, \ldots, \bar{x}_{n-m-1}$ 是醋变量的有限集合，共 $n-m$ 个醋变量。

2）$\bar{x}_{n-m}, \bar{x}_{n-m-1}, \ldots, \bar{x}_{n-1}$ 是油变量的有限集合，m 个油变量。

多变量多次多项式 $f_0, f_1, \ldots, f_{m-1}$ 的定义如下：

$$f(O_0, O_1, \ldots, O_{m-1}) = \sum \alpha_{ij} O_i V_j + \sum \beta_{ij} V_i V_j + \sum \gamma_i V_i + \sum \delta_i O_i + \eta$$

这里，$O_j, (V_i, V_j)$ 分别是油变量和醋变量，α_{ij}、β_{ij}、γ_i、δ_i 和 η 是多变量多次多次多项式的系数，被当作私钥使用。

多变量多次多项式包括 5 项，最高次数不超过两次。若将醋变量的数值代入多变量多次多项式，它将变换成关于油变量的一次多项式。

1）$O_i V_j$，油变量和醋变量的组合；

2）$V_i V_j$，醋变量和醋变量的组合；

3）V_i，醋变量；

4）O_i，油变量；

5）η，常数。

计算中心映射变换的步骤如下：

1）我们为 $n-m$ 个醋变量 $\bar{x}_0, \bar{x}_1, \ldots, \bar{x}_{n-m-1}$ 选择随机的数值；

2）我们将 $n-m$ 个醋变量 $\bar{x}_0, \bar{x}_1, \ldots, \bar{x}_{n-m-1}$ 的数值代入 m 个多变量多次多项式 $f_0, f_1, \ldots, f_{m-1}$ 中，并计算它们的系数；

3）$f_0, f_1, \ldots, f_{m-1}$ 被转化为只关于油变量 $\bar{x}_{n-m}, \bar{x}_{n-m+1}, \ldots, \bar{x}_{n-1}$ 的线性方程组，它的系数矩阵的规模是 $m \times m$，然后我们求解这个线性方程组。

到此为止，我们计算完成并获得所有变量 $\bar{x}_0, \ldots, \bar{x}_{n-1}$ 的数值。

最后，我们将 $n-m$ 个醋变量和 m 个油变量 $\bar{x}(\bar{x}_0, \bar{x}_1, \ldots, \bar{x}_{n-1})$ 代入可逆仿射变换，计算它的逆变换得 $x(x_0, x_1, \ldots, x_{n-1})$，即消息的散列值的非平衡油醋签名：

$$x = L^{-1}(\overline{x}_0, \overline{x}_1, \ldots, \overline{x}_{n-1})$$

可逆仿射的逆变换 L^{-1} 有如下形式：

$$x = A\overline{x} + b.$$

这里，A 是一个规模为 $n \times n$ 的矩阵，b 是一个维度为 n 的向量，A 和 b 都被当作私钥来运算。

所以，$x(x_0, x_1, \ldots, x_{n-1})$ 是 $y(y_0, y_1, \ldots, y_{m-1})$ 的签名。

4.3.10 Rainbow 签名算法

Rainbow 密码算法是一种签名算法，属于多变量公钥密码算法。它是由两位美国学者 Jintai Ding 和 Dieter Schmidt 在 2005 年共同提出的。它是油醋签名家族的一员，可以被看成多层的非平衡油醋方案。

我们假定 Rainbow 消息的散列值是 $y(y_0, y_1, \ldots, y_{m-1})$，它的长度是 m 字节，其中 $y_0, y_1, \ldots, y_{m-1}$ 是某个有限域的元素。而且，我们假定签名是 $x(x_0, x_1, \ldots, x_{n-1})$，它的长度是 n 字节，其中 $x_0, x_1, \ldots, x_{n-1}$ 是某个有限域的元素。

为了对消息的散列值 $y(y_0, y_1, \ldots, y_{m-1})$ 签名，我们需要进行如下计算：

$$F \circ L_2(x_0, x_1, \ldots, x_{n-1}) = L_1^{-1}(y_0, y_1, \ldots, y_{m-1})$$

这里 F 是一个中心映射变换，L_1 和 L_2 是两个可逆的仿射变换。

为了实现上述计算，我们首先要做可逆仿射 L_1 的逆变换：

$$\overline{y} = L_1^{-1}(y_0, y_1, \ldots, y_{m-1})$$

这里 $\overline{y}(\overline{y}_0, \overline{y}_1, \ldots, \overline{y}_{m-1})$ 是可逆仿射变换的结果，$\overline{y}$ 的长度是 m 字节。

可逆仿射 L_1 的逆变换 L_1^{-1} 有如下形式：

$$\overline{y} = Ay + b$$

这里 A 是一个规模为 $m \times m$ 的矩阵，b 是一个维度为 m 的向量，A 和 b 都被当作私钥来运算。

通过求解中心映射 F 的逆变换，我们获得 $\overline{x}(\overline{x}_0, \overline{x}_1, \ldots, \overline{x}_{m-1})$，$\overline{x}$ 的长度是 m 字节：

$$\overline{x} = F^{-1}(\overline{y}_0, \overline{y}_1, \ldots, \overline{y}_{m-1})$$

中心映射变换 F 由 m 个多变量多次多项式 $(f_0, f_1, \ldots, f_{m-1})$ 组成，它有如下形式：

$$F(\overline{x}_0, \overline{x}_1, \ldots, \overline{x}_{n-1}) = (f_0, f_1, \ldots, f_{m-1})$$

那么，我们将 $\overline{y}(\overline{y}_0, \overline{y}_1, \ldots, \overline{y}_{m-1})$ 代入中心映射的变换中，它变成如下形式：

$$\overline{y}(\overline{y}_0, \overline{y}_1, \ldots, \overline{y}_{m-1}) = f(f_0, f_1, \ldots, f_{m-1})$$

中心映射 F 是一个 k 层的油醋结构，由多个多变量多次多项式组成，即 $(f_0, f_1, \ldots, f_{m-1})$ 被分成 k 层，它的定义如下：

$$f(O_0, O_1, \ldots, O_{m-1}) = \sum \alpha_{ij} O_i V_j + \sum \beta_{ij} V_i V_j + \sum \gamma_i V_i + \sum \delta_i O_i + \eta$$

这里 $O_j, (V_i, V_j)$ 分别是油变量和醋变量，β_{ij}、γ_i、δ_i、和 η 是多变量多次多项式的系数，被当作私钥使用。

多变量多次多项式包括 5 项，最高次数不超过二次，若将醋变量的数值代入多变量多次多项式，它将变换成关于油变量的一次多项式。

1）$O_i V_j$，油变量和醋变量的组合；

2）$V_i V_j$，醋变量和醋变量的组合；

3）V_i，醋变量；

4）O_i，油变量；

5）η，常数。

计算中心映射变换的步骤如下：

1）我们为第一层的醋变量选择随机的数值；

2）我们将醋变量的数值代入中心映射变换 F 的第一层，并计算它们的系数；

3）它们被转化为关于第一层的油变量的线性方程组，然后我们求解这个线性方程组；

4）我们将第一层的醋变量和第一层的油变量作为第二层的醋变量代入中心映射变换 F 的第二层，并计算它们的系数；

5）它们被转化为关于第二层的油变量的线性方程组，然后我们求解这个线性方程组；

6）我们将第二层的醋变量和第二层的油变量作为第三层的醋变量代入中心映射变换 F 的第三层，并计算它们的系数；

7）它们被转化为关于第三层的油变量的线性方程组，然后我们求解这个线性方程组；

8）以下的计算类似，直到最后一层的所有元素计算完成，所有计算结束。

到此为止，$\bar{x}(\bar{x}_0, \bar{x}_1, \ldots, \bar{x}_{n-1})$ 的 n 个元素都计算完成，求得数值。

最后，我们将 $\bar{x}(\bar{x}_0, \bar{x}_1, \ldots, \bar{x}_{n-1})$ 的 n 个元素代入可逆仿射 L_2 的逆变换 L_2^{-1}，计算它的逆变换得 $x(x_0, x_1, \ldots, x_{n-1})$，即消息的散列值的 Rainbow 签名：

$$x = L_2^{-1}(\bar{x}_0, \bar{x}_1, \ldots, \bar{x}_{n-1})$$

可逆仿射的逆变换 L_2^{-1} 有如下形式：

$$x = C\bar{x} + d$$

这里，C 是一个规模为 $n \times n$ 的矩阵，d 是一个维度为 n 的向量，C 和 d 都被当作私钥来运算。

所以 $x(x_0, x_1, \ldots, x_{n-1})$ 是 $y(y_0, y_1, \ldots, y_{m-1})$ 的签名。

Rainbow 验证签名的算法相对比较简单，要验证签名 $x(x_0, x_1, \ldots, x_{n-1})$ 是否为 $y(y_0, y_1, \ldots, y_{m-1})$ 的签名信息，只需要将签名 $x(x_0, x_1, \ldots, x_{n-1})$ 代入公钥多项式组 $\bar{F}(x_0, x_1, \ldots, x_{n-1}) = L_1 \circ F \circ L_2(y_0, y_1, \ldots, y_{m-1})$ 中，验证所得的结果是否与 $y(y_0, y_1, \ldots, y_{m-1})$ 相

同即可。

Rainbow 算法的复杂度如表 4-2 所示。

表 4-2　　　　　　　　　　Rainbowbi 算法的复杂度

Rainbow 算法	复杂度
生成公私钥	$O(mn^3)$
生成签名	$O(n^3)$
验证签名	$O(m(n+1)^2)$

4.4　多变量公钥密码分析方法

近 20 年来，多变量公钥密码体制的方案层出不穷，但到目前为止，安全的方案非常少，大多数都是在提出不久便被攻破，针对多变量公钥密码体制的攻击的研究也成了多变量公钥密码体制研究的热点。当前主要的攻击方法可以分为 5 种。

1）暴力攻击也叫枚举攻击。
2）直接攻击，也就是求解多变量非线性方程。
3）线性化方程攻击。
4）秩攻击，又分为低秩攻击、高秩攻击和油醋分离攻击。
5）差分攻击。

这 5 种攻击方法单独使用并不一定有效，最常用的攻击手段是结合几种攻击方法进行的攻击。

4.4.1　暴力攻击

暴力攻击是指直接枚举整个解空间。对多变量公钥密码体制的暴力攻击分为两种：

1）对密文的攻击和对密钥的攻击。对密文的攻击又叫唯密文攻击，指的是通过枚举明文，计算得到密文之后，将其与目标密文进行比较，如果相等则攻击成功。

2）对密钥的攻击又叫唯密钥攻击，直接枚举密钥空间。由于暴力攻击的计算量太大，单纯使用暴力攻击效率太低，一般会先通过其他算法缩小搜索的空间，然后再使用暴力攻击。

4.4.2　直接攻击

直接攻击又叫求解多变量非线性方程。多变量公钥密码体制的公钥是一组多变量非线性多项式，将明文代入多变量非线性方程组即可获得密文，公钥和密文可以组成一组多变量非线性方程组。攻击者可以通过直接求解多变量非线性方程组来破解明文。当前求解非线性方程组比较有效的方法有 Gröbner 基算法和 XL 算法。

Gröbner 基算法是求解非线性方程组最有效的一种算法，它是对 1970 年提出的 Buchberger 算法的改进。Gröbner 基理论的本质是从多变量多项式环中随机选取一组理想的生成元，利用该生成元计算出一组具有好性质的生成元，然后利用这些好的性质来研究多变量多项式环的结构，求解方程的解。Gröbner 基算法是一种启发式算法，复杂度不确定，所以不能一定求出非线性方程组的解，当前最好的 Gröbner 基算法是由 Faugère 提出的 F4 算法和 F5 算法。

XL 算法是在 2000 年由 Courtois 等人首次提出的，算法的核心思想是通过引入变量将非线性方程组转化为线性方程组来进行求解。XL 算法的攻击复杂度为 $\left(\dfrac{n^{\sqrt{n}}}{\sqrt{n!}}\right)^w$，其中 w 为 3，改进后的 w 为 2.3766。

4.4.3 线性化方程攻击

线性化方程攻击算法最早是由 Patarin 于 1996 年提出，用于攻击 MI 体制。线性化方程攻击复杂度比较低，可以作为辅助攻击算法降低攻击的复杂度，再结合其他算法进行攻击。对于一个多变量非线性方程组，如果存在线性化方程，则一定存在恒等式 $\sum_{i=1}^{n}\sum_{j=1}^{m}a_{ij}x_iy_j + \sum_{i=1}^{n}b_ix_i + \sum_{j=1}^{m}c_jy_j + d = 0$。Patarin 敏锐地发现，线性化方程可以用来攻击 MI 公钥体制：对于 MI 的公钥多项式 $\bar{F}(x_1,x_2,\ldots,x_n) = (y_1,y_2,\ldots,y_n)$ 来说，可以很容易地发现 MI 的中心映射满足表达式 $XY^{q^\theta} - X^{q^{2\theta}}Y = 0$，即存在线性化方程。当我们将 $(y_1,y_2,\ldots,y_n)$ 代入恒等式，可得关于 $(x_1,x_2,\ldots,x_n)$ 的线性方程。我们通过将多组 $(y_1,y_2,\ldots,y_n)$ 代入恒等式可以获得 n 个关于 $(x_1,x_2,\ldots,x_n)$ 线性不相关的线性方程，这样就可以解出明文 $(x_1,x_2,\ldots,x_n)$。判断某个多变量公钥密码体制能否用线性化方程的方法攻击，有两种方法：

1）一种是通过对中心映射的表达式进行线性变换和构造来寻找线性化方程，例如 MI 公钥体制；

2）另一种是通过将多组密文和明文对代入恒等式，计算出恒等式的系数，从而获得关于线性化方程的恒等式。

4.4.4 秩攻击

秩攻击主要是指结合多变量公钥密码的中心映射的构造和对多变量公钥多项式的系

数矩阵的秩对多变量公钥密码体制进行分析，以得到中心映射的结构，从而可以进行签名的伪造和加密解密的中间人攻击。

当前比较有效的秩攻击算法可以分为以下 3 种：高秩攻击、低秩攻击和油醋分离攻击。

1）高秩攻击算法是由 Coppersmith 等人在 1997 年分析 Birational Permutation 签名体制的时候首次提出的，其算法复杂度为 $q^u(un^2+n^3/6)$。

2）低秩攻击算法最早由 Kipnis 和 Shamir 在 1999 年首次提出，用于分析 HFE 加密体制算法的安全性，其攻击算法的复杂度为 $q^r(m^2(nt/2-m/6)+mn^2t)$，其中 r 是中心映射矩阵的最小秩的大小，m 为中心映射方程的个数，n 为变量个数，并且 $t=\lceil m/n \rceil$。

3）油醋分离攻击算法是由 Kipnis 在 1999 年首次提出，用于对当时出现不久的油醋体制的多变量公钥密码算法进行攻击，攻击算法的复杂度为 $q^{2v-n-1}(n-v)^4$，其中 v 为醋变量的个数，n 为油变量和醋变量的个数之和。

4.4.5 差分攻击

差分攻击算法的基本思想是通过计算多变量公钥密码的中心映射多项式或公钥多项式的差分来消除扰动，再结合其他攻击方法对多变量公钥密码的私钥进行恢复。Dubois 等人在 2007 年用差分分析法和线性化方程结合的方法攻破了 SFLASH 签名体制，可以从其公钥中恢复私钥。Fouque 等人又在 2008 年的将差分分析法和 Gröbner 基算法结合起来攻破了 ℓ-IC 签名体制，从其公钥中恢复了等价私钥，可以用来伪造签名。差分攻击作为辅助攻击手段，通常结合其他多变量公钥密码的攻击方法，有很好的效果。

4.5 本章参考代码

4.5.1 Rainbow

1. C 代码

```
//Compose.cpp
#include "stdafx.h"
#include "rainbow.h"


extern int counter ;
int getchr( void ){return rand();}
```

第4章 多变量公钥密码技术

```cpp
    ofstream kout("key.txt",ios::out ) ;

    //密钥生成函数
    //sk 储存私钥, sklen 私钥的长度, OUT
    //pk 储存公钥, pklen 公钥的长度, OUT
    int keypair( unsigned char * sk, unsigned long * sklen,
                 unsigned char * pk, unsigned long * pklen) //sk:secret
keys;pk:public keys;
    {

const int ovn = 37 ;   // oil + vinegar variables
const int v1 = 10 ;    // first set of vinegar variables: o < v1 < v2 <...< ovn=n
const int un = 5 ; // number of entries into Slist, first entry always v1, last always ovn
//const int Slist[un] ={v1,20,24,27,ovn} ;

//      const int nv1 = ovn - v1 ;
//const int ovn1 = ovn+1 ;
//const int nv2 = 2 * nv1 ;
//const int ovn2 = 2 * ovn ;

    GFpow * a[nv1], * b[nv1], * c[nv1],  e[nv1] ; //中心映射的系数
    GFpow  S[nv1*nv1], s[nv1] ;//initialize affine transformation  L1 in S and s,L1
//的矩阵和偏移量;
    GFpow  T[ovn*ovn], t[ovn];//L2 的矩阵和偏移量;

    GFpow R[2*ovn*ovn] ;//用来求逆矩阵用的;

    //求公钥要用到的临时变量
    GFpow C[ovn1*ovn1] ;
    GFpow C1[ovn1*ovn1];
    GFpow L2[ovn1*ovn1] ;
    GFpow D0[nv1][ovn1*ovn1] ;
    GFpow D[nv1][ovn1*ovn1] ;
    GFpow temp ;


    int i,j,k, k0, ij,  eof,kj=0;
    const int rept = 1;
    int oil, vinegar ;

    // create L1 and check that L1 is not singular
    // initialize affine transformation  L1 in S and s
    // and also put s and S^-1 on sk

    for ( ij = 0 ; ij < nv1 ; ij++ ) {
            s[ij] = sk[ij] = getchr() % pow ;
```

```
    }//initalize s,randomly choose;
//关于并行，随机产生私钥的地方可以并行
    eof = 1 ;
    while ( eof ) {
        for ( i = 0 ; i < nv1 ; i ++ ) {
            for ( j = 0 ; j < nv1 ; j++ ) {
                    R[i*nv2+j] = S[i*nv1+j] = getchr() % pow ;//randomly choose S
            }

            for ( ; j < nv2 ; j++ ) {
                    R[i*nv2+j] = 0 ;
            }
            R[i*nv2+nv1+i] = 1 ;
        }
        eof = nv1 - gauss( R, nv1, nv2 ) ;

        if ( eof ) cout << "L1 was singular \n" ;
        else
        {
            kout<<"L1 的矩阵"<<endl;
            for ( i = 0 ; i < nv1 ; i ++ ) {
                    for ( j = 0 ; j < nv1 ; j++ ) {
                            kout<<setw(5)<<S[i*nv1+j] ;
                    }
                    kout<<endl;
            }
            kout<<"L1 的偏移量"<<endl;
            for ( kj = 0 ; kj < nv1 ; kj++ ) {
                kout<<s[kj]<<endl;
            }
            kout<<"L1 的矩阵的逆矩阵"<<endl;
            for ( i = 0 ; i < nv1 ; i++ ) {
                    for ( j = nv1 ; j < nv2; j++ ){
                            kout<<setw(5)<< R[i*nv2+j] ;
                    }
                    kout<<endl;
            }
        }
    }

    // store S^(-1) in secrete key sk
    for ( i = 0 ; i < nv1 ; i++ ) {
            for ( j = nv1 ; j < nv2; j++ ){
                    sk[ij++] = R[i*nv2+j].GFpowtochar() ;
            }
}//sk 的前 27（0-26）项装 L1 的偏移量，(27-755) 装 L1 矩阵的逆 (756-855) 装第一条 OV 映射的 a
//(856) 装第一条 OV 映射的 e
```

```
            k = 0 ;
            for( k0 = 0; k0 < un-1; k0++ ) {
                    oil = Slist[k0+1] - Slist[k0] ;
                    vinegar = Slist[k0] ;

                    for ( ; k < Slist[k0+1]-v1; k++ ) {

                            a[k] = new GFpow[oil*vinegar] ;
                            b[k] = new GFpow[vinegar*vinegar] ;
                            c[k] = new GFpow[vinegar+oil] ;

                            cout << "\nallocate when k= "<< k <<"   " <<oil*vinegar<<"
+ "<<vinegar*vinegar<<" + " <<vinegar + oil
            <<" oil="<<oil<< " vinegar = "<<vinegar;

                            kout << "\nallocate when k= "<< k <<"   " <<oil*vinegar<<"
+ "<<vinegar*vinegar<<" + " <<vinegar + oil
            <<" oil="<<oil<< " vinegar = "<<vinegar<<endl;
                            kout<<"第"<<k<<"条多项式(中心映射)"<<endl;

                            //k条式子之间没有数据上的联系，可以并行，而且每条的a,b,c也没什
                            //么关系，也可以并行
                            e[k] = getchr() % pow;
                            kout<<e[k];

                            if ( oil == 1 ) {
                                    for ( i = 0 ; i < vinegar; i++ ){
                                            a[k][i] = 0 ;
                                    }
                            }
                            else {
                                    for ( i =0 ; i < oil; i++)    {
                                            for( j=0; j < vinegar; j++) {
                                                    a[k][j*oil+i] = sk[ij++] =
getchr() % pow ;//generate the kth o_v map//如果用二维矩阵表示，则为v行，o列；
                                                    if(0 != a[k][j*oil+i])
                                                    {
                                                            kout<<"+"<<a[k][j
*oil+i]<<"*w_"<<j<<"*w_"<<(j+i);
                                                    }
                                            }
                                    }
                            }

                            sk[ij++] = e[k].GFpowtochar();
            // generate in ordered needed by inverse

                            for( i=0; i < vinegar; i++) {
```

4.5 本章参考代码

```
                                for ( j =0 ; j < vinegar; j++) {
                                        if( j < i )
                                                b[k][i*vinegar+j]=0 ;
                                        else
                                                b[k][i*vinegar+j]         =
sk[ij++] = getchr() % pow ;//up triangle

                                        if(0 != b[k][i*vinegar+j])
                                        {
                                                kout<<"+"<<b[k][i*vinegar
+j]<<"*w_"<<i<<"*w_"<<j;
                                        }

                                }
                        }

                        for ( i = 0 ; i < vinegar+oil; i++)
                        {
                                c[k][i] = sk[ij++] = getchr() % pow ;
                                if(0 != c[k][i])
                                {
                                        kout<<"+"<<c[k][i]<<"*w_"<<i;
                                }
                        }
                        if ( oil == 1) {
                                i--;
                                while ( c[k][i] == 0 ) {
                                        c[k][i] =  sk[ij-1] = getchr() % pow ;
                                }
                        }
                        kout<<endl;
                }
        }

        // check that  L2 is not singular
        // initialize affine transformation L2 in T and t

    //过程与L1 的产生一样，维数不同而已
        for ( i = 0 ; i < ovn ; i++ ) {
                sk[ij++] = getchr() % pow ;
                t[i] = GFpow( sk[ij-1] ) ;
        }

        eof = 1 ;
        while ( eof ) {
                for ( i = 0 ; i < ovn ; i ++ ) {
                        for ( j = 0 ; j < ovn ; j++ ) {
                                R[i*ovn2+j] = T[i*ovn+j] = getchr() % pow ;
```

第4章 多变量公钥密码技术

```
                                }
                                for ( ; j < ovn2 ; j++ ) {
                                        R[i*ovn2+j] = 0 ;
                                }
                                R[i*ovn2+ovn+i] = 1 ;
                }
                eof = ovn - gauss( R, ovn, ovn2 ) ;
                if ( eof ) cout << "L2 was singular \n" ;
                else
                {
                                kout<<"L2 的矩阵"<<endl;
                                for ( i = 0 ; i < ovn ; i ++ ) {
                                        for ( j = 0 ; j <ovn ; j++ ) {
                                                kout<<setw(5)<<T[i*ovn+j];
                                        }
                                        kout<<endl;
                                }
                                kout<<"L2 的偏移量"<<endl;
                                for (kj = 0 ; kj < ovn ; kj++ ) {
                                        kout<<t[kj]<<endl;
                                }
                                kout<<"L2 的矩阵的逆矩阵"<<endl;
                                for ( i = 0 ; i <ovn ; i++ ) {
                                        for ( j =ovn ; j < ovn2; j++ ){
                                                kout<<setw(5)<< R[i*ovn2+j] ;
                                        }
                                        kout<<endl;
                                }
                }
        }

        // store T^(-1) in secrete key sk
        for ( i = 0 ; i < ovn ; i++ ) {
                for ( j = ovn ; j < ovn2; j++ ){
                        sk[ij++] = R[i*ovn2+j].GFpowtochar() ;
                }
        }

        * sklen = ij ;


        // Now create public key
        // set up linear transformation L2 in homogeneous form
        // using matrices T and t

        k = 0 ;
        for ( i = 0 ; i < ovn; i++) {
                for ( j = 0 ; j < ovn ; j++ ) {
                        L2[k++] = T[ovn*i + j ] ;
```

```
                    }
                    L2[k++] = t[i] ;
        }
        for ( j = 0; j < ovn ; j++ ) L2[k++] = 0 ;
        L2[k] = 1 ;

        k = 0 ;
        for ( k0 = 0; k0 < un-1 ; k0++ ) {
                    oil = Slist[k0+1] - Slist[k0] ;
                    vinegar = Slist[k0] ;

                    for ( ; k < Slist[k0+1]-v1; k++ ) {

                                for ( i = 0 ; i < vinegar ; i++ ) {
                                            for ( j = 0 ; j < vinegar; j++ ) {
                                                        C[i*ovn1+j] = b[k][i*vinegar+j] ;
                                            }

                                            for (j=0 ; j < oil; j++ ) C[i*ovn1+j+vinegar] = a[k][i*oil+j] ;

                                }
        for (j=0 ; j < vinegar+oil ; j++ )  C[ovn*ovn1+j] = c[k][j] ;



                    C[ovn*ovn1+ovn] = e[k] ;



                    // form compostion with L2

                    for ( i = 0 ; i < ovn1 ; i++ ) {
                                for ( j = 0 ; j < ovn1; j++ ) {
                                            temp = 0 ;
                                            for ( ij = 0 ; ij < ovn1 ; ij++ ) {
                                                        temp += C[i*ovn1+ij]*L2[ij*ovn1+j] ;

                                            }

                                            C1[i*ovn1+j] =temp ;

                                }
                    }

                    for ( i = 0 ; i < ovn1*ovn1; i++) D0[k][i]=0 ;
```

```
                    for ( i = 0 ; i < ovn1 ; i++ ) {
                        for ( j = 0 ; j < ovn1; j++ ) {
                            temp = 0 ;
                            for ( ij = 0 ; ij < ovn1 ; ij++ ) {
                                temp += L2[ij*ovn1+i] * C1[ij*ovn1+j] ;
                            }
                            if ( i <= j ) D0[k][i*ovn1+j] =temp ;
                            else D0[k][j*ovn1+i] += temp ;
                        }
                    }
            }

// Composition with L1, uses matrices S and s directly

for ( k = 0 ; k < nv1 ; k++ ) {

    for ( i = 0 ; i < ovn1*ovn1 ; i++ ) D[k][i] = 0 ; // may not be needed,
                                                      // just in case
    for ( j = 0 ; j < nv1; j++ ) {
            for ( i = 0 ; i < ovn1*ovn1 ; i++ ) D[k][i]+=S[k*nv1+j]*D0[j][i] ;
    }

    D[k][--i] += s[k] ;       // affine part of L1

}


// write out public key
ij = 0 ;
for ( k = 0 ; k < nv1 ; k++ ) {
        kout<<"第"<<k<<"个公钥多项式"<<endl;
        kout<<D[k][ovn1*ovn1-1];

        for(i=0;i<ovn1-1;i++)
        {
                if(0!= D[k][i*ovn1+ovn1-1])
                        kout<<"+"<<D[k][i*ovn1+ovn1-1]<<"*x_"<<i;
        }
        for ( i = 0 ; i < ovn1 ; i++ ) {
                for ( j = i ; j < ovn1 ; j++ )
                {
                        pk[ij++]= D[k][i*ovn1+j].GFpowtochar() ;

                        if((0!= D[k][i*ovn1+j])&&(j != (ovn1-1)))
                        {
                                kout<<"+"<<D[k][i*ovn1+j]
                                        <<"*x_"<<i<<"*x_"<<j;
```

4.5 本章参考代码

```
                        }
                    }
                }
                kout<<endl;
    }

    * pklen = ij ;

    for ( k = 0 ; k < Slist[k0]-v1; k++ ) {
        delete [] a[k] ;
        delete [] b[k] ;
        delete [] c[k] ;
    }




    return(0) ;
}
```

```
//签名验证函数
//m 是消息,IN
//sm 是签名,     IN
//pk 是公钥,     IN
//temp 是还原的消息
//eof=0 表示验证签名成功, eof=-101 表示输入参数 sm 长度不符合规定, eof=-100 表示 temp !=m; OUT
int shortmessagesigned( unsigned char * m,unsigned long * mlen,
                       const unsigned char * sm, unsigned long smlen,
                       const unsigned char * pk, unsigned long pklen){




    int i,j,k, ij=0, eof=0 ;
    if (smlen != ovn) return (-101) ;
    GFpow temp, x[ovn1];

    for ( k = 0 ; k < ovn; k++ ) x[k] = GFpow(sm[k]);
    x[ovn] = 1;

    //每一个多项式之间可以并行
    for ( k = 0 ; k < nv1 ; k++ ) {
        temp = 0 ;
```

```
        //每一个多项式里的求和符号可以并行
        for ( i = 0 ; i < ovn1 ; i++ ) {
                for ( j = i ; j < ovn1 ; j++ ) {
                        temp += GFpow(pk[ij++])*x[i]*x[j] ;
                }
        }

        if (m[k] != temp.GFpowtochar() ) {
            //cout << "mismatch at k="<<k<< " m[k] = "<< GFpow(m[k])
            //     << "  temp="<< temp << endl ;
                eof = -100 ;
                m[k] = temp.GFpowtochar() ;
                break ;
            }
    }
    * mlen = k ;

    return(eof) ;
}




//sk 私钥，IN
//m 要签名的消息，IN
//sm 签名，OUT
//返回值为-1，签名失败；
int signedshortmessage( unsigned char * sm,unsigned long * smlen,
                        const unsigned char * m, unsigned long mlen,
                        const unsigned char * sk, unsigned long sklen)

//int Finverse( GFpow * x, GFpow * y, unsigned char * sk)
// given y[0],...,y[nv1-1]
// select vinegar variables x[0],...,x[v1-1] at random
// and solve for x[v1],..., x[ovn]
// matrix A is set up to hold resulting linear equations
{
    int i,j,k,k0,k1,ij, ij0, oil, vinegar, repeat, eof ;
    GFpow c0, A[nv1*(nv1+1)], B[ovn*ovn1], * xoil, y1[ovn];

    GFpow y[nv1] ;
    GFpow x[ovn] ;


    for (ij=0; ij < nv1 ; ij++) y[ij] = GFpow( m[ij] ) + GFpow( sk[ij] ) ;
    //消息加上 L1 的偏移量


    // compute first L1^{-1}
```

```
        k = 0 ;
        for ( i = 0 ; i < nv1 ; i++ )
        {
                y1[i] = 0 ;
                for ( j = 0 ; j < nv1 ; j++ ) {
                        y1[i] += GFpow(sk[ij++]) * y[j]  ;
                }

        }

        // next find F^{-1}


        ij0 = ij ;
        eof = 0 ;
        do {
                ij = ij0 ;     // in case we have to repeat
                repeat = 0;
                for ( j = 0 ; j < v1; j++ ) x[j] = getchr() % pow ;
                //randomly choose v1 variables


                for ( k0 = 0; k0 < un-1 ; k0++ ) {
                        oil = Slist[k0+1] - Slist[k0] ;
                        vinegar = Slist[k0] ;
                        xoil = x + Slist[k0] ;
                        if ( oil == 1 ) { // special case   mixed terms set to zero
                                // no difficulty in solving equation
                                // constant term caused by vinegar, subtract from y1[k]
                                k1 = Slist[k0] - v1 ;
                                c0 = GFpow( sk[ij++] )+ y1[k1] ; //e[k1]+y1[k1];
                                for ( i =0; i < vinegar; i++ ) {
                                        for ( j = i ; j < vinegar; j++ ) {
                                                c0 += GFpow(sk[ij++])* x[i]* x[j] ;
// b[k1][i*vinegar+j]
                                        }

                                }
                                for ( i =0; i < vinegar; i++ ) {
                                        c0 += GFpow(sk[ij++]) * x[i] ;
                                        // c[k1][i] * x[i] ;
                                }
                                x[vinegar] = c0 / GFpow( sk[ij++] ) ;   /// c0 /
c[k1][vinegar] ;
                        }
                        else {  // general case
                           for ( k = 0 ; k < oil ; k++ ) {
```

第4章　多变量公钥密码技术

```
                                        k1 = k + Slist[k0]-v1 ;  //表示第 k1 条式子
                                        for ( i =0 ; i < oil ; i++) {
                                                c0 = 0 ;
                                                for( j = 0 ; j < vinegar; j++ ) c0
+= GFpow(sk[ij++])*x[j] ; //a[k1][j*oil+i]*x[j]; //二次项的 x[i]的系数
                                                A[k*(oil+1)+i] = c0 ;//A 为增广矩阵
                                        }
                                        // constant term caused by vinegar, subtract
                                        // from y1[k]
                                        c0 = GFpow( sk[ij++] ) ;        //  e[k1] ;
                                        for ( i =0; i < vinegar; i++ ) {
                                                for ( j = i ; j < vinegar; j++ ) {
                                                        c0 += GFpow(sk[ij++]) *
x[i]* x[j] ;    // b[k1][i*vinegar+j]
                                                }
                                        }
                                        for ( i = 0 ; i < vinegar ; i++ ) {
                                                c0 += GFpow( sk[ij++] ) * x[i] ;
// c[k1][i] * x[i] ;
                                        }

                                        A[k*(oil+1)+oil] = y1[k1]+ c0 ;  //方程右边的向量
                                        // update equation for oil variables with
                                        // terms from vector c
                                        for ( i = 0 ; i < oil; i++ ) {
                                                A[k*(oil+1)+i] += GFpow(sk[ij++]) ;
//  c[k1][i+vinegar] ;//总的 x[i]的系数
                                        }
                                }
                                i = gauss( A, oil, oil+1 ) ;
                                if ( A[(oil-1)*(oil+2)] == 0  ) {
                                        eof++ ;
                                        if (eof < 100) {
                                                repeat = 1 ;
                                        }//这一段的目的是保证随机取的 v 变量能使 o 变量有解，当取了 100
                                         //次以后就放弃，签名不成功
                                        break ;
                                }
                                else {
                                        for ( i = 0 ; i< oil ; i++) x[i+vinegar] =A[i*(oil+1)+oil] ;
                                }
                        }  // general case oil > 1
                }
        } while( repeat ) ;

        if ( eof >= 100 ) {
                return ( -1 ) ;
        }
```

```
    // finally apply L2^{-1}
    k = 0 ;
    for ( i = 0 ; i < ovn ; i++ ) {
            y1[i] = GFpow(sk[ij++]) + x[i] ;
    }
    for ( i = 0 ; i < ovn ; i++ ) {
            x[i] = 0 ;
            for ( j = 0 ; j < ovn ; j++ ) {
                    x[i] += GFpow(sk[ij++]) * y1[j] ;
            }
    }


    for ( i = 0 ; i< ovn ; i++ ) sm[i] = x[i].GFpowtochar() ;

    * smlen = ovn ;

    return(repeat) ;
}
//Compose.h
#include "stdafx.h"


int keypair( unsigned char * sk, unsigned long * sklen,
                     unsigned char * pk, unsigned long * pklen);
int shortmessagesigned( unsigned char * m,unsigned long * mlen,
                                        const unsigned char * sm, unsigned long smlen,
                                        const unsigned char * pk, unsigned long pklen);
int signedshortmessage( unsigned char * sm,unsigned long * smlen,
                                        const unsigned char * m, unsigned long mlen,
                                        const unsigned char * sk, unsigned long sklen);

//Gauss.cpp
#include "stdafx.h"
#include "rainbow.h"

#ifdef OUTFILE
ofstream fout(OUTFILE,ios::out ) ; //Output Document


void display(GFpow *a, int const n, int const m, char * ch ) {
        // display in a in matrix form: n rows , m columns
        int i,j ;
        fout<<endl<<ch<<endl;
        for (i =0 ; i < n ; i++ ) {
                fout << endl ;
                for (j=0 ; j < m ; j++ ) if( a[i*m+j] == 0 ) fout<<". ";else fout
```

第4章 多变量公钥密码技术

```cpp
                << a[i*m+j]<<' ';
            }
            fout << endl ;
    }

    #endif

    int gauss( GFpow ab[], int const nequ, int const nvar ) {
            // nequ = number of equations,
            // nvar = number of variables
            // dimension: ab[nequ*nvar] ab[neq][nvar]

            int i, j, k, i0, i00, j0;
            int * ind ;

            int *ij = new int[nequ] ;


            for (i=0; i < nequ ; i++) ij[i]=i ;


            GFpow temp ;


    #ifdef DEBUG
            int b0,b1 ;

            fout << "starting gauss"<<endl ;

            for ( i = 0 ; i < nequ ; i++ ) {
                    b0 = 0 ;
                    for ( j = 0 ; j < nvar; j++) {
                            if (ab[i*nvar+j] != 0 ) {
                                    if (b0==0)  b0=1 ;
                                    else fout <<'+' ;
                                    fout << ab[i*nvar+j]<<"*u["<<j<<"]" ;
                            }
                    }
                    if (b0==0) fout <<"0,"<<endl ;
                    else fout <<','<<endl ;
            }
    #endif DEBUG


        ind = new int[nvar] ;


            i0 = j0 = 0 ;
            while ( (i0< nequ ) && ( j0 < nvar )) {
```

```
            i = i0 ;
            while ( ( ab[i*nvar+j0] == 0 ) && ( i < nequ ) ) i++ ;
            //找到第j0列第一个非0的元素所在的行i

            if ( i < nequ ) {
                    if ( i > i0 ) {  // interchange row i and i0

                            for ( j = j0 ; j < nvar ; j++ ) {
                                    temp = ab[i*nvar+j] ;
                                    ab[i*nvar+j] = ab[i0*nvar+j] ;
                                    ab[i0*nvar+j] = temp ;
                            }

                    }


                    temp = 1/ab[i0*nvar+j0] ;//归一

                    ab[i0*nvar+j0] = 1 ;
                    ind[i0] = j0 ;
                    for ( k = j0+1 ; k < nvar; k++ ) ab[i0*nvar+k] =
ab[i0*nvar+k]*temp ;

                    for ( j = i+1; j < nequ; j++ ) {//消元
                            if( (temp = ab[j*nvar+j0]) != 0 ){
                                    ab[j*nvar+j0] = 0 ;
                                    for ( k = j0+1; k < nvar ; k++ )
                                            ab[j*nvar+k] += ab[i0*nvar+k]*temp ;
                            }
                    }
                    i0++ ;
            }//if(i<nequ)
            j0++;

    }//while ( (i0< nequ ) && ( j0 < nvar ))


    i00 = i0 ;

    for ( i0--; i0>=0 ; i0-- ) {
            j0 = ind[i0] ;
            for ( i = i0-1 ; i >= 0 ; i-- ) {
                    temp = ab[i*nvar+j0] ;
                    ab[i*nvar+j0] = 0 ;
```

第4章 多变量公钥密码技术

```
                        for ( k = j0+1;  k < nvar ; k++ )
                                ab[i*nvar+k] = ab[i*nvar+k]+ab[i0*nvar+k]*temp ;
                }

        }

#ifdef DEBUG
        for ( i = 0 ; i < i00 ; i++ ) {
                b0 = b1 = 0 ;
                for ( j = 0 ; j < nvar; j++) {
                        if (ab[i*nvar+j] != 0 ) {
                                if (b0==0) {
                                        b0=1 ;
                                        if (ab[i*nvar+j]!=1) fout<<ab[i*nvar+j]<<'*';
                                        fout <<"u["<<j<<"]->" ;
                                }
                                else {
                                        if ( b1==1 ) fout << '+' ;
                                        fout <<ab[i*nvar+j]<< "*u["<<j<<"]" ;
                                        b1=1 ;
                                }
                        }
                }
                if (b1==0) fout <<"0,"<<endl ;
                else fout <<','<<endl ;
        }
#endif
        return ( i00 ) ;

}


//GFpow.cpp
#include "stdafx.h"
#include "GFpow.h"


void mult( int * c, int * a, int * b ) {
        /* multiplication of polynomials mod 2                    */
        /* c polynomial of degree 14,                             */
        /* x^7+s polynomials of degree 7, with x^7 not stored,    */
        /* Compute  c = c mod s                                   */
        int i,j ;
        for ( i = 0; i < 2*deg-1 ; i++ ) c[i] = 0 ;
        for ( i = 0; i < deg; i++ )
                if ( a[i]==1 ) for ( j = 0 ; j < deg ; j++ ) {
                        c[i+j] += b[j] ;
                        if ( c[i+j] == 2 ) c[i+j]=0 ;
                }
```

4.5 本章参考代码

```
                for( j = 2*deg-2 ; j >= deg ; j-- ) {
                        if ( c[j] == 1 ) {
                                c[j] = 0 ;
                                for ( i = 0 ; i < deg ; i++ ) {
                                        c[j-1-i] += fx[deg-1-i] ;
                                        if ( c[j-1-i] == 2 ) c[j-1-i] = 0 ;
                                }
                        }
                }
}

void binaryform( int * a, int n) {//Change an integer into polynomial (A coeffi
//cient array)
        int i ;
        for ( i=0 ; i < deg ; i++ ) {//get the last digit of n each time and put
//into a[i]
                a[i] = n & 1 ;
                n = n >> 1 ;
        }
}


int eval ( int * c ) {
        /* convert polynomial to integer between 0 to 256 */
        int i, a = 0 ;
        for ( i = deg-1; i >= 0; i-- )
                a = c[i]+ a + a ;
        return a ;
}

GFpow operator*( const GFpow & x, const GFpow & y ) {
        int j ;
        if (( x.a == 0 ) || ( y.a == 0 )) return (0) ;
        if ((j = lg[x.a] + lg[y.a] ) >= pow ) j -= powm1 ;
        return( ex[j] ) ;
}


GFpow operator/( const GFpow & x, const GFpow & y ) {
        int j ;
        if ( y.a==0 ) {
                cout << "Division by zero"<< endl ;
                abort() ;
        }
        if ( x.a == 0 ) return (0) ;

        if( (j = lg[x.a]-lg[y.a]) < 0 ) j += powm1 ;

        return ( ex[j] ) ;
```

```cpp
}


GFpow operator+( const GFpow & x, const GFpow & y ) {
      return (x.a^y.a) ;
}//加法就是按位异或



GFpow& GFpow:: operator+=( const GFpow & x ) {
      a= a^x.a ;
      return *this ;
}



GFpow& GFpow:: operator*=( const GFpow & x ) {
      int j ;
      if (a == 0 ) return *this;
      if ( x.a == 0 ) {a = 0 ; return *this ;}

      if ((j = lg[a] + lg[x.a] ) >= pow ) j -= powm1 ;
      a = ex[j] ;
      return *this ;
}



GFpow& GFpow:: operator/=( const GFpow & x ) {
      int j ;

      if ( x.a==0 ) {
                cout << "Division by zero in operator /="<< endl ;
                abort() ;
      }
      if ( a == 0 ) return *this ;

      if( (j = lg[a]-lg[x.a]) < 0 ) j += powm1 ;

      a = ex[j] ;
      return *this ;
}

ostream& operator<< (ostream& out, const GFpow& v){
        out << v.a ;
        return ( out ) ;
}
```

```cpp
istream& operator>> (istream& in, GFpow& v){
    in >> v.a ;
    return ( in ) ;
}

//GFpow.h
#ifndef GFPOW_H
#define GFPOW_H

#include <iostream>
#include <fstream>
#include <cstdlib>

using namespace std ;

const int deg = 8 ;      // degree of generating polynomial for finite field extension
const int pow = 256 ;    // 2^deg
const int powm1 = pow - 1 ;

// generating function for GF(2^8) was used to determine entries in arrays inv, lg, ex
const int short static fx[deg]= {1,0,1,1,0,0,1,0} ;   //1+x^2+x^3+x^6+x^8
// do not forget to call init() in main program when changing arithmitic

//static int short mul[pow][pow] ;
//static int short inv[pow], lg[pow], ex[pow];

//according to ex[i];
    static int short lg[pow]={0,0,1,23,2,46,24,83,3,106,47,147,25,52,84,69,4,92,107,
182,48,166,148,75,26,140,53,129,85,170,70,13,5,36,93,135,108,155,183,193,49,43,167,
163,149,152,76,202,27,230,141,115,54,205,130,18,86,98,171,240,71,79,14,189,6,212,37
,210,94,39,136,102,109,214,156,121,184,8,194,223,50,104,44,253,168,138,164,90,150,4
1,153,34,77,96,203,228,28,123,231,59,142,158,116,244,55,216,206,249,131,111,19,178,
87,225,99,220,172,196,241,175,72,10,80,66,15,186,190,199,7,222,213,120,38,101,211,2
09,95,227,40,33,137,89,103,252,110,177,215,248,157,243,122,58,185,198,9,65,195,174,
224,219,51,68,105,146,45,82,254,22,169,12,139,128,165,74,91,181,151,201,42,162,154,
192,35,134,78,188,97,239,204,17,229,114,29,61,124,235,232,233,60,234,143,125,159,23
6,117,30,245,62,56,246,217,63,207,118,250,31,132,160,112,237,20,144,179,126,88,251,
226,32,100,208,221,119,173,218,197,64,242,57,176,247,73,180,11,127,81,21,67,145,16,
113,187,238,191,133,200,161};
//x*f(x);
    static int short ex[pow]={1,2,4,8,16,32,64,128,77,154,121,242,169,31,62,124,248,
189,55,110,220,245,167,3,6,12,24,48,96,192,205,215,227,139,91,182,33,66,132,69,138,
89,178,41,82,164,5,10,20,40,80,160,13,26,52,104,208,237,151,99,198,193,207,211,235,
155,123,246,161,15,30,60,120,240,173,23,46,92,184,61,122,244,165,7,14,28,56,112,224
,141,87,174,17,34,68,136,93,186,57,114,228,133,71,142,81,162,9,18,36,72,144,109,218
,249,191,51,102,204,213,231,131,75,150,97,194,201,223,243,171,27,54,108,216,253,183
,35,70,140,85,170,25,50,100,200,221,247,163,11,22,44,88,176,45,90,180,37,74,148,101
,202,217,255,179,43,86,172,21,42,84,168,29,58,116,232,157,119,238,145,111,222,241,1
75,19,38,76,152,125,250,185,63,126,252,181,39,78,156,117,234,153,127,254,177,47,94,
```

第4章 多变量公钥密码技术

188,53,106,212,229,135,67,134,65,130,73,146,105,210,233,159,115,230,129,79,158,113,226,137,95,190,49,98,196,197,199,195,203,219,251,187,59,118,236,149,103,206,209,239,147,107,214,225,143,83,166,1};
  static int short inv[pow]={-1,1,166,196,83,135,98,116,143,44,229,36,49,94,58,125,225,43,22,240,212,141,18,241,190,51,47,223,29,28,152,236,214,159,179,131,11,228,120,207,106,65,224,17,9,142,222,26,95,12,191,25,177,80,201,219,168,148,14,124,76,238,118,123,107,41,233,164,255,105,231,90,163,89,114,183,60,239,193,227,53,176,134,4,112,204,174,172,162,73,71,230,111,202,13,48,137,216,6,117,249,186,170,242,254,69,40,64,194,221,203,92,84,205,74,182,7,99,62,122,38,206,119,63,59,15,155,208,147,139,178,35,210,180,82,5,217,96,146,129,213,21,45,8,247,184,138,128,57,169,253,234,30,237,209,126,198,244,215,33,188,250,88,72,67,232,2,197,56,149,102,243,87,175,86,173,81,52,130,34,133,211,115,75,145,246,101,248,160,251,24,50,226,78,108,220,3,167,156,245,218,54,93,110,85,113,121,39,127,154,132,181,20,140,32,158,97,136,200,55,195,109,46,27,42,16,192,79,37,10,91,70,165,66,151,252,31,153,61,77,19,23,103,171,157,199,185,144,187,100,161,189,235,150,104,161};

```
        class GFpow {
        private:
                int short a;

        public:
                GFpow() : a(0) {} ;
                GFpow( int x ): a(x % pow) {} ;
                GFpow( const GFpow & x ): a( x.a) {} ;
                unsigned char GFpow::GFpowtochar( ) {
                        return ((unsigned char)(a)) ;}



                friend GFpow operator+( const GFpow & x, const GFpow & y ) ;

                GFpow& operator+=( const GFpow & x ) ;
                GFpow& operator*=( const GFpow & x ) ;
                GFpow& operator/=( const GFpow & x ) ;

                friend int operator==(const GFpow& x, int z) { return ( x.a== z); }
                friend int operator!=(const GFpow& x, int z) { return ( x.a!= z); }
                friend int operator==(const GFpow& x, const GFpow& z) { return ( x.a== z.a); }
                friend int operator!=(const GFpow& x, const GFpow& z) { return ( x.a!= z.a); }

                friend GFpow operator*( const GFpow & x, const GFpow & y ) ;

                friend GFpow operator/( const GFpow & x, const GFpow & y ) ;

                friend ostream& operator<< (ostream& , const GFpow & ) ;
                friend istream& operator>> (istream& in,  GFpow& v ) ;
        } ;



        #endif /* GFPOW_H */
```

```cpp
//R.cpp
// R.cpp : 定义控制台应用程序的入口点
//

#include "stdafx.h"
#include "GFpow.h"
#include "sizes.h"
#include "Compose.h"
#include "rainbow.h"
//extern const int ovn =37 ;


int _tmain(int argc, _TCHAR* argv[])
{
        unsigned char sk[SECRETKEY_BYTES]; unsigned long sklen;
        unsigned char pk[PUBLICKEY_BYTES]; unsigned long pklen;
        unsigned char m[SHORTMESSAGE_BYTES]; unsigned long mlen = nv1 ;
        unsigned char sm[SIGNATURE_BYTES] ; unsigned long smlen = ovn ;
        int i, eof ;
        i=keypair( sk, & sklen, pk, & pklen ) ;
        cout << "\ndocument: " ;
        for ( i = 0 ; i < nv1; i++ ) {
                m[i] = rand() % pow ; // pow = 256;randomly generate message
                cout << GFpow(m[i])<<" ";
        }
        cout << endl ;
        eof = signedshortmessage( sm, & smlen, m, mlen, sk, sklen);//generate signature
        if ( eof < 0 ) cout <<"Could not sign message"<< endl;
        else {
                cout << "\nSignature: " ;
                for ( i =0 ; i < ovn; i++)
                        cout << GFpow(sm[i])<<" " ; // signed message
                eof = shortmessagesigned( m,& mlen, sm, smlen, pk, pklen);//verification
                if ( eof )
                        cout <<"\n reject signature\n";
                else
                        cout << "\nrecovered: " ;
        }
        for ( i = 0 ; i < signed long(mlen); i++ )
        {
                cout << GFpow( m[i] ) <<" ";//输出还原的消息
        }
return 0;
}

//rainbow.h
#include <cstdio>
#include <iostream>
```

第4章 多变量公钥密码技术

```cpp
#include <cstring>
#include <fstream>
#include <cstdlib>
#include <iomanip>
#include <ctime>

#include "GFpow.h"

using namespace std ;


// output file for Gauss can remove definition
#define OUTFILE  "temp.txt"
#define DEBUG



#ifdef OUTFILE
void display(GFpow * ap, int const n, int const m, char * ch =" ") ;
#endif

const int ovn = 37 ;   // oil + vinegar variables
const int v1 = 10 ;    // first set of vinegar variables: o < v1 < v2 <...< ovn=n
const int un = 5 ;// number of entries into Slist, first entry always v1, last always ovn
const int Slist[un] ={v1,20,24,27,ovn} ;

const int nv1 = ovn - v1 ;
const int ovn1 = ovn+1 ;
const int nv2 = 2 * nv1 ;
const int ovn2 = 2 * ovn ;

int gauss( GFpow ab[], int const , int const ) ;   //Definition in Gauss.cpp

//sizes.h
// header file as requested for e-bats
// exact values depend on parameters selected in rainbow.h
// choosing larger values here causes no harm
#define SECRETKEY_BYTES 13242
#define PUBLICKEY_BYTES 20007
#define SHAREDSECRET_BYTES 0
#define SHORTMESSAGE_BYTES 27
#define SIGNATURE_BYTES 37
```

2. VHDL 代码（签名生成）

```vhdl
--secretKeyParts.vhd
LIBRARY IEEE;
USE IEEE.STD_LOGIC_1164.ALL;
USE WORK.signedTypes.ALL;
```

```vhdl
    PACKAGE secretKeyParts IS
    CONSTANT y0 : matrix1 := (X"72",X"B1",X"46",X"3E",X"1B",X"63",X"61",X"9B",X"42",
X"7A",X"B5",X"0E",X"D5",X"3C",X"42",X"2E",X"62",X"0C",X"B0",X"8F",X"7E",X"3B",X"D8",
X"37");
    END secretKeyParts;

--secretKeyRom.vhd
-- megafunction wizard: %ROM: 1-PORT%
-- GENERATION: STANDARD
-- VERSION: WM1.0
-- MODULE: altsyncram

-- ============================================================
-- File Name: secretKeyRom.vhd
-- Megafunction Name(s):
--                  altsyncram
--
-- Simulation Library Files(s):
--                  altera_mf


LIBRARY ieee;
USE ieee.std_logic_1164.all;

LIBRARY altera_mf;
USE altera_mf.all;

ENTITY secretKeyRom IS
    PORT
    (
        address             : IN STD_LOGIC_VECTOR (14 DOWNTO 0);
        clock               : IN STD_LOGIC ;
        q                   : OUT STD_LOGIC_VECTOR (7 DOWNTO 0)
    );
END secretKeyRom;


ARCHITECTURE SYN OF secretkeyrom IS

    SIGNAL sub_wire0        : STD_LOGIC_VECTOR (7 DOWNTO 0);


    COMPONENT altsyncram
    GENERIC (
        address_aclr_a              : STRING;
        init_file                   : STRING;
        intended_device_family      : STRING;
        lpm_hint                    : STRING;
```

```vhdl
                lpm_type                        : STRING;
                numwords_a                      : NATURAL;
                operation_mode                  : STRING;
                outdata_aclr_a                  : STRING;
                outdata_reg_a                   : STRING;
                widthad_a                       : NATURAL;
                width_a                         : NATURAL;
                width_byteena_a                 : NATURAL
        );
        PORT (
                clock0          : IN STD_LOGIC ;
                address_a       : IN STD_LOGIC_VECTOR (14 DOWNTO 0);
                q_a             : OUT STD_LOGIC_VECTOR (7 DOWNTO 0)
        );
        END COMPONENT;

BEGIN
        q    <= sub_wire0(7 DOWNTO 0);

        altsyncram_component : altsyncram
        GENERIC MAP (
                address_aclr_a => "NONE",
                init_file => "./secretKeyRom.mif",
                intended_device_family => "Cyclone",
                lpm_hint => "ENABLE_RUNTIME_MOD=NO",
                lpm_type => "altsyncram",
                numwords_a => 17838,
                operation_mode => "ROM",
                outdata_aclr_a => "NONE",
                outdata_reg_a => "CLOCK0",
                widthad_a => 15,
                width_a => 8,
                width_byteena_a => 1
        )
        PORT MAP (
                clock0 => clock,
                address_a => address,
                q_a => sub_wire0
        );



END SYN;

-- ============================================================
-- CNX file retrieval info
-- ============================================================
-- Retrieval info: PRIVATE: ADDRESSSTALL_A NUMERIC "0"
-- Retrieval info: PRIVATE: AclrAddr NUMERIC "0"
```

```
-- Retrieval info: PRIVATE: AclrByte NUMERIC "0"
-- Retrieval info: PRIVATE: AclrOutput NUMERIC "0"
-- Retrieval info: PRIVATE: BYTE_ENABLE NUMERIC "0"
-- Retrieval info: PRIVATE: BYTE_SIZE NUMERIC "8"
-- Retrieval info: PRIVATE: BlankMemory NUMERIC "0"
-- Retrieval info: PRIVATE: CLOCK_ENABLE_INPUT_A NUMERIC "0"
-- Retrieval info: PRIVATE: CLOCK_ENABLE_OUTPUT_A NUMERIC "0"
-- Retrieval info: PRIVATE: Clken NUMERIC "0"
-- Retrieval info: PRIVATE: IMPLEMENT_IN_LES NUMERIC "0"
-- Retrieval info: PRIVATE: INIT_FILE_LAYOUT STRING "PORT_A"
-- Retrieval info: PRIVATE: INIT_TO_SIM_X NUMERIC "0"
-- Retrieval info: PRIVATE: INTENDED_DEVICE_FAMILY STRING "Stratix II"
-- Retrieval info: PRIVATE: JTAG_ENABLED NUMERIC "0"
-- Retrieval info: PRIVATE: JTAG_ID STRING "NONE"
-- Retrieval info: PRIVATE: MAXIMUM_DEPTH NUMERIC "0"
-- Retrieval info: PRIVATE: MIFfilename STRING "../../../multi144/signedMessage/
secretKeyRom.mif"
-- Retrieval info: PRIVATE: NUMWORDS_A NUMERIC "17838"
-- Retrieval info: PRIVATE: RAM_BLOCK_TYPE NUMERIC "0"
-- Retrieval info: PRIVATE: RegAddr NUMERIC "1"
-- Retrieval info: PRIVATE: RegOutput NUMERIC "1"
-- Retrieval info: PRIVATE: SYNTH_WRAPPER_GEN_POSTFIX STRING "0"
-- Retrieval info: PRIVATE: SingleClock NUMERIC "1"
-- Retrieval info: PRIVATE: UseDQRAM NUMERIC "0"
-- Retrieval info: PRIVATE: WidthAddr NUMERIC "15"
-- Retrieval info: PRIVATE: WidthData NUMERIC "8"
-- Retrieval info: PRIVATE: rden NUMERIC "0"
-- Retrieval info: CONSTANT: CLOCK_ENABLE_INPUT_A STRING "BYPASS"
-- Retrieval info: CONSTANT: CLOCK_ENABLE_OUTPUT_A STRING "BYPASS"
-- Retrieval info: CONSTANT: INIT_FILE STRING "../../../multi144/signedMessage/
secretKeyRom.mif"
-- Retrieval info: CONSTANT: INTENDED_DEVICE_FAMILY STRING "Stratix II"
-- Retrieval info: CONSTANT: LPM_HINT STRING "ENABLE_RUNTIME_MOD=NO"
-- Retrieval info: CONSTANT: LPM_TYPE STRING "altsyncram"
-- Retrieval info: CONSTANT: NUMWORDS_A NUMERIC "17838"
-- Retrieval info: CONSTANT: OPERATION_MODE STRING "ROM"
-- Retrieval info: CONSTANT: OUTDATA_ACLR_A STRING "NONE"
-- Retrieval info: CONSTANT: OUTDATA_REG_A STRING "CLOCK0"
-- Retrieval info: CONSTANT: WIDTHAD_A NUMERIC "15"
-- Retrieval info: CONSTANT: WIDTH_A NUMERIC "8"
-- Retrieval info: CONSTANT: WIDTH_BYTEENA_A NUMERIC "1"
-- Retrieval info: USED_PORT: address 0 0 15 0 INPUT NODEFVAL address[14..0]
-- Retrieval info: USED_PORT: clock 0 0 0 0 INPUT NODEFVAL clock
-- Retrieval info: USED_PORT: q 0 0 8 0 OUTPUT NODEFVAL q[7..0]
-- Retrieval info: CONNECT: @address_a 0 0 15 0 address 0 0 15 0
-- Retrieval info: CONNECT: q 0 0 8 0 @q_a 0 0 8 0
-- Retrieval info: CONNECT: @clock0 0 0 0 0 clock 0 0 0 0
-- Retrieval info: LIBRARY: altera_mf altera_mf.altera_mf_components.all
-- Retrieval info: GEN_FILE: TYPE_NORMAL secretKeyRom.vhd TRUE
```

```vhdl
-- Retrieval info: GEN_FILE: TYPE_NORMAL secretKeyRom.inc FALSE
-- Retrieval info: GEN_FILE: TYPE_NORMAL secretKeyRom.cmp FALSE
-- Retrieval info: GEN_FILE: TYPE_NORMAL secretKeyRom.bsf FALSE
-- Retrieval info: GEN_FILE: TYPE_NORMAL secretKeyRom_inst.vhd FALSE
-- Retrieval info: GEN_FILE: TYPE_NORMAL secretKeyRom_waveforms.html TRUE
-- Retrieval info: GEN_FILE: TYPE_NORMAL secretKeyRom_wave*.jpg FALSE
-- Retrieval info: LIB_FILE: altera_mf

--signedMessage.vhd
LIBRARY IEEE;
USE IEEE.STD_LOGIC_1164.ALL;
USE IEEE.STD_LOGIC_ARITH.ALL;
USE IEEE.STD_LOGIC_UNSIGNED.ALL;
USE WORK.secretKeyParts.ALL;
USE WORK.signedTypes.ALL;


ENTITY signedMessage IS
       PORT (clk,start: IN STD_LOGIC;
                       --signatrue: OUT matrix3;
                       SignedDone,SignedRight: OUT STD_LOGIC);
END signedMessage;

ARCHITECTURE rtl OF signedMessage IS

       SIGNAL sign_state: statetype;
       --rom component
       SIGNAL r: STD_LOGIC_VECTOR(7 DOWNTO 0);
       SIGNAL addr: STD_LOGIC_VECTOR(14 DOWNTO 0);
       ---mult component
       SIGNAL d,e,f: STD_LOGIC_VECTOR(7 DOWNTO 0);
       ---intermidiate_result
       SIGNAL i_result : matrix3;
       SIGNAL l1_inverse_result : matrix3;
       ----matrixA
       SIGNAL matrixA :  matrix2;
---------------------------------------------
--for guassian elimination
signal col:INT13:=0;--主元所在行
SIGNAL mainvalue:STD_LOGIC_VECTOR(7 DOWNTO 0);--主元
SIGNAL inversevalue:STD_LOGIC_VECTOR(7 DOWNTO 0);--逆元
SIGNAL midint:INT13;--存找到非零主元行
SIGNAL midvector:STD_LOGIC_VECTOR(7 DOWNTO 0);--临时使用
signal finish:bit;
signal gi:INT13;
signal ei:INT13;
signal ej:INT13;
signal ls:INT13:=0;
-----------------------------------------------------
```

```vhdl
        ----gauss
        SIGNAL xv : matrix3 := (X"67",X"20",X"86",X"51",X"E8",X"6B",X"7D",X"26",
X"7D",X"B6",X"BC",X"48",X"B8",X"5D",X"00",X"10",X"E3",X"90",X"DB",X"DD",X"C9",X"AA",
X"3C",X"18",X"D7",X"D6",X"03",X"BF",X"9E",X"6C",X"93",X"F0",X"A0",X"B7",X"28",X"8A",
X"52",X"ED",X"AB",X"C2",X"81",X"98");
        SIGNAL isSloved: STD_LOGIC;


        COMPONENT GF_mult
            PORT(a,b: IN STD_LOGIC_VECTOR(7 DOWNTO 0);
                    c: OUT STD_LOGIC_VECTOR(7 DOWNTO 0));
        END COMPONENT;
        -----rom

        --求逆部件
COMPONENT GF_inverse
        PORT (inB:IN STD_LOGIC_VECTOR(7 DOWNTO 0);
        BInvert: OUT STD_LOGIC_VECTOR(7 DOWNTO 0));
END COMPONENT;

        COMPONENT secretKeyRom IS
        PORT
        (
            address              : IN STD_LOGIC_VECTOR (14 DOWNTO 0);
            clock                : IN STD_LOGIC ;
            q                    : OUT STD_LOGIC_VECTOR (7 DOWNTO 0)
        );
        END COMPONENT;

BEGIN

---------乘法
M0:COMPONENT GF_mult PORT MAP (d, e, f);

I0:GF_inverse PORT MAP (mainvalue,inversevalue);--求逆元

------rom
R0:COMPONENT secretKeyRom PORT MAP (addr, clk, r);

        PROCESS(clk)
        VARIABLE li: INTEGER range 0 to 2; ---layer i
        VARIABLE fi: INTEGER range 0 to 12; ---function i
        VARIABLE o: array_int; ---oil
        VARIABLE v: array_int; ---vinager
        --VARIABLE count: INTEGER; ------count<100

        VARIABLE start_addr: STD_LOGIC_VECTOR(14 DOWNTO 0);
        VARIABLE steps : INTEGER range 0 to 31;
```

```vhdl
VARIABLE i: INTEGER range 0 to 42;
VARIABLE j: INTEGER range 0 to 42;

BEGIN
    IF (clk'event AND clk='1') THEN
        CASE sign_state IS
            WHEN idle =>

                if (start = '1') then
                    sign_state <= init;
                    SignedDone<='0';
                else
                    sign_state <= idle;
                end if;
-------------------------------------------------------------------------------
            WHEN tests =>
                if(l1_inverse_result(i)/=sig(i)) then
                    SignedRight<='0';
                    sign_state <= signFinish;
                END IF;
                i := i +1;
                if(i>41) then
                    sign_state <= signFinish;
                end if;


-------------------------------------------------------------------------------

            when signFinish =>
                SignedDone<='1';
                sign_state<=signStop;
            when signStop =>
                if (start = '0') then --reset
                    sign_state<=idle;
                    SignedDone<='0';
                    SignedRight<='0';
                end if;

            WHEN init =>
                --count := 0;
                SignedRight<='1';
                i := 0;
                steps := 0;
                start_addr := "0000000000000000";
                sign_state <= readData0;
            ------readData
            WHEN readData0 =>
                addr <= start_addr;
                start_addr := start_addr+"0000000000000001";
                sign_state <= readData1;
```

```vhdl
                    WHEN readData1 =>
                         sign_state <= readData2;
                    WHEN readData2 =>
                         IF (steps = 0) THEN
                              sign_state <= l1_offset_state;
                         ELSIF (steps = 1) THEN
                              sign_state <= l1_inverse_mult;
                         ELSIF (steps = 20) THEN
                              sign_state <= matrixA_a_mult;
                         ELSIF (steps = 21) THEN
                              sign_state <= matrixA_l_add;
                         ELSIF (steps = 22) THEN
                              sign_state <= matrixA_b_mult2;
                         ELSIF (steps = 23) THEN
                              sign_state <= matrixA_r_mult;
                         ELSIF (steps = 24) THEN
                              sign_state <= matrixA_n_l1_add;
                         ELSIF (steps = 30) THEN
                              sign_state <= l2_offset_state;
                         ELSIF (steps = 31) THEN
                              sign_state <= l2_inverse_mult;
                         END IF;

                    ------l1 offset
                    WHEN l1_offset_state =>
                         i_result(i) <= y0(i) XOR r;
                         i := i +1;
                         IF (i >= 24) THEN
                                   steps := 1;
                                   j := 0;
                                   i := 0;
                         END IF;
                         sign_state <= readData0;

                    WHEN l1_inverse_mult =>
                         d <= i_result(i);
                         e <= r;
                         sign_state <= l1_inverse_add;

                    WHEN l1_inverse_add =>
                         IF (i/=0) THEN
                                   l1_inverse_result(j) <= l1_inverse_result(j) XOR f;
                         ELSE
                                   l1_inverse_result(j) <= f;
                         END IF;
                         i := i +1;
                         IF (i = 24) THEN
```

```
                        j := j +1;
                    i := 0;
                END IF;
                IF (j >= 24) THEN
                        sign_state <= matrixA_init;
                ELSE
                        sign_state <= readData0;
                END IF;
----------------------------------------------------------------
        WHEN matrixA_init =>
            li := 0;
            o(0) := 12; o(1) := 12;
            v(0) := 18; v(1) := 30; v(2) := 42;
            sign_state <= matrixA_start;
----------------------------------------------------------------
----------------------------------------------------------------
        WHEN matrixA_start =>
            i := 0;
            j := 0;
            fi := 0;
            IF (li = 0) THEN
                    sign_state <= chooseRandom;
            ELSE
                        steps := 20;
                        sign_state <= readData0;
            END IF;

        WHEN chooseRandom =>
            steps := 20;
            sign_state <= readData0;
----------------------------------------------------------------
        WHEN matrixA_a_mult =>
            d <= xv(i);
            e <= r;
            sign_state <= matrixA_a_add;

        WHEN matrixA_a_add =>
            IF (i=0) THEN
                    matrixA(fi,j) <= f;
            ELSE
                    matrixA(fi,j) <= matrixA(fi,j) XOR f;
            END IF;
            i := i +1;
            IF (i = v(li)) THEN
                    j := j +1;
                    i := 0;
            END IF;
            IF (j >= o(li)) THEN
                    steps := 21;
```

```
                                i := 0;
                END IF;
                sign_state <= readData0;
        WHEN matrixA_l_add =>
                matrixA(fi,i) <= matrixA(fi,i) XOR r ;
                i := i+1;
                IF (i >= o(li)) THEN
                        sign_state <= matrixA_b_mult1;
                        steps := 22;
                        i := 0;
                        j := 0;
                ELSE
                        sign_state <= readData0;
                END IF;
```

```
        WHEN matrixA_b_mult1 =>
                d <= xv(i);
                e <= xv(j);
                sign_state <= readData0;
        WHEN matrixA_b_mult2 =>
                d <= f;
                e <= r;
                sign_state <= matrixA_b_add;
        WHEN matrixA_b_add =>
                IF (i = 0 and j = 0) THEN
                        matrixA(fi,o(li)) <= f ;
                ELSE
                        matrixA(fi,o(li)) <= matrixA(fi,o(li)) XOR f ;
                END IF;
                j := j +1;
                IF (j >= v(li)) THEN
                        i:= i +1;
                        j:= i;
                END IF;
                IF (i >= v(li)) THEN
                        steps := 23;
                        sign_state <= readData0;
                        i := 0;
                ELSE
                        sign_state <= matrixA_b_mult1;
                END IF;
```

```
        WHEN matrixA_r_mult =>
                d <= xv(i);
                e <= r;
                sign_state <= matrixA_r_add;
        WHEN matrixA_r_add =>
                matrixA(fi,o(li)) <= matrixA(fi,o(li)) XOR f;
                i := i +1;
```

```
                              IF (i >= v(li)) THEN
                                     steps := 24;
                              END IF;
                              sign_state <= readData0;
                       WHEN matrixA_n_l1_add =>
                              matrixA(fi,o(li))  <= matrixA(fi,o(li)) XOR r XOR
l1_inverse_result(li*12+fi);
                              IF (fi >= o(li)-1) THEN
                                     sign_state <= gaussA;
                              ELSE
                                     steps := 20;
                                     sign_state <= readData0;
                                     fi := fi +1;
                                     i := 0;
                                     j := 0;
                              END IF;


--------------------------------------------------------------------------------
--gauss
                       WHEN gaussA =>
                              col<=0;
              finish<='0';
              midvector<=matrixA(0,0);--主元
              gi<=1;
              if(matrixA(0,0)/="00000000") then
                   mainvalue<=matrixA(0,0);
                   sign_state <= normalizing1;

              else
                   --找主元
                   ls<=col+1;
                   sign_state <= findmainelement3;
              end if;
-----------------------------------------------------
                       WHEN gaussB =>
                              li := li + 1;
                              IF (li >= 2) THEN
                                     steps := 30;
                                     sign_state <= readData0;
--                                   sign_state <= tests;
                                     i := 0;
                              ELSE
                                     sign_state <= matrixA_start;
                              END IF;
--------------------------------------------------------------------------------
                       ------l2 offset
                       WHEN l2_offset_state =>
                              i_result(i) <= xv(i) XOR r;
```

```vhdl
                        i := i +1;
                        IF (i >= 42) THEN
                                steps := 31;
                                j := 0;
                                i := 0;
                        END IF;
                        sign_state <= readData0;
----------------------------------------------------------------------------
                WHEN l2_inverse_mult =>
                        d <= i_result(i);
                        e <= r;
                        sign_state <= l2_inverse_add;

                WHEN l2_inverse_add =>
                        IF (i=0) THEN
                                l1_inverse_result(j) <= f;
                        ELSE
                                l1_inverse_result(j) <= 
l1_inverse_result(j) XOR f;
                        END IF;
--                      l1_inverse_result(j) <= l1_inverse_result(j) XOR f;
                        i := i +1;
                        IF (i = 42) THEN
                                j := j +1;
                                i := 0;
                        END IF;
                        IF (j >= 42) THEN
                                sign_state <= tests;
                                i:=0;
                        ELSE
                                sign_state <= readData0;
                        END IF;
----------------------------------------------------------------------------


----------------------------------------------------------------------------
            --gaussian elimination
    ---代入求解
    WHEN outputresult2 =>
       --获得乘法结果
         matrixA(gi,n)<=f XOR matrixA(gi,n);
       --一列代入结束
       if(gi/=0) then
       --当列继续代入
         d<=matrixA(gi-1,col);
          --sign_state<=outputresult2;
          gi<=gi-1;
       else
       --一列代入结束
```

```
                gi<=col-1;
                if(li=0) then
                      xv(col-1+18)<=matrixA(col-1,n);
                elsif(li=1) then
                      xv(col-1+30)<=matrixA(col-1,n);
                end if;
         --结束代入
                if(col/=1) then
                      col<=col-1;
                      --送乘数和被乘数
                      d<=matrixA(col-2,col-1);
                      e<=matrixA(col-1,n);
                      --sign_state<=outputresult2;
                else
                      --结束代入
                      if(li=0) then
                            xv(18)<=f XOR matrixA(0,n);
                      elsif(li=1) then
                            xv(30)<=f XOR matrixA(0,n);
                      end if;
                      finish<='1';
                      sign_state<=gaussB;
                end if;
         end if;


------------------

              --换行
         WHEN findmainelement2 =>
              IF (midvector/="00000000") THEN
                   mainvalue<=midvector;--传出要求逆的元素
                   sign_state <= findmainelement1;
                   ls<=col;
              else
                   sign_state <= idle;
              end if;


                   --换行
         WHEN findmainelement1 =>
              if ((ls>=col) and (ls<=n)) then
                       matrixA(col,ls)<=matrixA(midint,ls);
                       matrixA(midint,ls)<=midvector;
              end if;
              ls<=ls+1;
              if(ls>=n) then
                      sign_state <= normalizing1;
              end if;
---------------------------------------------------------------
```

```
      --归一
WHEN normalizing1 =>
    d<=matrixA(col,col+1);
    e<=inversevalue;
    sign_state <= normalizing2;
-------
      --归一
WHEN normalizing2 =>
    matrixA(col,gi)<=f;
    --继续归一
    if(gi/=n) then
         d<=matrixA(col,gi+1);
         --sign_state <= normalizing2;
         gi<=gi+1;
     --本行归一结束
         elsif (col/=n-1) then
             e<=matrixA(col+1,col);
             d<=matrixA(col,col+1);
             sign_state<=elimiresult;
             ei<=col+1;
             ej<=col+1;
         --最后一行归一结束，不用消元
         else
             --送乘数和被乘数
             d<=matrixA(col-1,col);
             e<=f;
             sign_state<=outputresult2;
             if(li=0) then
                  xv(n-1+18)<=f;
             elsif(li=1) then
                  xv(n-1+30)<=f;
             end if;
             gi<=col-1;
    end if;

 ---消元
WHEN elimiresult=>
     matrixA(ei,ej)<=f XOR matrixA(ei,ej);
     if ((ej=col+1) and (ei=col+1)) then
          mainvalue<=f XOR matrixA(ei,ej);
     end if;

     --继续本行的消元
     if(ej/=n) then
         ej<=ej+1;
         --sign_state<=elimiresult;
         d<=matrixA(col,ej+1);
     else
     --本行消元结束，开始下一行
```

```
                    if (ei/=n-1) then
                        ej<=col+1;
                        ei<=ei+1;
                        --sign_state<=elimiresult;
                        e<=matrixA(ei+1,col);
                        d<=matrixA(col,col+1);
                     else
                         --结束消元
                        col<=col+1;
                        gi<=col+2;
                        --nonzero judging
                          if(matrixA(col+1,col+1)/="00000000") then
                           if(col=n-2) then
                                    d<=matrixA(col+1,col+2) xor f;
                           else
                                    d<=matrixA(col+1,col+2);
                           end if;
                           e<=inversevalue;
                           sign_state <= normalizing2;
                         else
                           --找主元
                           midvector<=matrixA(col+1,col+1);--主元
                           midint<=col+1;
                           sign_state <= findmainelement3;
                           ls<=col+1;
                          end if;
                        end if;
                    end if;

                    When findmainelement3=>
                                      IF (matrixA(i,col)/="00000000") THEN
                                 midvector<=matrixA(ls,col);
                                 midint<=ls;
                                 sign_state <= findmainelement2;
                             end if;

                             if ((ls>n-1) or (ls<col+1)) then
                                               sign_state <= idle;
                             END IF;

                             ls<=ls+1;
-------------------------------------------------------------------------------

                    WHEN OTHERS =>
                            sign_state <= idle;
                 END CASE;

         END IF;
    END PROCESS;
```

```vhdl
END rtl;

--signedTypes.vhd
LIBRARY IEEE;
USE IEEE.STD_LOGIC_1164.ALL;

PACKAGE signedTypes IS
      ------general types
      TYPE array_int IS ARRAY (0 TO 4) OF INTEGER;

      TYPE matrix1 IS ARRAY (0 TO 23) OF STD_LOGIC_VECTOR(7 DOWNTO 0);
      TYPE matrix2 IS ARRAY (0 TO 11, 0 TO 12) OF STD_LOGIC_VECTOR(7 DOWNTO 0);
      TYPE matrix3 IS ARRAY (0 TO 41) OF STD_LOGIC_VECTOR(7 DOWNTO 0);

      CONSTANT n:INTEGER:=12;
      SUBTYPE INT13 IS INTEGER range n downto 0;
      TYPE statetype IS (idle,init,tests,signFinish,signStop,readData0,readData1,
readData2,l1_offset_state,l1_inverse_mult,l1_inverse_add,
      chooseRandom,matrixA_start,matrixA_init,matrixA_a_mult,matrixA_a_add,matrixA_l_add,matrixA_l,matrixA_b_mult1,matrixA_b_mult2,
      matrixA_b_add,matrixA_r_mult,matrixA_n_l1_add,matrixA_r_add,matrixA_n_add,matrixA_l1_add,
      gaussA,gaussB,l2_offset_state,l2_inverse_mult,l2_inverse_add,normalizing1,normalizing2,
      findmainelement1,findmainelement2,findmainelement3,elimiresult,outputresult2);


      constant sig : matrix3 := (
           X"60",X"bd",X"0b",X"38",X"4f",X"1f",X"80",X"08",X"4e",X"27",
           X"7d",X"84",X"83",X"2d",X"50",X"4c",X"ff",X"ff",X"25",X"cd",
           X"64",X"8e",X"bd",X"a9",X"bf",X"d7",X"66",X"85",X"27",X"ab",
           X"34",X"c6",X"93",X"a5",X"3b",X"d2",X"0e",X"80",X"3e",X"07",
           X"5d",X"b8"
      );
END signedTypes;

--test.vhd
-- megafunction wizard: %ROM: 1-PORT%
-- GENERATION: STANDARD
-- VERSION: WM1.0
-- MODULE: altsyncram


-- ============================================================
-- File Name: test.vhd
-- Megafunction Name(s):
--                    altsyncram
--
-- Simulation Library Files(s):
--                    altera_mf
```

第4章 多变量公钥密码技术

```vhdl
-- ============================================================


LIBRARY ieee;
USE ieee.std_logic_1164.all;

LIBRARY altera_mf;
USE altera_mf.all;

ENTITY test IS
    PORT
    (
        address         : IN STD_LOGIC_VECTOR (8 DOWNTO 0);
        clock           : IN STD_LOGIC ;
        q               : OUT STD_LOGIC_VECTOR (998 DOWNTO 0)
    );
END test;


ARCHITECTURE SYN OF test IS

    SIGNAL sub_wire0    : STD_LOGIC_VECTOR (998 DOWNTO 0);



    COMPONENT altsyncram
    GENERIC (
        clock_enable_input_a            : STRING;
        clock_enable_output_a           : STRING;
        init_file           : STRING;
        intended_device_family          : STRING;
        lpm_hint        : STRING;
        lpm_type        : STRING;
        numwords_a          : NATURAL;
        operation_mode          : STRING;
        outdata_aclr_a          : STRING;
        outdata_reg_a           : STRING;
        widthad_a           : NATURAL;
        width_a         : NATURAL;
        width_byteena_a             : NATURAL
    );
    PORT (
            clock0      : IN STD_LOGIC ;
            address_a   : IN STD_LOGIC_VECTOR (8 DOWNTO 0);
            q_a     : OUT STD_LOGIC_VECTOR (998 DOWNTO 0)
    );
    END COMPONENT;

BEGIN
```

```vhdl
        q    <= sub_wire0(998 DOWNTO 0);

    altsyncram_component : altsyncram
    GENERIC MAP (
        clock_enable_input_a => "BYPASS",
        clock_enable_output_a => "BYPASS",
        init_file => "secretKeyRom.mif",
        intended_device_family => "Stratix II",
        lpm_hint => "ENABLE_RUNTIME_MOD=NO",
        lpm_type => "altsyncram",
        numwords_a => 512,
        operation_mode => "ROM",
        outdata_aclr_a => "NONE",
        outdata_reg_a => "CLOCK0",
        widthad_a => 9,
        width_a => 999,
        width_byteena_a => 1
    )
    PORT MAP (
        clock0 => clock,
        address_a => address,
        q_a => sub_wire0
    );

END SYN;


-- ============================================================
-- CNX file retrieval info
-- ============================================================
-- Retrieval info: PRIVATE: ADDRESSSTALL_A NUMERIC "0"
-- Retrieval info: PRIVATE: AclrAddr NUMERIC "0"
-- Retrieval info: PRIVATE: AclrByte NUMERIC "0"
-- Retrieval info: PRIVATE: AclrOutput NUMERIC "0"
-- Retrieval info: PRIVATE: BYTE_ENABLE NUMERIC "0"
-- Retrieval info: PRIVATE: BYTE_SIZE NUMERIC "9"
-- Retrieval info: PRIVATE: BlankMemory NUMERIC "0"
-- Retrieval info: PRIVATE: CLOCK_ENABLE_INPUT_A NUMERIC "0"
-- Retrieval info: PRIVATE: CLOCK_ENABLE_OUTPUT_A NUMERIC "0"
-- Retrieval info: PRIVATE: Clken NUMERIC "0"
-- Retrieval info: PRIVATE: IMPLEMENT_IN_LES NUMERIC "0"
-- Retrieval info: PRIVATE: INIT_FILE_LAYOUT STRING "PORT_A"
-- Retrieval info: PRIVATE: INIT_TO_SIM_X NUMERIC "0"
-- Retrieval info: PRIVATE: INTENDED_DEVICE_FAMILY STRING "Stratix II"
-- Retrieval info: PRIVATE: JTAG_ENABLED NUMERIC "0"
-- Retrieval info: PRIVATE: JTAG_ID STRING "NONE"
-- Retrieval info: PRIVATE: MAXIMUM_DEPTH NUMERIC "0"
-- Retrieval info: PRIVATE: MIFfilename STRING "secretKeyRom.mif"
-- Retrieval info: PRIVATE: NUMWORDS_A NUMERIC "512"
-- Retrieval info: PRIVATE: RAM_BLOCK_TYPE NUMERIC "0"
```

```
-- Retrieval info: PRIVATE: RegAddr NUMERIC "1"
-- Retrieval info: PRIVATE: RegOutput NUMERIC "1"
-- Retrieval info: PRIVATE: SYNTH_WRAPPER_GEN_POSTFIX STRING "0"
-- Retrieval info: PRIVATE: SingleClock NUMERIC "1"
-- Retrieval info: PRIVATE: UseDQRAM NUMERIC "0"
-- Retrieval info: PRIVATE: WidthAddr NUMERIC "9"
-- Retrieval info: PRIVATE: WidthData NUMERIC "999"
-- Retrieval info: PRIVATE: rden NUMERIC "0"
-- Retrieval info: CONSTANT: CLOCK_ENABLE_INPUT_A STRING "BYPASS"
-- Retrieval info: CONSTANT: CLOCK_ENABLE_OUTPUT_A STRING "BYPASS"
-- Retrieval info: CONSTANT: INIT_FILE STRING "secretKeyRom.mif"
-- Retrieval info: CONSTANT: INTENDED_DEVICE_FAMILY STRING "Stratix II"
-- Retrieval info: CONSTANT: LPM_HINT STRING "ENABLE_RUNTIME_MOD=NO"
-- Retrieval info: CONSTANT: LPM_TYPE STRING "altsyncram"
-- Retrieval info: CONSTANT: NUMWORDS_A NUMERIC "512"
-- Retrieval info: CONSTANT: OPERATION_MODE STRING "ROM"
-- Retrieval info: CONSTANT: OUTDATA_ACLR_A STRING "NONE"
-- Retrieval info: CONSTANT: OUTDATA_REG_A STRING "CLOCK0"
-- Retrieval info: CONSTANT: WIDTHAD_A NUMERIC "9"
-- Retrieval info: CONSTANT: WIDTH_A NUMERIC "999"
-- Retrieval info: CONSTANT: WIDTH_BYTEENA_A NUMERIC "1"
-- Retrieval info: USED_PORT: address 0 0 9 0 INPUT NODEFVAL address[8..0]
-- Retrieval info: USED_PORT: clock 0 0 0 0 INPUT NODEFVAL clock
-- Retrieval info: USED_PORT: q 0 0 999 0 OUTPUT NODEFVAL q[998..0]
-- Retrieval info: CONNECT: @address_a 0 0 9 0 address 0 0 9 0
-- Retrieval info: CONNECT: q 0 0 999 0 @q_a 0 0 999 0
-- Retrieval info: CONNECT: @clock0 0 0 0 0 clock 0 0 0 0
-- Retrieval info: LIBRARY: altera_mf altera_mf.altera_mf_components.all
-- Retrieval info: GEN_FILE: TYPE_NORMAL test.vhd TRUE
-- Retrieval info: GEN_FILE: TYPE_NORMAL test.inc FALSE
-- Retrieval info: GEN_FILE: TYPE_NORMAL test.cmp FALSE
-- Retrieval info: GEN_FILE: TYPE_NORMAL test.bsf FALSE
-- Retrieval info: GEN_FILE: TYPE_NORMAL test_inst.vhd FALSE
-- Retrieval info: GEN_FILE: TYPE_NORMAL test_waveforms.html TRUE
-- Retrieval info: GEN_FILE: TYPE_NORMAL test_wave*.jpg FALSE
-- Retrieval info: LIB_FILE: altera_mf
```

3. VHDL 代码（签名验证）

```
--add14.vhd
LIBRARY IEEE;
USE IEEE.STD_LOGIC_1164.ALL;

ENTITY fau IS
PORT (a,cin: IN STD_LOGIC;
cout ,s : OUT STD_LOGIC);
END fau;

ARCHITECTURE fau OF fau IS
```

```vhdl
BEGIN
s <= a AND cin;
cout <= a XOR cin ;--(a AND cin) OR (b AND cin);
END fau;

LIBRARY IEEE;
USE IEEE.STD_LOGIC_1164.ALL;

ENTITY add14 IS
        PORT (a: IN STD_LOGIC_VECTOR(14 DOWNTO 0);
                b:  IN STD_LOGIC;
                prod: OUT STD_LOGIC_VECTOR(14 DOWNTO 0));
END add14;

ARCHITECTURE structural OF add14 IS
        SIGNAL s: STD_LOGIC_VECTOR (15 DOWNTO 0);
BEGIN
        s(0) <= b;
        U1:entity work.fau PORT MAP (a(0), s(0), prod(0), s(1));
        U2:entity work.fau PORT MAP (a(1), s(1), prod(1), s(2));
        U3:entity work.fau PORT MAP (a(2), s(2), prod(2), s(3));
        U4:entity work.fau PORT MAP (a(3), s(3), prod(3), s(4));
        U5:entity work.fau PORT MAP (a(4), s(4), prod(4), s(5));
        U6:entity work.fau PORT MAP (a(5), s(5), prod(5), s(6));
        U7:entity work.fau PORT MAP (a(6), s(6), prod(6), s(7));

        U8:entity work.fau PORT MAP (a(7), s(7), prod(7), s(8));
        U9:entity work.fau PORT MAP (a(8), s(8), prod(8), s(9));
        U10:entity work.fau PORT MAP (a(9), s(9), prod(9), s(10));
        U11:entity work.fau PORT MAP (a(10), s(10), prod(10), s(11));
        U12:entity work.fau PORT MAP (a(11), s(11), prod(11), s(12));
        U13:entity work.fau PORT MAP (a(12), s(12), prod(12), s(13));
        U14:entity work.fau PORT MAP (a(13), s(13), prod(13), s(14));
        U15:entity work.fau PORT MAP (a(14), s(14), prod(14), s(15));
END structural;

-- GF_add.vhd
LIBRARY IEEE;
USE IEEE.STD_LOGIC_1164.ALL;

ENTITY GF_add IS
        PORT (a,b: IN STD_LOGIC_VECTOR(7 DOWNTO 0);
                c: OUT STD_LOGIC_VECTOR(7 DOWNTO 0));
END GF_add;

ARCHITECTURE rtl OF GF_add IS
BEGIN
        c<=a XOR b;
```

```vhdl
END rtl;

-- GF_inverse.vhd
LIBRARY IEEE;
USE IEEE.STD_LOGIC_1164.ALL;
USE WORK.RAINBOW_COMPONENTS.ALL;

ENTITY GF_inverse IS
      PORT (inB:IN STD_LOGIC_VECTOR(7 DOWNTO 0);
            BInvert: OUT STD_LOGIC_VECTOR(7 DOWNTO 0));
END GF_inverse;

ARCHITECTURE behav OF GF_inverse IS

COMPONENT GF_mult
PORT(a,b: IN STD_LOGIC_VECTOR(7 DOWNTO 0);
     c: OUT STD_LOGIC_VECTOR(7 DOWNTO 0));
END COMPONENT;

SIGNAL B2,B4,B8,B16,B32,B64,B128,B2B4,B8B16,B32B64,B2B4B8B16,B32B64B128: STD_LOGIC_VECTOR(7 DOWNTO 0);

BEGIN
B2(0)<= inB(0) XOR inB(4) XOR inB(5) XOR inB(6);
B2(1)<= inB(7);
B2(2)<= inB(1) XOR inB(4) XOR inB(7);
B2(3)<= inB(4) XOR inB(5) XOR inB(6) XOR inB(7);
B2(4)<= inB(2) XOR inB(5) XOR inB(7);
B2(5)<= inB(5) XOR inB(6) XOR inB(7);
B2(6)<= inB(3) XOR inB(4) XOR inB(5);
B2(7)<= inB(6);
B4(0)<= inB(0) XOR inB(2) XOR inB(3);
B4(1)<= inB(6);
B4(2)<= inB(2) XOR inB(5) XOR inB(6);
B4(3)<= inB(2) XOR inB(3) XOR inB(4) XOR inB(5);
B4(4)<= inB(1) XOR inB(4) XOR inB(5);
B4(5)<= inB(3) XOR inB(4) XOR inB(7);
B4(6)<= inB(2) XOR inB(4) XOR inB(5) XOR inB(7);
B4(7)<= inB(3) XOR inB(4) XOR inB(5);
B8(0)<= inB(0) XOR inB(1) XOR inB(4);
B8(1)<= inB(3) XOR inB(4) XOR inB(5);
B8(2)<= inB(1) XOR inB(3) XOR inB(6);
B8(3)<= inB(1) XOR inB(2) XOR inB(5);
B8(4)<= inB(2) XOR inB(6) XOR inB(7);
B8(5)<= inB(2) XOR inB(4);
B8(6)<= inB(1) XOR inB(2) XOR inB(4) XOR inB(7);
B8(7)<= inB(2) XOR inB(4) XOR inB(5) XOR inB(7);
B16(0)<= inB(0) XOR inB(2) XOR inB(4) XOR inB(6);
B16(1)<= inB(2) XOR inB(4) XOR inB(5) XOR inB(7);
```

```vhdl
B16(2)<= inB(3) XOR inB(6);
B16(3)<= inB(1) XOR inB(4) XOR inB(5) XOR inB(6) XOR inB(7);
B16(4)<= inB(1) XOR inB(3) XOR inB(5) XOR inB(6) XOR inB(7);
B16(5)<= inB(1) XOR inB(2) XOR inB(4) XOR inB(5);
B16(6)<= inB(1) XOR inB(2) XOR inB(4) XOR inB(5) XOR inB(6) XOR inB(7);
B16(7)<= inB(1) XOR inB(2) XOR inB(4) XOR inB(7);
B32(0)<= inB(0) XOR inB(1) XOR inB(2) XOR inB(3) XOR inB(4) XOR inB(5) XOR inB(6);
B32(1)<= inB(1) XOR inB(2) XOR inB(4) XOR inB(7);
B32(2)<= inB(3) XOR inB(6) XOR inB(7);
B32(3)<= inB(2) XOR inB(3) XOR inB(4) XOR inB(5) XOR inB(7);
B32(4)<= inB(3) XOR inB(5) XOR inB(6) XOR inB(7);
B32(5)<= inB(1) XOR inB(2) XOR inB(4) XOR inB(6);
B32(6)<= inB(1) XOR inB(2) XOR inB(3) XOR inB(5);
B32(7)<= inB(1) XOR inB(2) XOR inB(4) XOR inB(5) XOR inB(6) XOR inB(7);
B64(0)<= inB(0) XOR inB(1) XOR inB(2) XOR inB(3) XOR inB(5) XOR inB(6) XOR inB(7);
B64(1)<= inB(1) XOR inB(2) XOR inB(4) XOR inB(5) XOR inB(6) XOR inB(7);
B64(2)<= inB(3) XOR inB(7);
B64(3)<= inB(1) XOR inB(2) XOR inB(5) XOR inB(6);
B64(4)<= inB(3) XOR inB(5) XOR inB(6);
B64(5)<= inB(1) XOR inB(2) XOR inB(3) XOR inB(7);
B64(6)<= inB(1);
B64(7)<= inB(1) XOR inB(2) XOR inB(3) XOR inB(5);
B128(0)<= inB(0) XOR inB(1) XOR inB(3);
B128(1)<= inB(1) XOR inB(2) XOR inB(3) XOR inB(5);
B128(2)<= inB(4) XOR inB(5) XOR inB(7);
B128(3)<= inB(1) XOR inB(3) XOR inB(6) XOR inB(7);
B128(4)<= inB(3) XOR inB(5);
B128(5)<= inB(1) XOR inB(5) XOR inB(7);
B128(6)<= inB(7);
B128(7)<= inB(1);

        U0:GF_mult PORT MAP (B2,B4,B2B4);
        U1:GF_mult PORT MAP (B8,B16,B8B16);
        U2:GF_mult PORT MAP (B32,B64,B32B64);
        U3:GF_mult PORT MAP (B2B4,B8B16,B2B4B8B16);
        U4:GF_mult PORT MAP (B32B64,B128,B32B64B128);
        U5:GF_mult PORT MAP (B2B4B8B16,B32B64B128,BInvert);

END behav;

-- GF_mult.vhd
LIBRARY IEEE;
USE IEEE.STD_LOGIC_1164.ALL;
use IEEE.STD_LOGIC_ARITH.ALL;
use IEEE.STD_LOGIC_UNSIGNED.ALL;

ENTITY GF_mult IS
PORT(a,b:IN STD_LOGIC_VECTOR(7 DOWNTO 0);
    c:OUT STD_LOGIC_VECTOR(7 DOWNTO 0));
```

```
END GF_mult;

ARCHITECTURE rtl OF GF_mult IS
SIGNAL ly:STD_LOGIC_VECTOR(14 DOWNTO 0);
SIGNAL lym:STD_LOGIC_VECTOR(7 DOWNTO 0);
BEGIN
ly(14)<=a(7) AND b(7);
ly(13)<=(a(7) AND b(6)) XOR (a(6) AND b(7));
ly(12)<=(a(7) AND b(5))XOR(a(6) AND b(6))XOR(a(5) AND b(7));
ly(11)<=(a(7) AND b(4))XOR(a(6) AND b(5))XOR(a(5) AND b(6))XOR(a(4) AND b(7));
ly(10)<=(a(7) AND b(3))XOR(a(6) AND b(4))XOR(a(5) AND b(5))XOR(a(4) AND b(6))XOR(a(3) AND b(7));
ly(9)<=(a(7) AND b(2))XOR(a(6) AND b(3))XOR(a(5) AND b(4))XOR(a(4) AND b(5))XOR(a(3) AND b(6))XOR(a(2) AND b(7));
ly(8)<=(a(7) AND b(1))XOR(a(6) AND b(2))XOR(a(5) AND b(3))XOR(a(4) AND b(4))XOR(a(3) AND b(5))XOR(a(2) AND b(6))XOR(a(1) AND b(7));
ly(7)<=(a(7) AND b(0))XOR(a(6) AND b(1))XOR(a(5) AND b(2))XOR(a(4) AND b(3))XOR(a(3) AND b(4))XOR(a(2) AND b(5))XOR(a(1) AND b(6))XOR(a(0) AND b(7));
ly(6)<=(a(6) AND b(0))XOR(a(5) AND b(1))XOR(a(4) AND b(2))XOR(a(3) AND b(3))XOR(a(2) AND b(4))XOR(a(1) AND b(5))XOR(a(0) AND b(6));
ly(5)<=(a(5) AND b(0))XOR(a(4) AND b(1))XOR(a(3) AND b(2))XOR(a(2) AND b(3))XOR(a(1) AND b(4))XOR(a(0) AND b(5));
ly(4)<=(a(4) AND b(0))XOR(a(3) AND b(1))XOR(a(2) AND b(2))XOR(a(1) AND b(3))XOR(a(0) AND b(4));
ly(3)<=(a(3) AND b(0))XOR(a(2) AND b(1))XOR(a(1) AND b(2))XOR(a(0) AND b(3));
ly(2)<=(a(2) AND b(0))XOR(a(1) AND b(1))XOR(a(0) AND b(2));
ly(1)<=(a(1) AND b(0))XOR(a(0) AND b(1));
ly(0)<=(a(0) AND b(0));
lym(7)<=ly(7) XOR ly(9) XOR ly(11) XOR ly(12);
lym(6)<=ly(6) XOR ly(8) XOR ly(10) XOR ly(11);
lym(5)<=ly(5) XOR ly(10) XOR ly(11) XOR ly(12) XOR ly(14);
lym(4)<=ly(4) XOR ly(9) XOR ly(10) XOR ly(11) XOR ly(13) XOR ly(14);
lym(3)<=ly(3) XOR ly(8) XOR ly(9) XOR ly(10) XOR ly(12) XOR ly(13) XOR ly(14);
lym(2)<=ly(2) XOR ly(8) XOR ly(13) XOR ly(14);
lym(1)<=ly(1) XOR ly(9) XOR ly(11) XOR ly(13) XOR ly(14);
lym(0)<=ly(0) XOR ly(8) XOR ly(10) XOR ly(12) XOR ly(13);
c<=lym;
END rtl;

-- mult8.vhd

LIBRARY IEEE;
USE IEEE.STD_LOGIC_1164.ALL;

ENTITY and_2 IS
PORT (a,b: IN STD_LOGIC;
y: OUT STD_LOGIC);
END and_2;
```

```vhdl
ARCHITECTURE and_2 OF and_2 IS

BEGIN
y <= a AND b;
END and_2;

LIBRARY IEEE;
USE IEEE.STD_LOGIC_1164.ALL;

ENTITY fau IS
PORT (a,b,cin: IN STD_LOGIC;
s, cout: OUT STD_LOGIC);
END fau;

ARCHITECTURE fau OF fau IS

BEGIN
s <= a XOR b XOR cin;
cout <= (a AND b) OR (a AND cin) OR (b AND cin);
END fau;

LIBRARY IEEE;
USE IEEE.STD_LOGIC_1164.ALL;
USE WORK.RAINBOW_COMPONENTS.ALL;
ENTITY lower_row IS
PORT (sin, cin: IN STD_LOGIC_VECTOR (6 DOWNTO 0);
p: OUT STD_LOGIC_VECTOR (7 DOWNTO 0) );
END lower_row;

ARCHITECTURE structural OF lower_row IS
SIGNAL local: STD_LOGIC_VECTOR (6 DOWNTO 0);
BEGIN
local(0) <= '0';
U1: ENTITY WORK.fau PORT MAP(sin(0), cin(0), local(0), p(0), local(1));
U2: ENTITY WORK.fau PORT MAP(sin(1), cin(1), local(1), p(1), local(2));
U3: ENTITY WORK.fau PORT MAP(sin(2), cin(2), local(2), p(2), local(3));
U4: ENTITY WORK.fau PORT MAP(sin(3), cin(3), local(3), p(3), local(4));
U5: ENTITY WORK.fau PORT MAP(sin(4), cin(4), local(4), p(4), local(5));
U6: ENTITY WORK.fau PORT MAP(sin(5), cin(5), local(5), p(5), local(6));
U7: ENTITY WORK.fau PORT MAP(sin(6), cin(6), local(6), p(6), p(7));
END structural;

LIBRARY IEEE;
USE IEEE.STD_LOGIC_1164.ALL;

ENTITY mid_row IS
PORT (a: IN STD_LOGIC;
b: IN STD_LOGIC_VECTOR (7 DOWNTO 0);
sin, cin: IN STD_LOGIC_VECTOR (6 DOWNTO 0);
```

```vhdl
    sout, cout: OUT STD_LOGIC_VECTOR (6 DOWNTO 0);
    p: OUT STD_LOGIC );
END mid_row;

ARCHITECTURE structural OF mid_row IS
SIGNAL and_out: STD_LOGIC_VECTOR (6 DOWNTO 0);
BEGIN
    U1: ENTITY WORK.and_2 PORT MAP(a, b(7), sout(6));
    U2: ENTITY WORK.and_2 PORT MAP(a, b(6), and_out(6));
    U3: ENTITY WORK.and_2 PORT MAP(a, b(5), and_out(5));
    U4: ENTITY WORK.and_2 PORT MAP(a, b(4), and_out(4));
    U5: ENTITY WORK.and_2 PORT MAP(a, b(3), and_out(3));
    U6: ENTITY WORK.and_2 PORT MAP(a, b(2), and_out(2));
    U7: ENTITY WORK.and_2 PORT MAP(a, b(1), and_out(1));
    U8: ENTITY WORK.and_2 PORT MAP(a, b(0), and_out(0));

    U9:  ENTITY WORK.fau PORT MAP(sin(6), cin(6), and_out(6), sout(5), cout(6));
    U10: ENTITY WORK.fau PORT MAP(sin(5), cin(5), and_out(5), sout(4), cout(5));
    U11: ENTITY WORK.fau PORT MAP(sin(4), cin(4), and_out(4), sout(3), cout(4));
    U12: ENTITY WORK.fau PORT MAP(sin(3), cin(3), and_out(3), sout(2), cout(3));
    U13: ENTITY WORK.fau PORT MAP(sin(2), cin(2), and_out(2), sout(1), cout(2));
    U14: ENTITY WORK.fau PORT MAP(sin(1), cin(1), and_out(1), sout(0), cout(1));
    U15: ENTITY WORK.fau PORT MAP(sin(0), cin(0), and_out(0), p, cout(0));

END structural;

LIBRARY IEEE;
USE IEEE.STD_LOGIC_1164.ALL;

ENTITY top_row IS
PORT (a: IN STD_LOGIC;
    b: IN STD_LOGIC_VECTOR (7 DOWNTO 0);
    sout, cout: OUT STD_LOGIC_VECTOR (6 DOWNTO 0);
    p: OUT STD_LOGIC );
END top_row;

ARCHITECTURE structural OF top_row IS

BEGIN
    U1: ENTITY WORK.and_2 PORT MAP(a, b(7), sout(6));
    U2: ENTITY WORK.and_2 PORT MAP(a, b(6), sout(5));
    U3: ENTITY WORK.and_2 PORT MAP(a, b(5), sout(4));
    U4: ENTITY WORK.and_2 PORT MAP(a, b(4), sout(3));
    U5: ENTITY WORK.and_2 PORT MAP(a, b(3), sout(2));
    U6: ENTITY WORK.and_2 PORT MAP(a, b(2), sout(1));
    U7: ENTITY WORK.and_2 PORT MAP(a, b(1), sout(0));
    U8: ENTITY WORK.and_2 PORT MAP(a, b(0), p);
    cout(6) <= '0'; cout(5) <= '0'; cout(4) <= '0';cout(3) <= '0'; cout(2) <= '0'; cout(1) <= '0';cout(0) <= '0';
```

```vhdl
END structural;

LIBRARY IEEE;
USE IEEE.STD_LOGIC_1164.ALL;

ENTITY mult8 IS
PORT (a,b: IN STD_LOGIC_VECTOR(7 DOWNTO 0);
prod: OUT STD_LOGIC_VECTOR(15 DOWNTO 0));
END mult8;

ARCHITECTURE structural OF mult8 IS
TYPE matrix IS ARRAY (0 TO 7) OF
STD_LOGIC_VECTOR (6 DOWNTO 0);
SIGNAL s,c: matrix;
BEGIN
U1:ENTITY WORK.top_row PORT MAP (a(0), b, s(0), c(0), prod(0));
U2:ENTITY WORK.mid_row PORT MAP (a(1), b, s(0), c(0), s(1), c(1), prod(1));
U3:ENTITY WORK.mid_row PORT MAP (a(2), b, s(1), c(1), s(2), c(2), prod(2));
U4:ENTITY WORK.mid_row PORT MAP (a(3), b, s(2), c(2), s(3), c(3), prod(3));
U5:ENTITY WORK.mid_row PORT MAP (a(4), b, s(3), c(3), s(4), c(4), prod(4));
U6:ENTITY WORK.mid_row PORT MAP (a(5), b, s(4), c(4), s(5), c(5), prod(5));
U7:ENTITY WORK.mid_row PORT MAP (a(6), b, s(5), c(5), s(6), c(6), prod(6));
U8:ENTITY WORK.mid_row PORT MAP (a(7), b, s(6), c(6), s(7), c(7), prod(7));
U9:ENTITY WORK.lower_row PORT MAP (s(7), c(7), prod(15 DOWNTO 8));
END structural;

Library IEEE ;
use IEEE.std_logic_1164.all ;
use IEEE.std_logic_arith.all ;

entity divide_by_n is
  generic (data_width    : natural := 8 );
  port (
        data_in : in  UNSIGNED(data_width - 1 downto 0) ;
        load    : in  std_logic ;
        clk     : in  std_logic ;
        reset   : in  std_logic ;
        divide  : out std_logic
        );
end divide_by_n ;

architecture rtl of divide_by_n is
  signal count_reg : UNSIGNED(data_width - 1 downto 0) ;
  constant max_count : UNSIGNED(data_width - 1 downto 0) := (others => '1') ;
  begin
  cont_it : process(clk,reset)
       begin
         if (reset = '1') then
           count_reg <= (others => '0') ;
```

```vhdl
            elsif (clk = '1' and clk'event) then
              if (load = '1') then
                  count_reg <= data_in ;
              else
                  count_reg <= count_reg + "01" ;
              end if ;
            end if;
         end process ;
   divide <= '1' when count_reg = max_count else '0' ;
end RTL ;

Library IEEE ;
use IEEE.std_logic_1164.all ;
use IEEE.std_logic_arith.all ;

entity dlatrg is
   generic (data_width    : natural := 16 );
   port (
         data_in  : in  UNSIGNED(data_width - 1 downto 0) ;
         clk      : in  std_logic ;
         reset    : in  std_logic ;
         data_out : out UNSIGNED(data_width - 1 downto 0)
        );
end dlatrg ;

architecture rtl of dlatrg is
  begin
  latch_it : process(data_in,clk,reset)
  begin
       if (reset = '1') then
          data_out <= (others => '0') ;
       elsif (clk = '1') then
          data_out <= data_in ;
       end if;
   end process ;
end RTL ;

Library IEEE ;
use IEEE.std_logic_1164.all ;
use IEEE.std_logic_arith.all ;

entity lfsr is
   generic (data_width    : natural := 8 );
   port (
         clk      : in  std_logic ;
         reset    : in  std_logic ;
         data_out : out UNSIGNED(data_width - 1 downto 0)
        );
end lfsr ;
```

4.5 本章参考代码

```vhdl
architecture rtl of lfsr is
  signal feedback : std_logic ;
  signal lfsr_reg : UNSIGNED(data_width - 1 downto 0) ;
  begin
    feedback <= lfsr_reg(7) xor lfsr_reg(0) ;
  latch_it : process(clk,reset)
      begin
        if (reset = '1') then
        lfsr_reg <= (others => '0') ;
      elsif (clk = '1' and clk'event) then
        lfsr_reg <= lfsr_reg(lfsr_reg'high - 1 downto 0) & feedback ;
      end if;
    end process ;
  data_out <= lfsr_reg ;
end RTL ;

Library IEEE ;
use IEEE.std_logic_1164.all ;
use IEEE.std_logic_arith.all ;

entity priority_encoder is
  generic (data_width    : natural := 25 ;
           address_width : natural := 5 ) ;
  port (
        data    : in  UNSIGNED(data_width - 1 downto 0) ;
        address : out UNSIGNED(address_width - 1 downto 0) ;
        none    : out STD_LOGIC
       );
end priority_encoder ;

architecture rtl of priority_encoder is
  attribute SYNTHESIS_RETURN : STRING ;

  FUNCTION to_stdlogic (arg1:BOOLEAN)  RETURN STD_LOGIC IS
     BEGIN
     IF(arg1) THEN
        RETURN('1') ;
     ELSE
        RETURN('0') ;
     END IF ;
  END ;

   function to_UNSIGNED(ARG: INTEGER; SIZE: INTEGER) return UNSIGNED is
       variable result: UNSIGNED(SIZE-1 downto 0);
       variable temp: integer;
       attribute SYNTHESIS_RETURN of result:variable is "FEED_THROUGH" ;
    begin
       temp := ARG;
```

```vhdl
            for i in 0 to SIZE-1 loop
                if (temp mod 2) = 1 then
                    result(i) := '1';
                else
                    result(i) := '0';
                end if;
                if temp > 0 then
                    temp := temp / 2;
                else
                    temp := (temp - 1) / 2;
                end if;
            end loop;
            return result;
      end;

    constant zero : UNSIGNED(data_width downto 1) := (others => '0') ;
    begin
PRIO :  process(data)
            variable temp_address : UNSIGNED(address_width - 1 downto 0) ;
            begin
            temp_address := (others => '0') ;
            for i in data_width - 1 downto 0 loop
              if (data(i) = '1') then
                 temp_address := to_unsigned(i,address_width) ;
                 exit ;
                 end if ;
            end loop ;
            address <= temp_address ;
            none <= to_stdlogic(data = zero) ;
          end process ;
end RTL ;


Library IEEE ;
use IEEE.std_logic_1164.all ;
use IEEE.std_logic_arith.all ;
use IEEE.std_logic_unsigned.all ;

entity ram is
     generic (data_width    : natural := 8 ;
              address_width : natural := 8);
     port (
          data_in : in  UNSIGNED(data_width - 1 downto 0) ;
          address : in  UNSIGNED(address_width - 1 downto 0) ;
          we      : in  std_logic ;
            clk   : in  std_logic;
          data_out : out UNSIGNED(data_width - 1 downto 0)
          );
end ram ;
```

4.5 本章参考代码

```vhdl
architecture rtl of ram is
    type mem_type is array (2**address_width downto 0) of UNSIGNED(data_width - 1 downto 0) ;
    signal mem : mem_type ;
    signal addr_reg : unsigned (address_width -1 downto 0);

    begin
      data_out <= mem(conv_integer(addr_reg)) ;
      I0 : process
            begin
          wait until clk'event and clk = '1';
          if (we = '1') then
            mem(conv_integer(address)) <= data_in ;
          end if ;
             addr_reg <= address;
       end process ;
end RTL ;

Library IEEE ;
use IEEE.std_logic_1164.all ;
use IEEE.std_logic_arith.all ;

entity tbuf is
     generic (data_width   : natural := 16 );
     port (
         data_in : in  UNSIGNED(data_width - 1 downto 0) ;
         en      : in  std_logic ;
         data_out : out UNSIGNED(data_width - 1 downto 0)
        );
end tbuf ;

architecture rtl of tbuf is
  begin
   three_state :  process(data_in,en)
        begin
         if (en = '1') then
             data_out <= data_in ;
         else
             data_out <= (others => 'Z') ;
          end if;
        end process ;
end RTL ;

Library IEEE ;
use IEEE.std_logic_1164.all ;
use IEEE.std_logic_arith.all ;

entity pseudorandom is
     generic (data_width   : natural := 8 );
```

```vhdl
        port (
            seed  : in  UNSIGNED (24 downto 0) ;
            init  : in  UNSIGNED (4 downto 0) ;
            load  : in  std_logic ;
            clk   : in  std_logic ;
            reset : in  std_logic ;
            read  : in  std_logic ;
            write : in  std_logic ;
            rand  : out UNSIGNED (7 downto 0) ;
            none  : out std_logic
        );
end pseudorandom ;

architecture rtl of pseudorandom is
  signal latch_seed : UNSIGNED(24 downto 0) ;
  signal encoder_address : UNSIGNED(4 downto 0) ;
  signal random_data : UNSIGNED(7 downto 0) ;
  signal write_enable : std_logic ;
  signal ram_data : UNSIGNED(7 downto 0) ;
  begin
    I0 : entity work.dlatrg(rtl)
            generic map (25)
            port map (seed,read,reset,latch_seed) ;
    I1 : entity work.priority_encoder(rtl)
            generic map (25,5)
            port map (latch_seed,encoder_address,none) ;
    I2 : entity work.ram(rtl)
            generic map (8,5)
            port map (random_data,encoder_address,write_enable,clk,ram_data) ;
    I3 : entity work.tbuf(rtl)
            generic map (8)
            port map (ram_data,write,rand) ;
    I4 : entity work.lfsr(rtl)
            generic map (8)
            port map (clk,reset,random_data) ;
    I5 : entity work.divide_by_n(rtl)
            generic map (5)
            port map (init,load,clk,reset,write_enable) ;
end rtl ;

-- Rainbow.vhd

LIBRARY IEEE;
USE IEEE.STD_LOGIC_1164.ALL;
use IEEE.numeric_std.all;
USE IEEE.STD_LOGIC_UNSIGNED.ALL;
--调用运算程序包,支持并行数据的直接相加并自动转换
USE WORK.RAINBOW_COMPONENTS.ALL;
USE WORK.RAINBOW_PARAMETER.ALL;
```

```vhdl
ENTITY CountK IS
        generic (k_num : Integer RANGE 0 TO 27 );

        PORT (clock,CountKstart: IN STD_LOGIC;
              PublicKey :IN STD_LOGIC_VECTOR(7 DOWNTO 0);
              Ksigned_accept: OUT STD_LOGIC
                );
END CountK;

ARCHITECTURE behave OF CountK IS

type ctl_state is (idle,counting,ijloop,finish, init ,wait1,stop);-- wait1, wait2,get);

SIGNAL mulina,mulinb,mulAmBre,mulAmB,mulPK,mulresult,sum:FINITE_FIELD;

signal mstate:ctl_state;

BEGIN

        GF_multiplier1: GF_mult
              PORT MAP(
                              a=>mulina,
                              b=>mulinb,
                              c=>mulAmBre);

        GF_multiplier2: GF_mult
              PORT MAP(
                              a=>mulAmB,
                              b=>mulPK,
                              c=>mulresult);

        countK:PROCESS(clock)

        VARIABLE i:INTEGER RANGE 0 TO ovn;
        VARIABLE j:INTEGER RANGE 0 TO ovn;

        BEGIN
        IF(clock'EVENT AND clock='1')THEN
              case mstate is
              when idle =>
                      mulina<="00000000";
                      mulinb<="00000000";
                      sum<="00000000";

                      if (CountKstart = '1') then
                              mstate <= wait1;
                      else
```

```
                    mstate <= idle;
            end if;
    when wait1 =>
            mstate <= init;
    when init =>                    --init
            mstate <= counting;
            Ksigned_accept<='0';
            i:=0;
            j:=0;
            sum<="00000000";

            mulina<=sig(i);--x(i);
            mulinb<=sig(j);--x(j);

    when counting =>
            --if(done='1') then
            mulAmB<=mulAmBre;
            mulPK<=PublicKey;

            j:=j+1;
            mulina<=sig(i);--x(i);
            mulinb<=sig(j);--x(j);
            mstate<=ijloop;
            --end if;
    when ijloop =>
            sum <= sum XOR mulresult;

            mulAmB<=mulAmBre;
            mulPK<=PublicKey;

            if(j/=ovn)then
                    j:=j+1;
                    --mstate<=counting;
            else
                    if(i<ovn)then       --j=ovn
                            i:=i+1;
                            j:=i;
                            --mstate<=counting;
                    else                        --j=ovn i=ovn
                            mstate<=finish;
                    end if;
            end if;
            mulina<=sig(i);--x(i);
            mulinb<=sig(j);--x(j);

    when finish =>

            IF( (sum XOR mulresult) = doc(k_num)) THEN
```

```vhdl
                                    Ksigned_accept<='1';
                            ELSE
                                    Ksigned_accept<='0';
                    END IF;

                        mstate<=stop;
                when stop =>
                    if (CountKstart = '0') then --reset
                            mstate<=idle;
                    end if;
                when others =>
                        mstate<=idle;
                end case;
        END IF;

        END PROCESS;

END behave;


LIBRARY IEEE;
USE IEEE.STD_LOGIC_1164.ALL;
use IEEE.numeric_std.all;
USE IEEE.STD_LOGIC_UNSIGNED.ALL;
--调用运算程序包,支持并行数据的直接相加并自动转换
USE WORK.RAINBOW_COMPONENTS.ALL;
USE WORK.RAINBOW_PARAMETER.ALL;

ENTITY Rainbow IS
        PORT (clk,start: IN STD_LOGIC;
                -- T_mstate ,T_mstate1_1,T_mstate2_1: OUT STD_LOGIC_VECTOR (5 DOWNTO 0);
                -- T_PK_NUM,T_doc1,T_doc2,T_sum1_1,T_sum2_1: OUT FINITE_FIELD;
                -- T_PK_addr : OUT STD_LOGIC_VECTOR (14 DOWNTO 0);
                signed_accept,done: OUT STD_LOGIC
                );
END Rainbow;

ARCHITECTURE behave OF Rainbow IS

COMPONENT Rom216_PublicKey
        PORT
        (
                address                 : IN STD_LOGIC_VECTOR (9 DOWNTO 0);
                clock                   : IN STD_LOGIC ;
                q                       : OUT STD_LOGIC_VECTOR (215 DOWNTO 0)
        );
END COMPONENT;

type ctl_state is (idle,add_addr,finish, init ,stop);-- wait1,wait2,get);
```

```vhdl
    SIGNAL PK_addr:STD_LOGIC_VECTOR (9 DOWNTO 0);

    SIGNAL PK_num:STD_LOGIC_VECTOR (215 DOWNTO 0);

    signal mstate:ctl_state;

    signal
    signed_accept0,signed_accept1,signed_accept2,signed_accept3,signed_accept4,signed_accept5,signed_accept6,
    signed_accept7,signed_accept8,signed_accept9,signed_accept10,signed_accept11,signed_accept12,signed_accept13,
    signed_accept14,signed_accept15,signed_accept16,signed_accept17,signed_accept18,signed_accept19,signed_accept20,
    signed_accept21,signed_accept22,signed_accept23,signed_accept24,signed_accept25,signed_accept26
    : STD_LOGIC;

    BEGIN

            PublicKey :Rom216_PublicKey
                    PORT MAP(
                                    address=>PK_addr,
                                    clock=>clk,
                                    q=>PK_num);


        K0: entity work.CountK generic map(0)
            port  map  (clock=>clk,CountKstart=>start,PublicKey=>PK_num(215  downto 208),Ksigned_accept=>signed_accept0) ;
        K1: entity work.CountK generic map(1)
            port  map  (clock=>clk,CountKstart=>start,PublicKey=>PK_num(207  downto 200),Ksigned_accept=>signed_accept1) ;
        K2: entity work.CountK generic map(2)
            port  map  (clock=>clk,CountKstart=>start,PublicKey=>PK_num(199  downto 192),Ksigned_accept=>signed_accept2) ;
        K3: entity work.CountK generic map(3)
            port  map  (clock=>clk,CountKstart=>start,PublicKey=>PK_num(191  downto 184),Ksigned_accept=>signed_accept3) ;
        K4: entity work.CountK generic map(4)
            port  map  (clock=>clk,CountKstart=>start,PublicKey=>PK_num(183  downto 176),Ksigned_accept=>signed_accept4) ;
        K5: entity work.CountK generic map(5)
            port  map  (clock=>clk,CountKstart=>start,PublicKey=>PK_num(175  downto 168),Ksigned_accept=>signed_accept5) ;
        K6: entity work.CountK generic map(6)
            port  map  (clock=>clk,CountKstart=>start,PublicKey=>PK_num(167  downto 160),Ksigned_accept=>signed_accept6) ;
        K7: entity work.CountK generic map(7)
            port  map  (clock=>clk,CountKstart=>start,PublicKey=>PK_num(159  downto
```

152),Ksigned_accept=>signed_accept7) ;
       K8: entity work.CountK generic map(8)
           port map (clock=>clk,CountKstart=>start,PublicKey=>PK_num(151 downto
144),Ksigned_accept=>signed_accept8) ;
       K9: entity work.CountK generic map(9)
           port map (clock=>clk,CountKstart=>start,PublicKey=>PK_num(143 downto
136),Ksigned_accept=>signed_accept9) ;
       K10: entity work.CountK generic map(10)
           port map (clock=>clk,CountKstart=>start,PublicKey=>PK_num(135 downto
128),Ksigned_accept=>signed_accept10) ;
       K11: entity work.CountK generic map(11)
           port map (clock=>clk,CountKstart=>start,PublicKey=>PK_num(127 downto
120),Ksigned_accept=>signed_accept11) ;
       K12: entity work.CountK generic map(12)
           port map (clock=>clk,CountKstart=>start,PublicKey=>PK_num(119 downto
112),Ksigned_accept=>signed_accept12) ;
       K13: entity work.CountK generic map(13)
           port map (clock=>clk,CountKstart=>start,PublicKey=>PK_num(111 downto
104),Ksigned_accept=>signed_accept13) ;

       K14: entity work.CountK generic map(14)
           port map (clock=>clk,CountKstart=>start,PublicKey=>PK_num(103 downto
96),Ksigned_accept=>signed_accept14) ;
       K15: entity work.CountK generic map(15)
           port map (clock=>clk,CountKstart=>start,PublicKey=>PK_num(95 downto
88),Ksigned_accept=>signed_accept15) ;
       K16: entity work.CountK generic map(16)
           port map (clock=>clk,CountKstart=>start,PublicKey=>PK_num(87 downto
80),Ksigned_accept=>signed_accept16) ;
       K17: entity work.CountK generic map(17)
           port map (clock=>clk,CountKstart=>start,PublicKey=>PK_num(79 downto
72),Ksigned_accept=>signed_accept17) ;
       K18: entity work.CountK generic map(18)
           port map (clock=>clk,CountKstart=>start,PublicKey=>PK_num(71 downto
64),Ksigned_accept=>signed_accept18) ;
       K19: entity work.CountK generic map(19)
           port map (clock=>clk,CountKstart=>start,PublicKey=>PK_num(63 downto
56),Ksigned_accept=>signed_accept19) ;
       K20: entity work.CountK generic map(20)
           port map (clock=>clk,CountKstart=>start,PublicKey=>PK_num(55 downto
48),Ksigned_accept=>signed_accept20) ;
       K21: entity work.CountK generic map(21)
           port map (clock=>clk,CountKstart=>start,PublicKey=>PK_num(47 downto
40),Ksigned_accept=>signed_accept21) ;
       K22: entity work.CountK generic map(22)
           port map (clock=>clk,CountKstart=>start,PublicKey=>PK_num(39 downto
32),Ksigned_accept=>signed_accept22) ;
       K23: entity work.CountK generic map(23)
           port map (clock=>clk,CountKstart=>start,PublicKey=>PK_num(31 downto

第4章 多变量公钥密码技术

```vhdl
24),Ksigned_accept=>signed_accept23) ;
          K24: entity work.CountK generic map(24)
                port    map    (clock=>clk,CountKstart=>start,PublicKey=>PK_num(23   downto
16),Ksigned_accept=>signed_accept24) ;
          K25: entity work.CountK generic map(25)
                port    map    (clock=>clk,CountKstart=>start,PublicKey=>PK_num(15   downto
8),Ksigned_accept=>signed_accept25) ;
          K26: entity work.CountK generic map(26)
                port    map    (clock=>clk,CountKstart=>start,PublicKey=>PK_num(7    downto
0),Ksigned_accept=>signed_accept26) ;

          signed_accept<=
          signed_accept0 AND signed_accept1 AND signed_accept2  AND signed_accept3
AND signed_accept4 AND
          signed_accept5 AND signed_accept6 AND signed_accept7  AND signed_accept8
AND signed_accept9 AND
          signed_accept10    AND    signed_accept11    AND    signed_accept12    AND
signed_accept13 AND
          signed_accept14 AND signed_accept15 AND signed_accept16 AND signed_accept17
AND signed_accept18 AND
          signed_accept19 AND signed_accept20 AND signed_accept21 AND signed_accept22
AND signed_accept23 AND
          signed_accept24 AND signed_accept25 AND signed_accept26;



          main:PROCESS(clk)

          VARIABLE i:INTEGER RANGE 0 TO 741;
          --VARIABLE j:INTEGER RANGE 0 TO ovn;

          BEGIN
          IF(clk'EVENT AND clk='1')THEN
                  case mstate is
                  when idle =>
                  --      T_mstate<="000000";

                          if (start = '1') then
                                  done<='0';
                                  mstate <= add_addr;
                                  i:=0;
                          else
                                  mstate <= idle;
                          end if;
                          PK_addr<="0000000000";

                  when add_addr =>
                          -- T_mstate<="000010";
                          if (i=741) then
```

```vhdl
                            mstate<=finish;
                    end if;

                    i:=i+1;

                    PK_addr <= PK_addr + "000000001";

                when finish =>
                    --T_mstate<="000100";
                    done<='1';
                    mstate<=stop;
                when stop =>
                    if (start = '0') then --reset
                        mstate<=idle;
                    end if;
                when others =>
                    mstate<=idle;
                end case;
    END IF;

    END PROCESS;



END behave;

-- rainbow_components.vhd
LIBRARY IEEE;
USE IEEE.STD_LOGIC_1164.ALL;
use IEEE.std_logic_arith.all ;

PACKAGE rainbow_components IS

COMPONENT GF_mult IS
       PORT(a,b:IN STD_LOGIC_VECTOR(7 DOWNTO 0);
              c:OUT STD_LOGIC_VECTOR(7 DOWNTO 0));
END COMPONENT;


END rainbow_components;

-- rainbow_parameter.vhd
LIBRARY IEEE;
USE IEEE.STD_LOGIC_1164.ALL;
use IEEE.std_logic_arith.all ;

PACKAGE rainbow_parameter IS

CONSTANT ovn:INTEGER := 37 ;
```

```vhdl
    CONSTANT v1:INTEGER := 10 ;
    CONSTANT nv1:INTEGER := ovn - v1 ;
    CONSTANT ovn1:INTEGER := ovn+1 ;

    SUBTYPE FINITE_FIELD            IS STD_LOGIC_VECTOR(7 DOWNTO 0);
    TYPE DOCUMENT_TYPE                    IS ARRAY(0 TO nv1-1) OF FINITE_FIELD;
    TYPE SIGNATURE_TYPE             IS ARRAY(0 TO ovn) OF FINITE_FIELD;
    --TYPE VERIFY_SIGNATURE_TYPE              IS ARRAY(0 TO ovn) OF FINITE_FIELD;

    constant doc : DOCUMENT_TYPE := (
     X"4d", X"8b", X"2f", X"df", X"e7", X"b5", X"0c", X"72", X"ed", X"a5",
     X"72", X"e1", X"ff", X"91", X"8b", X"3a", X"46", X"09", X"7c", X"a2",
     X"a0", X"87", X"a2", X"8f", X"fd", X"47", X"1b"
    );

    constant sig : SIGNATURE_TYPE := (
     X"c5",X"8f",X"d5",X"31",X"81",X"67",X"1c",X"73",X"9e",X"90",
     X"6a",X"b3",X"79",X"70",X"47",X"29",X"b3",X"07",X"c2",X"1e",
     X"06",X"11",X"65",X"95",X"03",X"4b",X"94",X"c8",X"bf",X"a7",
     X"83",X"9c",X"00",X"99",X"7f",X"6f",X"1b",X"01"
    );

    END rainbow_parameter;

    -- Rom_PublicKey.vhd
    -- megafunction wizard: %ROM: 1-PORT%
    -- GENERATION: STANDARD
    -- VERSION: WM1.0
    -- MODULE: altsyncram


    -- ============================================================
    -- File Name: Rom_PublicKey.vhd
    -- Megafunction Name(s):
    --                  altsyncram
    --
    -- Simulation Library Files(s):
    --                  altera_mf
    -- ============================================================
    -- ************************************************************
    -- THIS IS A WIZARD-GENERATED FILE. DO NOT EDIT THIS FILE!
    --
    -- 8.0 Build 215 05/29/2008 SJ Full Version
    -- ************************************************************


    LIBRARY ieee;
    USE ieee.std_logic_1164.all;

    LIBRARY altera_mf;
```

```vhdl
USE altera_mf.all;

ENTITY Rom_PublicKey IS
    PORT
    (
        address        : IN STD_LOGIC_VECTOR (14 DOWNTO 0);
        clock          : IN STD_LOGIC ;
        q              : OUT STD_LOGIC_VECTOR (7 DOWNTO 0)
    );
END Rom_PublicKey;


ARCHITECTURE SYN OF rom_publickey IS

    SIGNAL sub_wire0    : STD_LOGIC_VECTOR (7 DOWNTO 0);


    COMPONENT altsyncram
    GENERIC (
        address_aclr_a            : STRING;
        init_file                 : STRING;
        intended_device_family    : STRING;
        lpm_hint                  : STRING;
        lpm_type                  : STRING;
        numwords_a                : NATURAL;
        operation_mode            : STRING;
        outdata_aclr_a            : STRING;
        outdata_reg_a             : STRING;
        widthad_a                 : NATURAL;
        width_a                   : NATURAL;
        width_byteena_a           : NATURAL
    );
    PORT (
            clock0       : IN STD_LOGIC ;
            address_a    : IN STD_LOGIC_VECTOR (14 DOWNTO 0);
            q_a          : OUT STD_LOGIC_VECTOR (7 DOWNTO 0)
    );
    END COMPONENT;

BEGIN
    q    <= sub_wire0(7 DOWNTO 0);

    altsyncram_component : altsyncram
    GENERIC MAP (
        address_aclr_a => "NONE",
        init_file => "PublicKeyData.mif",
        intended_device_family => "Cyclone",
        lpm_hint => "ENABLE_RUNTIME_MOD=NO",
```

```
                lpm_type => "altsyncram",
                numwords_a => 24576,
                operation_mode => "ROM",
                outdata_aclr_a => "NONE",
                outdata_reg_a => "CLOCK0",
                widthad_a => 15,
                width_a => 8,
                width_byteena_a => 1
        )
        PORT MAP (
                clock0 => clock,
                address_a => address,
                q_a => sub_wire0
        );


END SYN;


-- ============================================================
-- CNX file retrieval info
-- ============================================================
-- Retrieval info: PRIVATE: ADDRESSSTALL_A NUMERIC "0"
-- Retrieval info: PRIVATE: AclrAddr NUMERIC "0"
-- Retrieval info: PRIVATE: AclrByte NUMERIC "0"
-- Retrieval info: PRIVATE: AclrOutput NUMERIC "0"
-- Retrieval info: PRIVATE: BYTE_ENABLE NUMERIC "0"
-- Retrieval info: PRIVATE: BYTE_SIZE NUMERIC "8"
-- Retrieval info: PRIVATE: BlankMemory NUMERIC "0"
-- Retrieval info: PRIVATE: CLOCK_ENABLE_INPUT_A NUMERIC "0"
-- Retrieval info: PRIVATE: CLOCK_ENABLE_OUTPUT_A NUMERIC "0"
-- Retrieval info: PRIVATE: Clken NUMERIC "0"
-- Retrieval info: PRIVATE: IMPLEMENT_IN_LES NUMERIC "0"
-- Retrieval info: PRIVATE: INIT_FILE_LAYOUT STRING "PORT_A"
-- Retrieval info: PRIVATE: INIT_TO_SIM_X NUMERIC "0"
-- Retrieval info: PRIVATE: INTENDED_DEVICE_FAMILY STRING "Cyclone"
-- Retrieval info: PRIVATE: JTAG_ENABLED NUMERIC "0"
-- Retrieval info: PRIVATE: JTAG_ID STRING "NONE"
-- Retrieval info: PRIVATE: MAXIMUM_DEPTH NUMERIC "0"
-- Retrieval info: PRIVATE: MIFfilename STRING "PublicKeyData.mif"
-- Retrieval info: PRIVATE: NUMWORDS_A NUMERIC "24576"
-- Retrieval info: PRIVATE: RAM_BLOCK_TYPE NUMERIC "0"
-- Retrieval info: PRIVATE: RegAddr NUMERIC "1"
-- Retrieval info: PRIVATE: RegOutput NUMERIC "1"
-- Retrieval info: PRIVATE: SYNTH_WRAPPER_GEN_POSTFIX STRING "0"
-- Retrieval info: PRIVATE: SingleClock NUMERIC "1"
-- Retrieval info: PRIVATE: UseDQRAM NUMERIC "0"
-- Retrieval info: PRIVATE: WidthAddr NUMERIC "15"
-- Retrieval info: PRIVATE: WidthData NUMERIC "8"
```

```
-- Retrieval info: PRIVATE: rden NUMERIC "0"
-- Retrieval info: CONSTANT: ADDRESS_ACLR_A STRING "NONE"
-- Retrieval info: CONSTANT: INIT_FILE STRING "PublicKeyData.mif"
-- Retrieval info: CONSTANT: INTENDED_DEVICE_FAMILY STRING "Cyclone"
-- Retrieval info: CONSTANT: LPM_HINT STRING "ENABLE_RUNTIME_MOD=NO"
-- Retrieval info: CONSTANT: LPM_TYPE STRING "altsyncram"
-- Retrieval info: CONSTANT: NUMWORDS_A NUMERIC "24576"
-- Retrieval info: CONSTANT: OPERATION_MODE STRING "ROM"
-- Retrieval info: CONSTANT: OUTDATA_ACLR_A STRING "NONE"
-- Retrieval info: CONSTANT: OUTDATA_REG_A STRING "CLOCK0"
-- Retrieval info: CONSTANT: WIDTHAD_A NUMERIC "15"
-- Retrieval info: CONSTANT: WIDTH_A NUMERIC "8"
-- Retrieval info: CONSTANT: WIDTH_BYTEENA_A NUMERIC "1"
-- Retrieval info: USED_PORT: address 0 0 15 0 INPUT NODEFVAL address[14..0]
-- Retrieval info: USED_PORT: clock 0 0 0 0 INPUT NODEFVAL clock
-- Retrieval info: USED_PORT: q 0 0 8 0 OUTPUT NODEFVAL q[7..0]
-- Retrieval info: CONNECT: @address_a 0 0 15 0 address 0 0 15 0
-- Retrieval info: CONNECT: q 0 0 8 0 @q_a 0 0 8 0
-- Retrieval info: CONNECT: @clock0 0 0 0 0 clock 0 0 0 0
-- Retrieval info: LIBRARY: altera_mf altera_mf.altera_mf_components.all
-- Retrieval info: GEN_FILE: TYPE_NORMAL Rom_PublicKey.vhd TRUE
-- Retrieval info: GEN_FILE: TYPE_NORMAL Rom_PublicKey.inc FALSE
-- Retrieval info: GEN_FILE: TYPE_NORMAL Rom_PublicKey.cmp TRUE
-- Retrieval info: GEN_FILE: TYPE_NORMAL Rom_PublicKey.bsf TRUE
-- Retrieval info: GEN_FILE: TYPE_NORMAL Rom_PublicKey_inst.vhd FALSE
-- Retrieval info: GEN_FILE: TYPE_NORMAL Rom_PublicKey_waveforms.html TRUE
-- Retrieval info: GEN_FILE: TYPE_NORMAL Rom_PublicKey_wave*.jpg FALSE
-- Retrieval info: LIB_FILE: altera_mf

-- Rom_TestData.vhd
-- megafunction wizard: %ROM: 1-PORT%
-- GENERATION: STANDARD
-- VERSION: WM1.0
-- MODULE: altsyncram

-- ============================================================
-- File Name: Rom_TestData.vhd
-- Megafunction Name(s):
--                      altsyncram
--
-- Simulation Library Files(s):
--                      altera_mf
-- ============================================================


LIBRARY ieee;
USE ieee.std_logic_1164.all;

LIBRARY altera_mf;
```

```
    USE altera_mf.all;

ENTITY Rom_TestData IS
    PORT
    (
        address         : IN STD_LOGIC_VECTOR (5 DOWNTO 0);
        clock           : IN STD_LOGIC ;
        q               : OUT STD_LOGIC_VECTOR (7 DOWNTO 0)
    );
END Rom_TestData;


ARCHITECTURE SYN OF rom_testdata IS

    SIGNAL sub_wire0        : STD_LOGIC_VECTOR (7 DOWNTO 0);



    COMPONENT altsyncram
    GENERIC (
        address_aclr_a              : STRING;
        init_file                   : STRING;
        intended_device_family              : STRING;
        lpm_hint                : STRING;
        lpm_type                : STRING;
        numwords_a              : NATURAL;
        operation_mode              : STRING;
        outdata_aclr_a              : STRING;
        outdata_reg_a               : STRING;
        widthad_a               : NATURAL;
        width_a             : NATURAL;
        width_byteena_a             : NATURAL
    );
    PORT (
            clock0      : IN STD_LOGIC ;
            address_a   : IN STD_LOGIC_VECTOR (5 DOWNTO 0);
            q_a         : OUT STD_LOGIC_VECTOR (7 DOWNTO 0)
    );
    END COMPONENT;

BEGIN
        q    <= sub_wire0(7 DOWNTO 0);

    altsyncram_component : altsyncram
    GENERIC MAP (
        address_aclr_a => "NONE",
        init_file => "TestData.mif",
        intended_device_family => "Cyclone",
        lpm_hint => "ENABLE_RUNTIME_MOD=NO",
```

```
                lpm_type => "altsyncram",
                numwords_a => 64,
                operation_mode => "ROM",
                outdata_aclr_a => "NONE",
                outdata_reg_a => "CLOCK0",
                widthad_a => 6,
                width_a => 8,
                width_byteena_a => 1
        )
        PORT MAP (
                clock0 => clock,
                address_a => address,
                q_a => sub_wire0
        );


END SYN;

-- ============================================================
-- CNX file retrieval info
-- ============================================================
-- Retrieval info: PRIVATE: ADDRESSSTALL_A NUMERIC "0"
-- Retrieval info: PRIVATE: AclrAddr NUMERIC "0"
-- Retrieval info: PRIVATE: AclrByte NUMERIC "0"
-- Retrieval info: PRIVATE: AclrOutput NUMERIC "0"
-- Retrieval info: PRIVATE: BYTE_ENABLE NUMERIC "0"
-- Retrieval info: PRIVATE: BYTE_SIZE NUMERIC "8"
-- Retrieval info: PRIVATE: BlankMemory NUMERIC "0"
-- Retrieval info: PRIVATE: CLOCK_ENABLE_INPUT_A NUMERIC "0"
-- Retrieval info: PRIVATE: CLOCK_ENABLE_OUTPUT_A NUMERIC "0"
-- Retrieval info: PRIVATE: Clken NUMERIC "0"
-- Retrieval info: PRIVATE: IMPLEMENT_IN_LES NUMERIC "0"
-- Retrieval info: PRIVATE: INIT_FILE_LAYOUT STRING "PORT_A"
-- Retrieval info: PRIVATE: INIT_TO_SIM_X NUMERIC "0"
-- Retrieval info: PRIVATE: INTENDED_DEVICE_FAMILY STRING "Cyclone"
-- Retrieval info: PRIVATE: JTAG_ENABLED NUMERIC "0"
-- Retrieval info: PRIVATE: JTAG_ID STRING "NONE"
-- Retrieval info: PRIVATE: MAXIMUM_DEPTH NUMERIC "0"
-- Retrieval info: PRIVATE: MIFfilename STRING "TestData.mif"
-- Retrieval info: PRIVATE: NUMWORDS_A NUMERIC "64"
-- Retrieval info: PRIVATE: RAM_BLOCK_TYPE NUMERIC "0"
-- Retrieval info: PRIVATE: RegAddr NUMERIC "1"
-- Retrieval info: PRIVATE: RegOutput NUMERIC "1"
-- Retrieval info: PRIVATE: SYNTH_WRAPPER_GEN_POSTFIX STRING "0"
-- Retrieval info: PRIVATE: SingleClock NUMERIC "1"
-- Retrieval info: PRIVATE: UseDQRAM NUMERIC "0"
-- Retrieval info: PRIVATE: WidthAddr NUMERIC "6"
-- Retrieval info: PRIVATE: WidthData NUMERIC "8"
```

```vhdl
-- Retrieval info: PRIVATE: rden NUMERIC "0"
-- Retrieval info: CONSTANT: ADDRESS_ACLR_A STRING "NONE"
-- Retrieval info: CONSTANT: INIT_FILE STRING "TestData.mif"
-- Retrieval info: CONSTANT: INTENDED_DEVICE_FAMILY STRING "Cyclone"
-- Retrieval info: CONSTANT: LPM_HINT STRING "ENABLE_RUNTIME_MOD=NO"
-- Retrieval info: CONSTANT: LPM_TYPE STRING "altsyncram"
-- Retrieval info: CONSTANT: NUMWORDS_A NUMERIC "64"
-- Retrieval info: CONSTANT: OPERATION_MODE STRING "ROM"
-- Retrieval info: CONSTANT: OUTDATA_ACLR_A STRING "NONE"
-- Retrieval info: CONSTANT: OUTDATA_REG_A STRING "CLOCK0"
-- Retrieval info: CONSTANT: WIDTHAD_A NUMERIC "6"
-- Retrieval info: CONSTANT: WIDTH_A NUMERIC "8"
-- Retrieval info: CONSTANT: WIDTH_BYTEENA_A NUMERIC "1"
-- Retrieval info: USED_PORT: address 0 0 6 0 INPUT NODEFVAL address[5..0]
-- Retrieval info: USED_PORT: clock 0 0 0 0 INPUT NODEFVAL clock
-- Retrieval info: USED_PORT: q 0 0 8 0 OUTPUT NODEFVAL q[7..0]
-- Retrieval info: CONNECT: @address_a 0 0 6 0 address 0 0 6 0
-- Retrieval info: CONNECT: q 0 0 8 0 @q_a 0 0 8 0
-- Retrieval info: CONNECT: @clock0 0 0 0 0 clock 0 0 0 0
-- Retrieval info: LIBRARY: altera_mf altera_mf.altera_mf_components.all
-- Retrieval info: GEN_FILE: TYPE_NORMAL Rom_TestData.vhd TRUE
-- Retrieval info: GEN_FILE: TYPE_NORMAL Rom_TestData.inc FALSE
-- Retrieval info: GEN_FILE: TYPE_NORMAL Rom_TestData.cmp TRUE
-- Retrieval info: GEN_FILE: TYPE_NORMAL Rom_TestData.bsf TRUE
-- Retrieval info: GEN_FILE: TYPE_NORMAL Rom_TestData_inst.vhd FALSE
-- Retrieval info: GEN_FILE: TYPE_NORMAL Rom_TestData_waveforms.html TRUE
-- Retrieval info: GEN_FILE: TYPE_NORMAL Rom_TestData_wave*.jpg FALSE
-- Retrieval info: LIB_FILE: altera_mf

-- Rom112_PublicKey.vhd
-- megafunction wizard: %ROM: 1-PORT%
-- GENERATION: STANDARD
-- VERSION: WM1.0
-- MODULE: altsyncram

-- ============================================================
-- File Name: Rom112_PublicKey.vhd
-- Megafunction Name(s):
--                  altsyncram
--
-- Simulation Library Files(s):
--                  altera_mf
-- ============================================================


LIBRARY ieee;
USE ieee.std_logic_1164.all;

LIBRARY altera_mf;
```

```vhdl
USE altera_mf.all;

ENTITY Rom112_PublicKey IS
    PORT
    (
        address         : IN STD_LOGIC_VECTOR (10 DOWNTO 0);
        clock           : IN STD_LOGIC ;
        q               : OUT STD_LOGIC_VECTOR (111 DOWNTO 0)
    );
END Rom112_PublicKey;


ARCHITECTURE SYN OF rom112_publickey IS

    SIGNAL sub_wire0    : STD_LOGIC_VECTOR (111 DOWNTO 0);



    COMPONENT altsyncram
    GENERIC (
        address_aclr_a          : STRING;
        init_file               : STRING;
        intended_device_family  : STRING;
        lpm_hint                : STRING;
        lpm_type                : STRING;
        numwords_a              : NATURAL;
        operation_mode          : STRING;
        outdata_aclr_a          : STRING;
        outdata_reg_a           : STRING;
        widthad_a               : NATURAL;
        width_a                 : NATURAL;
        width_byteena_a         : NATURAL
    );
    PORT (
        clock0      : IN STD_LOGIC ;
        address_a   : IN STD_LOGIC_VECTOR (10 DOWNTO 0);
        q_a         : OUT STD_LOGIC_VECTOR (111 DOWNTO 0)
    );
    END COMPONENT;

BEGIN
    q    <= sub_wire0(111 DOWNTO 0);

    altsyncram_component : altsyncram
    GENERIC MAP (
        address_aclr_a => "NONE",
        init_file => "Rom112_PublicKey.mif",
        intended_device_family => "Cyclone",
        lpm_hint => "ENABLE_RUNTIME_MOD=NO",
```

```
                lpm_type => "altsyncram",
                numwords_a => 1482,
                operation_mode => "ROM",
                outdata_aclr_a => "NONE",
                outdata_reg_a => "CLOCK0",
                widthad_a => 11,
                width_a => 112,
                width_byteena_a => 1
        )
        PORT MAP (
                clock0 => clock,
                address_a => address,
                q_a => sub_wire0
        );



END SYN;

-- ============================================================
-- CNX file retrieval info
-- ============================================================
-- Retrieval info: PRIVATE: ADDRESSSTALL_A NUMERIC "0"
-- Retrieval info: PRIVATE: AclrAddr NUMERIC "0"
-- Retrieval info: PRIVATE: AclrByte NUMERIC "0"
-- Retrieval info: PRIVATE: AclrOutput NUMERIC "0"
-- Retrieval info: PRIVATE: BYTE_ENABLE NUMERIC "0"
-- Retrieval info: PRIVATE: BYTE_SIZE NUMERIC "8"
-- Retrieval info: PRIVATE: BlankMemory NUMERIC "0"
-- Retrieval info: PRIVATE: CLOCK_ENABLE_INPUT_A NUMERIC "0"
-- Retrieval info: PRIVATE: CLOCK_ENABLE_OUTPUT_A NUMERIC "0"
-- Retrieval info: PRIVATE: Clken NUMERIC "0"
-- Retrieval info: PRIVATE: IMPLEMENT_IN_LES NUMERIC "0"
-- Retrieval info: PRIVATE: INIT_FILE_LAYOUT STRING "PORT_A"
-- Retrieval info: PRIVATE: INIT_TO_SIM_X NUMERIC "0"
-- Retrieval info: PRIVATE: INTENDED_DEVICE_FAMILY STRING "Cyclone"
-- Retrieval info: PRIVATE: JTAG_ENABLED NUMERIC "0"
-- Retrieval info: PRIVATE: JTAG_ID STRING "NONE"
-- Retrieval info: PRIVATE: MAXIMUM_DEPTH NUMERIC "0"
-- Retrieval info: PRIVATE: MIFfilename STRING "Rom112_PublicKey.mif"
-- Retrieval info: PRIVATE: NUMWORDS_A NUMERIC "1482"
-- Retrieval info: PRIVATE: RAM_BLOCK_TYPE NUMERIC "0"
-- Retrieval info: PRIVATE: RegAddr NUMERIC "1"
-- Retrieval info: PRIVATE: RegOutput NUMERIC "1"
-- Retrieval info: PRIVATE: SYNTH_WRAPPER_GEN_POSTFIX STRING "0"
-- Retrieval info: PRIVATE: SingleClock NUMERIC "1"
-- Retrieval info: PRIVATE: UseDQRAM NUMERIC "0"
-- Retrieval info: PRIVATE: WidthAddr NUMERIC "11"
-- Retrieval info: PRIVATE: WidthData NUMERIC "112"
```

```
-- Retrieval info: PRIVATE: rden NUMERIC "0"
-- Retrieval info: CONSTANT: ADDRESS_ACLR_A STRING "NONE"
-- Retrieval info: CONSTANT: INIT_FILE STRING "Rom112_PublicKey.mif"
-- Retrieval info: CONSTANT: INTENDED_DEVICE_FAMILY STRING "Cyclone"
-- Retrieval info: CONSTANT: LPM_HINT STRING "ENABLE_RUNTIME_MOD=NO"
-- Retrieval info: CONSTANT: LPM_TYPE STRING "altsyncram"
-- Retrieval info: CONSTANT: NUMWORDS_A NUMERIC "1482"
-- Retrieval info: CONSTANT: OPERATION_MODE STRING "ROM"
-- Retrieval info: CONSTANT: OUTDATA_ACLR_A STRING "NONE"
-- Retrieval info: CONSTANT: OUTDATA_REG_A STRING "CLOCK0"
-- Retrieval info: CONSTANT: WIDTHAD_A NUMERIC "11"
-- Retrieval info: CONSTANT: WIDTH_A NUMERIC "112"
-- Retrieval info: CONSTANT: WIDTH_BYTEENA_A NUMERIC "1"
-- Retrieval info: USED_PORT: address 0 0 11 0 INPUT NODEFVAL address[10..0]
-- Retrieval info: USED_PORT: clock 0 0 0 0 INPUT NODEFVAL clock
-- Retrieval info: USED_PORT: q 0 0 112 0 OUTPUT NODEFVAL q[111..0]
-- Retrieval info: CONNECT: @address_a 0 0 11 0 address 0 0 11 0
-- Retrieval info: CONNECT: q 0 0 112 0 @q_a 0 0 112 0
-- Retrieval info: CONNECT: @clock0 0 0 0 0 clock 0 0 0 0
-- Retrieval info: LIBRARY: altera_mf altera_mf.altera_mf_components.all
-- Retrieval info: GEN_FILE: TYPE_NORMAL Rom112_PublicKey.vhd TRUE
-- Retrieval info: GEN_FILE: TYPE_NORMAL Rom112_PublicKey.inc FALSE
-- Retrieval info: GEN_FILE: TYPE_NORMAL Rom112_PublicKey.cmp TRUE
-- Retrieval info: GEN_FILE: TYPE_NORMAL Rom112_PublicKey.bsf TRUE
-- Retrieval info: GEN_FILE: TYPE_NORMAL Rom112_PublicKey_inst.vhd FALSE
-- Retrieval info: GEN_FILE: TYPE_NORMAL Rom112_PublicKey_waveforms.html TRUE
-- Retrieval info: GEN_FILE: TYPE_NORMAL Rom112_PublicKey_wave*.jpg FALSE
-- Retrieval info: LIB_FILE: altera_mf

-- Rom216_PublicKey.vhd
-- megafunction wizard: %ROM: 1-PORT%
-- GENERATION: STANDARD
-- VERSION: WM1.0
-- MODULE: altsyncram

-- ============================================================
-- File Name: Rom216_PublicKey.vhd
-- Megafunction Name(s):
--                  altsyncram
--
-- Simulation Library Files(s):
--                  altera_mf
-- ============================================================
LIBRARY ieee;
USE ieee.std_logic_1164.all;

LIBRARY altera_mf;
USE altera_mf.all;
```

```vhdl
ENTITY Rom216_PublicKey IS
    PORT
    (
        address        : IN STD_LOGIC_VECTOR (9 DOWNTO 0);
        clock          : IN STD_LOGIC ;
        q              : OUT STD_LOGIC_VECTOR (215 DOWNTO 0)
    );
END Rom216_PublicKey;


ARCHITECTURE SYN OF rom216_publickey IS

    SIGNAL sub_wire0    : STD_LOGIC_VECTOR (215 DOWNTO 0);



    COMPONENT altsyncram
    GENERIC (
        address_aclr_a          : STRING;
        init_file               : STRING;
        intended_device_family  : STRING;
        lpm_hint                : STRING;
        lpm_type                : STRING;
        numwords_a              : NATURAL;
        operation_mode          : STRING;
        outdata_aclr_a          : STRING;
        outdata_reg_a           : STRING;
        widthad_a               : NATURAL;
        width_a                 : NATURAL;
        width_byteena_a         : NATURAL
    );
    PORT (
            clock0      : IN STD_LOGIC ;
            address_a   : IN STD_LOGIC_VECTOR (9 DOWNTO 0);
            q_a         : OUT STD_LOGIC_VECTOR (215 DOWNTO 0)
    );
    END COMPONENT;

BEGIN
    q    <= sub_wire0(215 DOWNTO 0);

    altsyncram_component : altsyncram
    GENERIC MAP (
        address_aclr_a => "NONE",
        init_file => "Rom216_PublicKey.mif",
        intended_device_family => "Cyclone",
        lpm_hint => "ENABLE_RUNTIME_MOD=NO",
        lpm_type => "altsyncram",
        numwords_a => 741,
```

```
                    operation_mode => "ROM",
                    outdata_aclr_a => "NONE",
                    outdata_reg_a => "CLOCK0",
                    widthad_a => 10,
                    width_a => 216,
                    width_byteena_a => 1
            )
            PORT MAP (
                    clock0 => clock,
                    address_a => address,
                    q_a => sub_wire0
            );



END SYN;


-- ============================================================
-- CNX file retrieval info
-- ============================================================
-- Retrieval info: PRIVATE: ADDRESSSTALL_A NUMERIC "0"
-- Retrieval info: PRIVATE: AclrAddr NUMERIC "0"
-- Retrieval info: PRIVATE: AclrByte NUMERIC "0"
-- Retrieval info: PRIVATE: AclrOutput NUMERIC "0"
-- Retrieval info: PRIVATE: BYTE_ENABLE NUMERIC "0"
-- Retrieval info: PRIVATE: BYTE_SIZE NUMERIC "8"
-- Retrieval info: PRIVATE: BlankMemory NUMERIC "0"
-- Retrieval info: PRIVATE: CLOCK_ENABLE_INPUT_A NUMERIC "0"
-- Retrieval info: PRIVATE: CLOCK_ENABLE_OUTPUT_A NUMERIC "0"
-- Retrieval info: PRIVATE: Clken NUMERIC "0"
-- Retrieval info: PRIVATE: IMPLEMENT_IN_LES NUMERIC "0"
-- Retrieval info: PRIVATE: INIT_FILE_LAYOUT STRING "PORT_A"
-- Retrieval info: PRIVATE: INIT_TO_SIM_X NUMERIC "0"
-- Retrieval info: PRIVATE: INTENDED_DEVICE_FAMILY STRING "Cyclone"
-- Retrieval info: PRIVATE: JTAG_ENABLED NUMERIC "0"
-- Retrieval info: PRIVATE: JTAG_ID STRING "NONE"
-- Retrieval info: PRIVATE: MAXIMUM_DEPTH NUMERIC "0"
-- Retrieval info: PRIVATE: MIFfilename STRING "Rom216_PublicKey.mif"
-- Retrieval info: PRIVATE: NUMWORDS_A NUMERIC "741"
-- Retrieval info: PRIVATE: RAM_BLOCK_TYPE NUMERIC "0"
-- Retrieval info: PRIVATE: RegAddr NUMERIC "1"
-- Retrieval info: PRIVATE: RegOutput NUMERIC "1"
-- Retrieval info: PRIVATE: SYNTH_WRAPPER_GEN_POSTFIX STRING "0"
-- Retrieval info: PRIVATE: SingleClock NUMERIC "1"
-- Retrieval info: PRIVATE: UseDQRAM NUMERIC "0"
-- Retrieval info: PRIVATE: WidthAddr NUMERIC "10"
-- Retrieval info: PRIVATE: WidthData NUMERIC "216"
-- Retrieval info: PRIVATE: rden NUMERIC "0"
-- Retrieval info: CONSTANT: ADDRESS_ACLR_A STRING "NONE"
```

```
-- Retrieval info: CONSTANT: INIT_FILE STRING "Rom216_PublicKey.mif"
-- Retrieval info: CONSTANT: INTENDED_DEVICE_FAMILY STRING "Cyclone"
-- Retrieval info: CONSTANT: LPM_HINT STRING "ENABLE_RUNTIME_MOD=NO"
-- Retrieval info: CONSTANT: LPM_TYPE STRING "altsyncram"
-- Retrieval info: CONSTANT: NUMWORDS_A NUMERIC "741"
-- Retrieval info: CONSTANT: OPERATION_MODE STRING "ROM"
-- Retrieval info: CONSTANT: OUTDATA_ACLR_A STRING "NONE"
-- Retrieval info: CONSTANT: OUTDATA_REG_A STRING "CLOCK0"
-- Retrieval info: CONSTANT: WIDTHAD_A NUMERIC "10"
-- Retrieval info: CONSTANT: WIDTH_A NUMERIC "216"
-- Retrieval info: CONSTANT: WIDTH_BYTEENA_A NUMERIC "1"
-- Retrieval info: USED_PORT: address 0 0 10 0 INPUT NODEFVAL address[9..0]
-- Retrieval info: USED_PORT: clock 0 0 0 0 INPUT NODEFVAL clock
-- Retrieval info: USED_PORT: q 0 0 216 0 OUTPUT NODEFVAL q[215..0]
-- Retrieval info: CONNECT: @address_a 0 0 10 0 address 0 0 10 0
-- Retrieval info: CONNECT: q 0 0 216 0 @q_a 0 0 216 0
-- Retrieval info: CONNECT: @clock0 0 0 0 0 clock 0 0 0 0
-- Retrieval info: LIBRARY: altera_mf altera_mf.altera_mf_components.all
-- Retrieval info: GEN_FILE: TYPE_NORMAL Rom216_PublicKey.vhd TRUE
-- Retrieval info: GEN_FILE: TYPE_NORMAL Rom216_PublicKey.inc FALSE
-- Retrieval info: GEN_FILE: TYPE_NORMAL Rom216_PublicKey.cmp TRUE
-- Retrieval info: GEN_FILE: TYPE_NORMAL Rom216_PublicKey.bsf TRUE
-- Retrieval info: GEN_FILE: TYPE_NORMAL Rom216_PublicKey_inst.vhd FALSE
-- Retrieval info: GEN_FILE: TYPE_NORMAL Rom216_PublicKey_waveforms.html TRUE
-- Retrieval info: GEN_FILE: TYPE_NORMAL Rom216_PublicKey_wave*.jpg FALSE
-- Retrieval info: LIB_FILE: altera_mf
```

4.5.2 HFE

以下是 HFE 的代码示例。

```
// HFE
// r1 and r2 determine the maximum degrees of the quadratic and
// linear parts of the hidden function. The coefficients are
// chosen at random.
// At the end of the program all possible plaintexts are considered
// and for each plaintext we determine how many solutions exists for
// the given function.

clear ;

SetSeed(1297481417,9091) ;    // values used for book
print GetSeed() ;
q := 2^2 ;
G<a>:=GF(q);
r1 := 1 ;
r2 := 2 ;
```

```
n:=4;

P1<[x]> := PolynomialRing( G, n+1 ) ;

Q1<y> := PolynomialRing(G);
g := IrreduciblePolynomial(G,n);
c1:=Coefficients(g) ;

print "Irreducible polynomial g=",g , IsIrreducible(g);

R0    := [g] ;
Q2<y> := quo < Q1| R0 > ;
Q0<X> := PolynomialRing(Q2,1) ;

print"c1",c1;

g := P1!0 ;
for i in [1..n+1] do
   g +:= c1[i]*x[n+1]^(i-1) ;    // x[n+1] takes on place of y in big field
end for ;

   print g;
R1:= [g] ;

for i in [1..n] do
   R1 := Append(R1,x[i]^q+x[i]);
end for ;

P0<[x]> := quo < P1|R1 > ;

// linear transformation L2
repeat
   L22 := Matrix(P1,n,n,[<i,j,Random(G)>: i,j in [1..n]]) ;
until  Determinant( L22 ) ne 0 ;
L21 := Matrix(P1,n,1,[Random(G) : i in [1..n]] ) ;

//L21 := Matrix(P1,n,1,[0:i in [1..n]] ) ;   // zero matrix for testing
//L22 := DiagonalMatrix(P1,n,[1:i in [1..n]] ) ;  //Identity Matrix

xVector :=  Matrix(P1,n,1,[x[i]: i in [1..n]]) ;

xV2 := L22 * xVector + L21  ;

print "L22=",L22," L21=",L21;

yMatrix := Matrix(P1,1,n,[x[n+1]^i: i in [0..n-1]]) ;

XX := yMatrix*xV2 ;
XX := P0!XX[1][1] ;                    // make it scalar
```

第4章 多变量公钥密码技术

```
// Construct hidden equations

a1 := Random(G) ;
f  := Q0!0 ;
f1 := P0!0 ;

for i in [1..r1] do
    a1 := Random(G) ;
    f  +:= a1 * X^(q^(i-1)) ;
    f1 +:= a1 * XX^(q^(i-1)) ;
end for ;
for i in [1..r2] do
    for j in [1..i] do
        a1 := Random(G) ;
        f  +:= a1 * X^(q^(i-1)+q^(j-1));
        f1 +:= a1 * XX^(q^(i-1)+q^(j-1));
        print "f=",f ;
        print "f1=",f1 ;

    end for ;
end for ;

f1c := Coefficients(f1, x[n+1]) ;
while ( #f1c ne n ) do
   Append(~f1c,0 ) ;
end while ;

polys := Matrix(P1,n,1, f1c ) ;

repeat
   L11 := Matrix(P1,n,n,[<i,j,Random(G)>: i,j in [1..n]]) ;
until  Determinant( L11 ) ne 0 ;

L12 := Matrix(P1,n,1,[Random(G) : i in [1..n]] ) ;

//L11 := DiagonalMatrix(P1,n,[1:i in [1..n]] ) ;  //Identity Matrix
//L12 := Matrix(P1,n,1,[0:i in [1..n]] ) ;   // zero matrix for testing

pol:= L11*polys + L12;

print "L11=",L11," L12=",L12;

count := [0: i in [1..n^2]];

d := [0: i in [1..n]];

field :=[a: a in G] ;
```

4.5 本章参考代码

```
for ij in [1..q^n] do  // loop to do all cases

    plaintext := [] ;
    for i in [1..n] do
       plaintext[i] := field[d[i]+1] ;
    end for ;

    print "Plaintext:", plaintext ;
    plaintext[n+1] := 0 ;

    ciphertext := Matrix(P1,n,1,[Evaluate( pol[i][1], plaintext ): i in [1..n]]) ;

    // decrypting
  y2 := ciphertext + L12;
  L11inv := L11^(-1) ;
  y1 := L11inv * y2 ;

    y0 :=  P0!y1[1][1] ;
    for i in [2..n] do
        y0 +:= P0!y1[i][1] * x[n+1]^(i-1) ;
    end for ;

    f0 := f ;

c0 := Coefficients(y0,x[n+1]) ;
for i in [1..#c0] do
   f0 +:= Q0! c0[i]* y^(i-1) ;
end for ;

s0 := Factorization( f0 ) ;

// extract linear factors

solfin := [] ;

for  i in [1.. #s0] do
    sq := s0[i] ;
    if LeadingTotalDegree( sq[1] ) eq 1 then
        Append( ~solfin, sq[1] ) ;
    end if ;
end for ;

  count[#solfin] +:= 1 ;

    print "Number of Solutions: ", #solfin ;
    p0 :=0 ;

    for j in [1..#solfin] do
        s1 := Coefficient(solfin[j],X,0) ;
        c1 := Coefficients(Q2 ! s1 ) ;
```

```
            while ( #c1 ne n ) do
                Append(~c1,0 ) ;
            end while ;

        // convert to matrix ;
        yb := Matrix( P1,n,1,c1 ) ;
        L22inv := L22^(-1) ;
        yc := yb + L21 ;
        yd := L22inv * yc ;
        plain := Transpose(yd) ;
        b0 := true ;
        for i in [1..n] do
            if plain[1][i] ne plaintext[i] then b0 := false ; end if ;
        end for ;

        if b0 then p0 +:= 1 ; end if ;
    end for ;
        if p0 ne 1 then print "number of correct :", p0 ; end if ;

        b0 := 1 ;
        j := 1 ;

        for j in [1..n] do
            d[j] +:= 1 ;
            if d[j] ne q then
                break ;
            else
                d[j] := 0 ;
            end if ;
        end for ;
    end for ;      // loop all cases

    print "count=",count ;
```

4.6 本章小结

本章首先介绍了多变量公钥密码的起源和发展,然后描述了多变量公钥密码系统的分类、代表性多变量公钥密码算法以及分析方法,最后给出了算法的软硬件实现参考代码。

4.7 本章参考文献

[1] H. Ong, Claus-Peter Schnorr. "Signatures through approximate representation by quadratic

forms." Pages 117–131 in: David Chaum (editor). Advances in cryptology, proceedings of CRYPTO '83. Plenum Press.

[2] H. Ong, Claus-Peter Schnorr, Adi Shamir. "Efficient signature schemes based on polynomial equations." Pages 37-46 in: G. R. Blakley, David Chaum (editors). Advances in cryptology, proceedings of CRYPTO '84, Santa Barbara, California, USA, August 19-22, 1984, proceedings. Lecture Notes in Computer Science 196. Springer. ISBN 3-540-15658-5.

[3] Harriet J. Fell, Whitfield Diffie. "Analysis of a public key approach based on polynomial substitution." Pages 340-349 in: Hugh C. Williams (editor). Advances in cryptology—CRYPTO '85, Santa Barbara, California, USA, August 18-22, 1985, proceedings.

[4] John M. Pollard, Claus-Peter Schnorr. "An efficient solution of the congruence $x^2 + ky^2 = m \pmod{n}$." IEEE Transactions on Information Theory 33, 702-709.

[5] Tsutomu Matsumoto, Hideki Imai. "Public quadratic polynomial-tuples for efficient signature-verification and message-encryption." MR 90d:94008. Pages 419-453 in: Christoph G. Günther (editor). Advances in cryptology—EUROCRYPT 1988, proceedings of the workshop on the theory and application of cryptographic techniques held in Davos, May 25-27, 1988. Lecture Notes in Computer Science 330. ISBN 3-540-50251-3. MR 90a:94002.

[6] Jacques Patarin. "Asymmetric cryptography with a hidden monomial and a candidate algorithm for =~ 64 bits asymmetric signatures." MR 99b:94040. Pages 45-60 in: Neal Koblitz (editor). Advances in cryptology—CRYPTO '96, proceedings of the 16th annual international cryptology conference held at the University of California, Santa Barbara, CA, August 18-22, 1996. Lecture Notes in Computer Science 1109. Springer. ISBN 3-540-61512-1. MR 98f:94001.

[7] Jacques Patarin. "Hidden Fields Equations (HFE) and Isomorphisms of Polynomials (IP): two new families of asymmetric algorithms." Pages 33-48 in: Ueli Maurer (editor). Advances in Cryptology—EUROCRYPT '96. International conference on the theory and application of cryptographic techniques, Saragossa, Spain, May 12-16, 1996, proceedings. Lecture Notes in Computer Science 1070. Springer. ISBN 978-3-540-61186-8.

[8] Jacques Patarin, Louis Goubin, Nicolas T. Courtois. "C*-+ and HM: variations around two schemes of T. Matsumoto and H. Imai." Pages 35-49 in: Kazuo Ohta, Dingyi Pei (editors). Advances in cryptology—ASIACRYPT'98. Proceedings of the International Conference on the Theory and Application of Cryptology and Information Security held in Beijing, October 18-22, 1998. Lecture Notes in Computer Science 1514. Springer. ISBN 3-540-65109-8.

[9] Aviad Kipnis, Adi Shamir. "Cryptanalysis of the HFE public key cryptosystem by linearization." Pages 19-30 in: Michael J. Wiener (editor). Advances in cryptology—CRYPTO '99, 19th annual international cryptology conference, Santa Barbara, California, USA, August

15-19, proceedings. Lecture Notes in Computer Science 1666. Springer. ISBN 3-540-66347-9.

[10] Jacques Patarin. "Cryptanalysis of the Matsumoto and Imai public key scheme of Eurocrypt '88." Designs, Codes and Cryptography 20, 175-209. Earlier version: 1995. MR 98d:94022. Pages 248-261 in: Don Coppersmith (editor). Advances in Cryptology—CRYPTO 1995, proceedings of the 15th annual international cryptology conference held at the University of California, Santa Barbara, CA, August 27-31, 1995. Lecture Notes in Computer Science 963. Springer. ISBN 3-540-60221-6. MR 97k:94002.

[11] Nicolas T. Courtois. "The security of hidden field equations (HFE)." MR 1907103. Pages 266-281 in: David Naccache (editor). Topics in cryptology—CT-RSA 2001, proceedings of the cryptographers' track at the RSA conference held in San Francisco, CA, April 8-12, 2001. Lecture Notes in Computer Science 2020. Springer. ISBN 3-540-41898-9. MR 2003a:94039.

[12] Jacques Patarin, Nicolas T. Courtois, Louis Goubin. "QUARTZ, 128-bit long digital signatures." MR 1907104. Pages 282-297 in: David Naccache (editor). Topics in cryptology—CT-RSA 2001, proceedings of the cryptographers' track at the RSA conference held in San Francisco, CA, April 8-12, 2001. Lecture Notes in Computer Science 2020. Springer. ISBN 3-540-41898-9. MR 2003a:94039.

[13] Jacques Patarin, Nicolas T. Courtois, Louis Goubin. "FLASH, a fast multivariate signature algorithm." MR 1907105. Pages 298-307 in: David Naccache (editor). Topics in cryptology—CT-RSA 2001, proceedings of the cryptographers' track at the RSA conference held in San Francisco, CA, April 8-12, 2001. Lecture Notes in Computer Science 2020. Springer. ISBN 3-540-41898-9. MR 2003a:94039.

[14] Nicolas T. Courtois, Magnus Daum, Patrick Felke. "On the security of HFE, HFEv- and Quartz." MR 2007g:94044. Pages 337-350 in: Yvo G. Desmedt (editor). Public key cryptography—PKC 2003, proceedings of the 6th international workshop on practice and theory in public key cryptography held in Miami, FL, January 6-8, 2003. Lecture Notes in Computer Science 2567. Springer. ISBN 3-540-00324-X. MR 2006d:94071.

[15] Nicolas T. Courtois. "Generic attacks and the security of Quartz." MR 2006i:94045. Pages 351-364 in: Yvo G. Desmedt (editor). Public key cryptography—PKC 2003, proceedings of the 6th international workshop on practice and theory in public key cryptography held in Miami, FL, January 6-8, 2003. Lecture Notes in Computer Science 2567. Springer. ISBN 3-540-00324-X. MR 2006d:94071.

[16] Jean-Charles Faugère., Antoine Joux. "Algebraic cryptanalysis of Hidden Field Equations (HFE) using Gröbner bases." MR 2005e:94140. Pages 44-60 in: Dan Boneh (editor). Advances in cryptology—CRYPTO 2003, proceedings of the 23rd annual international cryptology conference held in Santa Barbara, CA, August 17-21, 2003. Lecture Notes in Computer

Science 2729. Springer. ISBN 3-540-40674-3. MR 2005d:94151.

[17] Jintai Ding, Dieter Schmidt. "Cryptanalysis of SFlash v3." http://eprint.iacr.org/2004/103.

[18] Nicolas T. Courtois. "Algebraic attacks over GF(2^k), application to HFE Challenge 2 and Sflash-v2." MR 2005e:94132. Pages 201-217 in: Feng Bao, Robert Deng, Jianying Zhou (editors). Public key cryptography—PKC 2004, proceedings of the 7th international workshop on theory and practice in public key cryptography held in Singapore, March 1-4, 2004. Lecture Notes in Computer Science 2947. Springer. ISBN 3-540-21018-0. MR 2005d:94155.

[19] Jintai Ding. "A new variant of the Matsumoto-Imai cryptosystem through perturbation." Pages 305-318 in: Feng Bao, Robert Deng, Jianying Zhou (editors). Public key cryptography—PKC 2004, proceedings of the 7th international workshop on theory and practice in public key cryptography held in Singapore, March 1-4, 2004. Lecture Notes in Computer Science 2947. Springer. ISBN 3-540-21018-0. MR 2005d:94155.

[20] Jintai Ding, Jason E. Gower, Dieter Schmidt, Christopher Wolf, Zhijun Yin. "Complexity estimates for the F_4 attack on the perturbed Matsumoto-Imai cryptosystem." MR 2007f:94036. Pages 262-277 in: Nigel P. Smart (editor). Cryptography and coding, 10th IMA international conference, Cirencester, UK, December 19-21, 2005, proceedings. Lecture Notes in Computer Science 3796. Springer. ISBN 3-540-30276-X. MR 2006m:94081.

[21] Jintai Ding, Dieter Schmidt. "Cryptanalysis of HFEv and internal perturbation of HFE." MR 2006j:94061. Pages 288-301 in: Serge Vaudenay (editor). Public key cryptography—PKC 2005: proceedings of the 8th international workshop on theory and practice in public key cryptography held in Les Diablerets, January 23-26, 2005. Lecture Notes in Computer Science 3386. Springer. ISBN 3-540-24454-9. MR 2006d:94072.

[22] Pierre-Alain Fouque, Louis Granboulan, Jacques Stern. "Differential cryptanalysis for multivariate schemes." Pages 341-353 in: Ronald Cramer (editor). Advances in cryptology—EUROCRYPT 2005. Proceedings of the 24th annual international conference on the theory and applications of cryptographic techniques held in Aarhus, May 22-26, 2005. Lecture Notes in Computer Science 3494. Springer. ISBN 3-540-25910-4. MR 2008e:94035.

[23] Vivien Dubois, Louis Granboulan, Jacques Stern. "An efficient provable distinguisher for HFE." MR 2008b:94065. Pages 156-167 in: Michele Bugliesi, Bart Preneel, Vladimiro Sassone, Ingo Wegener (editors). Automata, languages and programming. Part II. Proceedings of the 33rd International Colloquium (ICALP 2006) held in Venice, July 10-14, 2006. Lecture Notes in Computer Science 4052. Springer. ISBN 978-3-540-35907-4.

[24] Louis Granboulan, Antoine Joux, Jacques Stern. "Inverting HFE is quasipolynomial." Pages 345-356 in: Cynthia Dwork (editor). Advances in cryptology—CRYPTO 2006. Proceedings of the 26th annual international cryptology conference held in Santa Barbara, CA,

August 20-24, 2006. Lecture Notes in Computer Science 4117. Springer. ISBN 978-3-540-37432-9. MR 2422188.

[25] Aline Gouget, Jacques Patarin. "Probabilistic multivariate cryptography." Pages 1-18 in: Phong Q. Nguyen (editor). Progress in cryptology—VIETCRYPT 2006, first international conference on cryptology in Vietnam, Hanoi, Vietnam, September 25-28, 2006, revised selected papers. Lecture Notes in Computer Science 4341. Springer. ISBN 3-540-68799-8.

[26] Jintai Ding, Jason E. Gower. "Inoculating multivariate schemes against differential attacks." MR 2423196. Pages 290-301 in: Moti Yung, Yevgeniy Dodis, Aggelos Kiayias, Tal Malkin (editors). Public key cryptography—PKC 2006. Proceedings of the 9th International Conference on Theory and Practice of Public-Key Cryptography held in New York, April 24-26, 2006. Lecture Notes in Computer Science 3958. Springer. MR 2009a:94034.

[27] Adama Diene, Jintai Ding, Jason E. Gower, Timothy J. Hodges, Zhijun Yin. "Dimension of the linearization equations of the Matsumoto-Imai cryptosystems." MR 2423694. Pages 242-251 in: Oyvind Ytrehus, Coding and cryptography, revised selected papers from the international workshop (WCC 2005) held in Bergen, March 14-18, 2005. Lecture Notes in Computer Science 3969. Springer. ISBN 3-540-35481-6. MR 2428233.

[28] Lih-Chung Wang, Fei-Hwang Chang. "Revision of tractable rational map cryptosystem." Version 20061227:165851. http://eprint.iacr.org/2004/046.

[29] Lih-Chung Wang, Bo-Yin Yang, Yuh-Hua Hu, Feipei Lai. "A 'medium-field' multivariate public-key encryption scheme." Pages 132-149 in: David Pointcheval (editor). Topics in cryptology—CT-RSA 2006, proceedings of the cryptographers' track at the RSA conference held in San Jose, CA, February 13-17, 2006. Lecture Notes in Computer Science 3860. Springer. ISBN 3-540-31033-9. MR 2007b:94003.

[30] Jintai Ding, Christopher Wolf, Bo-Yin Yang. "l-invertible cycles for Multivariate Quadratic (MQ) public key cryptography." MR 2404125. Pages 266-281 in: Tatsuaki Okamoto, Xiaoyun Wang (editors). Public key cryptography—PKC 2007, proceedings of the 10th international conference on practice and theory in public-key cryptography held at Tsinghua University, Beijing, April 16-20, 2007. Lecture Notes in Computer Science 4450. Springer. ISBN 3-540-71676-9. MR 2404107.

[31] Jintai Ding, Bo-Yin Yang, Chen-Mou Cheng, Owen Chen, Vivien Dubois. "Breaking the symmetry: a way to resist the new differential attack." http://eprint.iacr.org/2007/366.

[32] Vivien Dubois, Louis Granboulan, Jacques Stern. "Cryptanalysis of HFE with internal perturbation." MR 2404124. Pages 249-265 in: Tatsuaki Okamoto, Xiaoyun Wang (editors). Public key cryptography—PKC 2007, proceedings of the 10th international conference on practice and theory in public-key cryptography held at Tsinghua University, Beijing, April

16-20, 2007. Lecture Notes in Computer Science 4450. Springer. ISBN 3-540-71676-9. MR 2404107.

[33] Vivien Dubois, Pierre-Alain Fouque, Jacques Stern. "Cryptanalysis of SFLASH with slightly modified parameters." Pages 264-275 in: Moni Naor (editor). Advances in Cryptology—EUROCRYPT 2007. 26th annual international conference on the theory and applications of cryptographic techniques, Barcelona, Spain, May 20-24, 2007, proceedings. Lecture Notes in Computer Science 4515. Springer. ISBN 978-3-540-72539-8.

[34] Vivien Dubois, Pierre-Alain Fouque, Adi Shamir, Jacques Stern. "Practical cryptanalysis of SFLASH." MR 2419591. Pages 1-12 in: Alfred Menezes (editor). Advances in cryptology—CRYPTO 2007, proceedings of the 27th annual international cryptology conference held in Santa Barbara, CA, August 19-23, 2007. Lecture Notes in Computer Science 4622. Springer. ISBN 978-3-540-74142-8.

[35] Xin Jiang, Jintai Ding, Lei Hu. "Kipnis-Shamir attack on HFE revisited." Pages 399-411 in: Dingyi Pei, Moti Yung, Dongdai Lin, Chuankun Wu (editors). Information security and cryptology, third SKLOIS conference, Inscrypt 2007, Xining, China, August 31-September 5, 2007, revised selected papers. Lecture Notes in Computer Science 4990. Springer. ISBN 978-3-540-79498-1.

[36] Jintai Ding, Dieter Schmidt, Fabian Werner. "Algebraic attack on HFE revisited." Pages 215-227 in: Tzong-Chen Wu, Chin-Laung Lei, Vincent Rijmen, Der-Tsai Lee (editors). Information security: 11th international conference, ISC 2008, Taipei, Taiwan, September 15-18, 2008, proceedings. Lecture Notes in Computer Science 5222. Springer. ISBN 978-3-540-85884-3.

[37] Pierre-Alain Fouque, Gilles Macario-Rat, Ludovic Perret, Jacques Stern. "Total break of the l-IC signature scheme." Pages 1-17 in: Ronald Cramer (editor). Public key cryptography—PKC 2008, 11th international workshop on practice and theory in public-key cryptography, Barcelona, Spain, March 9-12, 2008, proceedings. Lecture Notes in Computer Science 4939. Springer. ISBN 978-3-540-78439-5.

[38] Shigeo Tsujii, Toshiya Itoh, Atsushi Fujioka, Kaoru Kurosawa, Tsutomu Matsumoto. "A public-key cryptosystem based on the difficulty of solving a system of nonlinear equations." Systems and Computers in Japan 19, 10-18.

[39] Adi Shamir. "Efficient signature schemes based on birational permutations." Pages 1-12 in: Douglas R. Stinson (editor). Advances in Cryptology—CRYPTO '93, 13th annual international cryptology conference, Santa Barbara, California, USA, August 22-26, 1993, proceedings. Lecture Notes in Computer Science 773. Springer. ISBN 3-540-57766-1.

[40] Don Coppersmith, Jacques Stern, Serge Vaudenay. "The security of the birational

permutation signature schemes." Journal of Cryptology 10, 207-221. MR 99e:94033.

[41] Aviad Kipnis, Adi Shamir. "Cryptanalysis of the oil and vinegar signature scheme." Pages 257-266 in: Hugo Krawczyk (editor). Advances in cryptology—CRYPTO '98, 18th annual international cryptology conference, Santa Barbara, California, USA, August 23-27, 1998, proceedings. Lecture Notes in Computer Science 1462. Springer. ISBN 3-540-64892-5.

[42] Aviad Kipnis, Jacques Patarin, Louis Goubin. "Unbalanced oil and vinegar signature schemes." MR 1717470. Pages 206-222 in: Jacques Stern (editor). Advances in cryptology—EUROCRYPT '99, proceedings of the 17th international conference on the theory and application of cryptographic techniques held in Prague, May 2-6, 1999. Lecture Notes in Computer Science 1592. Springer. ISBN 3-540-65889-0. MR 2000i:94001.

[43] T. Moh. "A public key system with signature and master key functions." Communications in Algebra 27, 2207-2222.

[44] Louis Goubin, Nicolas T. Courtois. "Cryptanalysis of the TTM cryptosystem." MR 2002j:94037. Pages 44-57 in: Tatsuaki Okamoto (editor). Advances in cryptology— ASIACRYPT 2000, proceedings of the 6th annual international conference on the theory and application of cryptology and information security held in Kyoto, December 3-7, 2000. Lecture Notes in Computer Science 1976. Springer. ISBN 3-540-41404-5. MR 2002d:94046.

[45] T. Moh, Jiun-Ming Chen. "On the Goubin-Courtois attack on TTM." http://eprint.iacr.org/2001/072.

[46] Jiun-Ming Chen, Bo-Yin Yang. "A more secure and efficacious TTS signature scheme." Pages 320-338 in: Lecture Notes in Computer Science 2971. Springer. ISBN 3-540-21376-7. MR 2005d:94153.

[47] Jintai Ding, Dieter Schmidt. "The new implementation schemes of the TTM cryptosystem are not secure." MR 2005d:94100. Pages 113-127 in: Keqin Feng, Harald Niederreiter, Chaoping Xing (editors). Coding, cryptography and combinatorics. Progress in Computer Science and Applied Logic 23. Birkhauser. ISBN 3-7643-2429-5.

[48] Jintai Ding, Timothy Hodges. "Cryptanalysis of an implementation scheme of the tamed transformation method cryptosystem." Journal of Algebra and its Applications 3, 273-282. MR 2005f:94093.

[49] Masao Kasahara, Ryuichi Sakai. "A construction of public key cryptosystem for realizing ciphertext of size 100 bit and digital signature scheme." IEICE Transactions on Fundamentals 87-A, 102-109. http://search.ieice.org/bin/summary.php?id=e87-a_1_102&category= D&year=2004&lang=E&abst=.

[50] Bo-Yin Yang, Jiun-Ming Chen, Yen-Hung Chen. "TTS: high-speed signatures on a low-cost smart card." Pages 371-385 in: Marc Joye, Jean-Jacques Quisquater (editors).

Cryptographic hardware and embedded systems—CHES 2004, 6th international workshop, Cambridge, MA, USA, August 11-13, 2004, proceedings. Lecture Notes in Computer Science 3156. Springer. ISBN 3-540-22666-4.

[51] Lih-Chung Wang, Fei-Hwang Chang. "Tractable rational map cryptosystem." Version 20040221:212731. http://eprint.iacr.org/2004/046.

[52] Bo-Yin Yang, Jiun-Ming Chen. "TTS: rank attacks in tame-like multivariate PKCs." http://eprint.iacr.org/2004/061.

[53] Masao Kasahara, Ryuichi Sakai. "A construction of public-key cryptosystem based on singular simultaneous equations." IEICE Transactions on Fundamentals 88-A, 74-80. http://search.ieice.org/bin/summary.php?id=e88-a_1_74&category=D&year=2005&lang=E&abst=.

[54] Christopher Wolf, Bart Preneel. "Large superfluous keys in multivariate quadratic asymmetric systems." Pages 275-287 in: Serge Vaudenay (editor). Public key cryptography—PKC 2005: proceedings of the 8th international workshop on theory and practice in public key cryptography held in Les Diablerets, January 23-26, 2005. Lecture Notes in Computer Science 3386. Springer. ISBN 3-540-24454-9. MR 2006d:94072.

[55] An Braeken, Christopher Wolf, Bart Preneel. "A study of the security of unbalanced oil and vinegar signature schemes." MR 2006h:94169. Pages 29-43 in: Alfred Menezes (editor). Topics in cryptology—CT-RSA 2005. Proceedings of the cryptographers' track at the RSA conference held in San Francisco, CA, February 14-18, 2005. Lecture Notes in Computer Science 3376. Springer. MR 2006d:94073. ISBN 3-540-24399-2.

[56] Lih-Chung Wang, Yuh-Hua Hu, Feipei Lai, Chun-yen Chou, Bo-Yin Yang. "Tractable rational map signature." Pages 244-257 in: Serge Vaudenay (editor). Public key cryptography—PKC 2005: proceedings of the 8th international workshop on theory and practice in public key cryptography held in Les Diablerets, January 23-26, 2005. Lecture Notes in Computer Science 3386. Springer. ISBN 3-540-24454-9. MR 2006d:94072.

[57] Bo-Yin Yang, Jiun-Ming Chen. "Building secure tame-like multivariate public-key cryptosystems: the new TTS." Pages 518-531 in: Colin Boyd, Juan Manuel González Nieto (editors). Information security and privacy, 10th Australasian conference, ACISP 2005, Brisbane, Australia, July 4-6, 2005, proceedings. Lecture Notes in Computer Science 3574. Springer. ISBn 3-540-26547-3.

[58] Jintai Ding, Dieter Schmidt. "Rainbow, a new multivariable polynomial signature scheme." Pages 164-175 in: John Ioannidis, Angelos D. Keromytis, Moti Yung (editors). Applied cryptography and network security, third international conference, ACNS 2005, New York, NY, USA, June 7-10, 2005, proceedings. Lecture Notes in Computer Science 3531.

Springer. ISBN 3-540-26223-7.

[59] Olivier Billet, Henri Gilbert. "Cryptanalysis of Rainbow." Pages 336-347 in: Roberto De Prisco, Moti Yung (editors). Security and cryptography for networks, 5th international conference, SCN 2006, Maiori, Italy, September 6-8, 2006, proceedings.. Lecture Notes in Computer Science 4116. Springer. ISBN 3-540-38080-9.

[60] Xuyun Nie, Lei Hu, Jianyu Li, Crystal Updegrove, Jintai Ding. "Breaking a new instance of TTM cryptosystems." Pages 210-225 in: Jianying Zhou, Moti Yung, Feng Bao (editors). Applied cryptography and network security, 4th international conference, ACNS 2006, Singapore, June 6-9, 2006, proceedings. Lecture Notes in Computer Science 3989. Springer. ISBN 3-540-34703-8.

[61] Christopher Wolf, An Braeken, Bart Preneel. "On the security of stepwise triangular systems." Designs, Codes and Cryptography 40, 285-302. Older version: 2004. "Efficient cryptanalysis of RSE(2)PKC and RSSE(2)PKC." Pages 294-309 in: SCN 2004. Lecture Notes in Computer Science 3352. Springer.

[62] Jintai Ding, Dieter Schmidt, Zhijun Yin. "Cryptanalysis of the new TTS scheme in CHES 2004." International Journal of Information Security 5, 231-240.

[63] T. Moh. "The recent attack of Nie et al on TTM is faulty." http://eprint.iacr.org/ 2006/417

[64] Jintai Ding, Lei Hu, Xuyun Nie, Jianyu Li, John Wanger. "High order linearization equation (HOLE) attack on multivariate public key cryptosystems." MR 2404123. Pages 230-247 in: Tatsuaki Okamoto, Xiaoyun Wang (editors). Public key cryptography—PKC 2007, proceedings of the 10th international conference on practice and theory in public-key cryptography held at Tsinghua University, Beijing, April 16-20, 2007. Lecture Notes in Computer Science 4450. Springer. ISBN 3-540-71676-9. MR 2404107.

[65] Xuyun Nie, Lei Hu, Jintai Ding, Jianyu Li, John Wagner. "Cryptanalysis of the TRMC-4 public key cryptosystem." Pages 104-115 in: Jonathan Katz, Moti Yung (editors). Applied Cryptography and Network Security, 5th international conference, ACNS 2007, Zhuhai, China, June 5-8, 2007, proceedings. Lecture Notes in Computer Science 4521. Springer. ISBN 978-3-540-72737-8.

[66] T. Moh. "Two new examples of TTM." http://eprint.iacr.org/2007/144.

[67] Jintai Ding, Vivien Dubois, Bo-Yin Yang, Chia-Hsin Owen Chen, Chen-Mou Cheng. "Could SFLASH be repaired?" Pages 691-701 in: Luca Aceto, Ivan Damgård, Leslie Ann Goldberg, Magnús M. Halldórsson, Anna Ingólfsdóttir, Igor Walukiewicz (editors). Automata, languages and programming, 35th international colloquium, ICALP 2008, Reykjavik, Iceland, July 7-11, 2008, proceedings, part II, track B: logic, semantics, and theory of programming; track C: security and cryptography foundations. Lecture Notes in Computer Science 5126.

Springer. ISBN 978-3-540-70582-6.

[68] Jintai Ding, Bo-Yin Yang, Chia-Hsin Owen Chen, Ming-Shing Chen, Chen-Mou Cheng. "New differential-algebraic attacks and reparametrization of Rainbow." Pages 242-257 in: Steven M. Bellovin, Rosario Gennaro, Angelos D. Keromytis, Moti Yung (editors). Applied cryptography and network security, 6th international conference, ACNS 2008, New York, NY, USA, June 3-6, 2008, proceedings. Lecture Notes in Computer Science 5037. Springer. ISBN 978-3-540-68913-3.

[69] Jean-Charles Faugère, Françoise Levy-dit-Vehel, Ludovic Perret. "Cryptanalysis of MinRank." Pages 280-296 in: David Wagner (editor). Advances in cryptology—CRYPTO 2008, 28th annual international cryptology conference, Santa Barbara, CA, USA, August 17-21, 2008, proceedings. Lecture Notes in Computer Science 5157. Springer. ISBN 978-3-540-85173-8.

[70] Luk Bettale, Jean-Charles Faugère, Ludovic Perret: "Cryptanalysis of the TRMS signature scheme of PKC'05." Pages 143-155 in: Serge Vaudenay (editor). Progress in cryptology—AFRICACRYPT 2008, first international conference on cryptology in Africa, Casablanca, Morocco, June 11-14, 2008, proceedings. Lecture Notes in Computer Science 5023. Springer. ISBN 978-3-540-68159-5.

[71] Jintai Ding, John Wagner. "Cryptanalysis of rational multivariate public key cryptosystems." Pages 124-136 in: Johannes Buchmann, Jintai Ding (editors). Post-quantum cryptography, second international workshop, PQCrypto 2008, Cincinnati, OH, USA, October 17-19, 2008, proceedings. Lecture Notes in Computer Science 5299, Springer.

[72] John Baena, Crystal Clough, Jintai Ding. "Square-vinegar signature scheme." Pages 17-30 in: Johannes Buchmann, Jintai Ding (editors). Post-quantum cryptography, second international workshop, PQCrypto 2008, Cincinnati, OH, USA, October 17-19, 2008, proceedings. Lecture Notes in Computer Science 5299, Springer.

[73] Mehdi-Laurent Akkar, Nicolas T. Courtois, Romain Duteuil, Louis Goubin. "A fast and secure implementation of Sflash." MR 2006i:94034. Pages 267-278 in: Yvo G. Desmedt (editor). Public key cryptography—PKC 2003, proceedings of the 6th international workshop on practice and theory in public key cryptography held in Miami, FL, January 6-8, 2003. Lecture Notes in Computer Science 2567. Springer. ISBN 3-540-00324-X. MR 2006d:94071.

[74] Christopher Wolf. "Efficient public key generation for multivariate cryptosystems." http://eprint.iacr.org/2003/089.

[75] Bo-Yin Yang, Chen-Mou Cheng, Bor-Rong Chen, Jiun-Ming Chen. "Implementing minimized multivariate PKC on low-resource embedded systems." Pages 73-88 in: John A. Clark, Richard F. Paige, Fiona Polack, Phillip J. Brooke (editors). Security in pervasive

computing, third international conference, SPC 2006, York, UK, April 18-21, 2006, proceedings. Lecture Notes in Computer Science 3934. Springer. ISBN 3-540-33376-2.

[76] Côme Berbain, Olivier Billet, Henri Gilbert. "Efficient implementations of multivariate quadratic systems." Pages 174-187 in: Eli Biham, Amr M. Youssef (editors). Selected areas in cryptography, 13th international workshop, SAC 2006, Montreal, Canada, August 17-18, 2006, revised selected papers. Lecture Notes in Computer Science 4356. Springer. ISBN 978-3-540-74461-0.

[77] Sundar Balasubramanian, Andrey Bogdanov, Andy Rupp, Jintai Ding, Harold W. Carter. "Fast multivariate signature generation in hardware: the case of Rainbow." ASAP 2008. IEEE.

[78] Bruno Buchberger. "Ein Algorithmus zum Auffinden der Basiselemente des Restklassenringes nach einem nulldimensionalen Polynomideal." Ph.D. thesis, University of Innsbruck.

[79] Daniel Lazard. "Gröbner-bases, Gaussian elimination and resolution of systems of algebraic equations." MR 86m:13002. Pages 146-156 in: J. A. van Hulzen (editor). Computer algebra: proceedings of the European computer algebra conference (EUROCAL) held in London, March 28-30, 1983. Lecture Notes in Computer Science 162. Springer. ISBN 3-540-12868-9. MR 86f:68004.

[80] Jean-Charles Faugère, Patrizia M. Gianni, Daniel Lazard, Teo Mora. "Efficient computation of zero-dimensional Gröbner bases by change of ordering." Journal of Symbolic Computation 16, 329-344.

[81] Jean-Charles Faugère. "A new efficient algorithm for computing Gröbner bases (F4)." Journal of Pure and Applied Algebra 139, 61-88.

[82] Nicolas T. Courtois, Alexander Klimov, Jacques Patarin, Adi Shamir. "Efficient algorithms for solving overdefined systems of multivariate polynomial equations." MR 1772028. Pages 392-407 in: Bart Preneel (editor). Advances in cryptology—EUROCRYPT 2000, proceedings of the 19th international annual conference on the theory and application of cryptographic techniques held in Bruges, May 14-18, 2000. Lecture Notes in Computer Science 1807. Springer. ISBN 3-540-67517-5. MR 2001b:94028.

[83] Jean-Charles Faugère. "A new efficient algorithm for computing Gröbner bases without reduction to zero (F5)." Pages 75-83 in: Marc Giusti (chair). Proceedings of the 2002 international symposium on symbolic and algebraic computation. ISBN 1-58113-484-3. ACM Press.

[84] Nicolas T. Courtois, Louis Goubin, Willi Meier, Jean-Daniel Tacier. "Solving underdefined systems of multivariate quadratic equations." Pages 211-227 in: David Naccache, Pascal Paillier (editors). Public key cryptography, proceedings of the 5th international workshop on

practice and theory in public key cryptosystems (PKC 2002) held in Paris, February 12-14, 2002. Lecture Notes in Computer Science 2274. Springer. ISBN 3-540-43168-3. MR 2005b:94044.

[85] Nicolas T. Courtois, Jacques Patarin. "About the XL algorithm over GF(2)." MR 2080135. Pages 141-157 in: Marc Joye (editor). Topics in cryptology—CT-RSA 2003, the cryptographers' track at the RSA conference 2003, San Francisco, CA, USA, April 13-17, 2003, proceedings. Lecture Notes in Computer Science 2612. Springer. ISBN 3-540-00847-0. MR 2005b:94045.

[86] Bo-Yin Yang, Jiun-Ming Chen. "Theoretical analysis of XL over small fields." Pages 277-288 in: Huaxiong Wang, Josef Pieprzyk, Vijay Varadharajan (editors). Information security and privacy, 9th Australasian conference, ACISP 2004, Sydney, Australia, July 13-15, 2004, proceedings. Lecture Notes in Computer Science 3108. Springer. ISBN 978-3-540-22379-5.

[87] Bo-Yin Yang, Jiun-Ming Chen, Nicolas T. Courtois. "On asymptotic security estimates in XL and Gröbner bases-related algebraic cryptanalysis." Pages 401-413 in: Javier López, Sihan Qing, Eiji Okamoto (editors). Information and communications security, 6th international conference, ICICS 2004, Malaga, Spain, October 27-29, 2004, proceedings. Lecture Notes in Computer Science 3269. Springer. ISBN 978-3-540-23563-7.

[88] Claus Diem. "The XL-algorithm and a conjecture from commutative algebra." MR 2006m:12011. Pages 323-337 in: Pil Joong Lee (editor). Advances in cryptology—ASIACRYPT 2004, proceedings of the 10th international conference on the theory and application of cryptology and information security held on Jeju Island, December 5-9, 2004. Lecture Notes in Computer Science 3329. Springer. ISBN 3-540-23975-8. MR 2006b:94042.

[89] Gwénolé Ars, Jean-Charles Faugère, Hideki Imai, Mitsuru Kawazoe, Makoto Sugita. "Comparison between XL and Gröbner Basis algorithms." Pages 338-353 in: Pil Joong Lee (editor). Advances in cryptology—ASIACRYPT 2004, proceedings of the 10th international conference on the theory and application of cryptology and information security held on Jeju Island, December 5-9, 2004. Lecture Notes in Computer Science 3329. Springer. ISBN 3-540-23975-8. MR 2006b:94042.

[90] Magali Bardet, Jean-Charles Faugère, Bruno Salvy. "On the complexity of Gröbner basis computation of semi-regular overdetermined algebraic equations." http://www-calfor.lip6.fr/ICPSS/papers/43BF/43bf.htm. Pages 71-74 in: Jean-Charles Faugère, Fabrice Rouillier (editors). Proceedings of the international conference on polynomial system solving.

[91] Bo-Yin Yang, Jiun-Ming Chen. "All in the XL family: theory and practice." MR 2006k:13060. Pages 67-86 in: Choonsik Park, Seongtaek Chee (editors). Information security and cryptology—ICISC 2004, revised selected papers from the 7th International Conference held in Seoul, December 2-3, 2004. Lecture Notes in Computer Science 3506. Springer. ISBN

3-540-26226-1. MR 2006j:94101.

[92] Magali Bardet, Jean-Charles Faugère, Bruno Salvy, Bo-Yin Yang. "Asymptotic expansion of the index of regularity of quadratic semi-regular polynomial systems." http://www-spiral.lip6.fr/~bardet/Publis/bardet_et_all_MEGA05.pdf.

[93] Bo-Yin Yang, Chia-Hsin Owen Chen, Jiun-Ming Chen. "The limit of XL implemented with sparse matrices." Pages 215-225 in: http://postquantum.cr.yp.to/pqcrypto 2006record.pdf

[94] Jintai Ding, Johannes Buchmann, Mohamed Saied Emam Mohamed, Wael Said Abd Elmageed Mohamed, Ralf-Philipp Weinmann. "MutantXL." http://www.cdc.informatik.tu-darmstadt.de/reports/reports/MutantXL_Algorithm.pdf.

[95] Mohamed Saied Emam Mohamed, Wael Said Abd Elmageed Mohamed, Jintai Ding, Johannes Buchmann. "Solving polynomial equations over GF(2) using an improved mutant strategy." Pages 203-215 in: Johannes Buchmann, Jintai Ding (editors). Post-quantum cryptography, second international workshop, PQCrypto 2008, Cincinnati, OH, USA, October 17-19, 2008, proceedings. Lecture Notes in Computer Science 5299, Springer.

[96] Jacques Patarin, Louis Goubin, Nicolas T. Courtois. "Improved algorithms for isomorphisms of polynomials." Pages 184-200 in: Kaisa Nyberg (editor), Advances in cryptology—EUROCRYPT '98, international conference on the theory and application of cryptographic techniques, Espoo, Finland, May 31-June 4, 1998, proceedings. Lecture Notes in Computer Science 1403. Springer. ISBN 978-3-540-64518-4.

[97] Willi Geiselmann, Willi Meier, Rainer Steinwandt. "An attack on the Isomorphisms of Polynomials problem with one secret." http://eprint.iacr.org/2002/143.

[98] Françoise Levy-dit-Vehel, Ludovic Perret. "Polynomial equivalence problems and applications to multivariate cryptosystems." MR 2005e:94175. Pages 235-251 in: Thomas Johansson, Subhamoy Maitra (editors). Proceedings in cryptology—INDOCRYPT 2003. Proceedings of the 4th international conference on cryptology in India held in New Delhi, December 8-10, 2003. Lecture Notes in Computer Science 2904. Springer. ISBN 3-540-20609-4. MR 2005d:94154.

[99] Ludovic Perret. "A fast cryptanalysis of the isomorphism of polynomials with one secret problem." Pages 354-370 in: Ronald Cramer (editor). Advances in cryptology—EUROCRYPT 2005. Proceedings of the 24th annual international conference on the theory and applications of cryptographic techniques held in Aarhus, May 22-26, 2005. Lecture Notes in Computer Science 3494. Springer. ISBN 3-540-25910-4. MR 2008e:94035.

[100] Jean-Charles Faugère, Ludovic Perret. "Polynomial equivalence problems: algorithmic and theoretical aspects." Pages 30-47 in: Serge Vaudenay (editor). Advances in Cryptology—EUROCRYPT 2006, 25th annual international conference on the theory and

applications of cryptographic techniques, St. Petersburg, Russia, May 28-June 1, 2006, proceedings. Lecture Notes in Computer Science 4004. Springer. ISBN 3-540-34546-9.

[101] Jean-Charles Faugère, Ludovic Perret. "Cryptanalysis of 2R^- schemes." MR 2422172. Pages 357-372 in: Cynthia Dwork (editor). Advances in cryptology—CRYPTO 2006. Proceedings of the 26th annual international cryptology conference held in Santa Barbara, CA, August 20-24, 2006. Lecture Notes in Computer Science 4117. Springer. ISBN 978-3-540-37432-9. MR 2422188.

[102] H. Ong, Claus-Peter Schnorr. "Signatures through approximate representation by quadratic forms." Pages 117-131 in: David Chaum (editor). Advances in cryptology, proceedings of CRYPTO '83. Plenum Press.

[103] H. Ong, Claus-Peter Schnorr, Adi Shamir. "Efficient signature schemes based on polynomial equations." Pages 37-46 in: G. R. Blakley, David Chaum (editors). Advances in cryptology, proceedings of CRYPTO '84, Santa Barbara, California, USA, August 19-22, 1984, proceedings. Lecture Notes in Computer Science 196. Springer. ISBN 3-540-15658-5.

[104] Harriet J. Fell, Whitfield Diffie. "Analysis of a public key approach based on polynomial substitution." Pages 340-349 in: Hugh C. Williams (editor). Advances in cryptology—CRYPTO '85, Santa Barbara, California, USA, August 18-22, 1985, proceedings.

[105] John M. Pollard, Claus-Peter Schnorr. "An efficient solution of the congruence $x^2 + ky^2 = m \pmod{n}$." IEEE Transactions on Information Theory 33, 702-709.

[106] Tsutomu Matsumoto, Hideki Imai. "Public quadratic polynomial-tuples for efficient signature-verification and message-encryption." MR 90d:94008. Pages 419-453 in: Christoph G. Günther (editor). Advances in cryptology—EUROCRYPT 1988, proceedings of the workshop on the theory and application of cryptographic techniques held in Davos, May 25-27, 1988. Lecture Notes in Computer Science 330. ISBN 3-540-50251-3. MR 90a:94002.

[107] Jacques Patarin. "Asymmetric cryptography with a hidden monomial and a candidate algorithm for =~ 64 bits asymmetric signatures." MR 99b:94040. Pages 45-60 in: Neal Koblitz (editor). Advances in cryptology—CRYPTO '96, proceedings of the 16th annual international cryptology conference held at the University of California, Santa Barbara, CA, August 18-22, 1996. Lecture Notes in Computer Science 1109. Springer. ISBN 3-540-61512-1. MR 98f:94001.

[108] Jacques Patarin. "Hidden Fields Equations (HFE) and Isomorphisms of Polynomials (IP): two new families of asymmetric algorithms." Pages 33-48 in: Ueli Maurer (editor). Advances in Cryptology—EUROCRYPT '96. International conference on the theory and application of cryptographic techniques, Saragossa, Spain, May 12-16, 1996, proceedings. Lecture Notes in Computer Science 1070. Springer. ISBN 978-3-540-61186-8.

[109] Jacques Patarin, Louis Goubin, Nicolas T. Courtois. "C*-+ and HM: variations around two schemes of T. Matsumoto and H. Imai." Pages 35-49 in: Kazuo Ohta, Dingyi Pei (editors). Advances in cryptology—ASIACRYPT'98. Proceedings of the International Conference on the Theory and Application of Cryptology and Information Security held in Beijing, October 18-22, 1998. Lecture Notes in Computer Science 1514. Springer. ISBN 3-540-65109-8.

[110] Aviad Kipnis, Adi Shamir. "Cryptanalysis of the HFE public key cryptosystem by linearization." Pages 19-30 in: Michael J. Wiener (editor). Advances in cryptology—CRYPTO '99, 19th annual international cryptology conference, Santa Barbara, California, USA, August 15-19, proceedings. Lecture Notes in Computer Science 1666. Springer. ISBN 3-540-66347-9.

[111] Jacques Patarin. "Cryptanalysis of the Matsumoto and Imai public key scheme of Eurocrypt '88." Designs, Codes and Cryptography 20, 175-209. Earlier version: 1995. MR 98d:94022. Pages 248-261 in: Don Coppersmith (editor). Advances in Cryptology—CRYPTO 1995, proceedings of the 15th annual international cryptology conference held at the University of California, Santa Barbara, CA, August 27-31, 1995. Lecture Notes in Computer Science 963. Springer. ISBN 3-540-60221-6. MR 97k:94002.

[112] Nicolas T. Courtois. "The security of hidden field equations (HFE)." MR 1907103. Pages 266-281 in: David Naccache (editor). Topics in cryptology—CT-RSA 2001, proceedings of the cryptographers' track at the RSA conference held in San Francisco, CA, April 8-12, 2001. Lecture Notes in Computer Science 2020. Springer. ISBN 3-540-41898-9. MR 2003a:94039.

[113] Jacques Patarin, Nicolas T. Courtois, Louis Goubin. "QUARTZ, 128-bit long digital signatures." MR 1907104. Pages 282-297 in: David Naccache (editor). Topics in cryptology—CT-RSA 2001, proceedings of the cryptographers' track at the RSA conference held in San Francisco, CA, April 8-12, 2001. Lecture Notes in Computer Science 2020. Springer. ISBN 3-540-41898-9. MR 2003a:94039.

[114] Jacques Patarin, Nicolas T. Courtois, Louis Goubin. "FLASH, a fast multivariate signature algorithm." MR 1907105. Pages 298-307 in: David Naccache (editor). Topics in cryptology—CT-RSA 2001, proceedings of the cryptographers' track at the RSA conference held in San Francisco, CA, April 8-12, 2001. Lecture Notes in Computer Science 2020. Springer. ISBN 3-540-41898-9. MR 2003a:94039.

[115] Nicolas T. Courtois, Magnus Daum, Patrick Felke. "On the security of HFE, HFEv- and Quartz." MR 2007g:94044. Pages 337-350 in: Yvo G. Desmedt (editor). Public key cryptography—PKC 2003, proceedings of the 6th international workshop on practice and theory in public key cryptography held in Miami, FL, January 6-8, 2003. Lecture Notes in Computer Science 2567. Springer. ISBN 3-540-00324-X. MR 2006d:94071.

[116] Nicolas T. Courtois. "Generic attacks and the security of Quartz." MR 2006i:94045. Pages 351-364 in: Yvo G. Desmedt (editor). Public key cryptography—PKC 2003, proceedings of the 6th international workshop on practice and theory in public key cryptography held in Miami, FL, January 6-8, 2003. Lecture Notes in Computer Science 2567. Springer. ISBN 3-540-00324-X. MR 2006d:94071.

[117] Jean-Charles Faugère., Antoine Joux. "Algebraic cryptanalysis of Hidden Field Equations (HFE) using Gröbner bases." MR 2005e:94140. Pages 44-60 in: Dan Boneh (editor). Advances in cryptology—CRYPTO 2003, proceedings of the 23rd annual international cryptology conference held in Santa Barbara, CA, August 17-21, 2003. Lecture Notes in Computer Science 2729. Springer. ISBN 3-540-40674-3. MR 2005d:94151.

[118] Jintai Ding, Dieter Schmidt. "Cryptanalysis of SFlash v3." http://eprint.iacr.org/2004/103.

[119] Nicolas T. Courtois. "Algebraic attacks over GF(2^k), application to HFE Challenge 2 and Sflash-v2." MR 2005e:94132. Pages 201-217 in: Feng Bao, Robert Deng, Jianying Zhou (editors). Public key cryptography—PKC 2004, proceedings of the 7th international workshop on theory and practice in public key cryptography held in Singapore, March 1-4, 2004. Lecture Notes in Computer Science 2947. Springer. ISBN 3-540-21018-0. MR 2005d:94155.

[120] Jintai Ding. "A new variant of the Matsumoto-Imai cryptosystem through perturbation." Pages 305-318 in: Feng Bao, Robert Deng, Jianying Zhou (editors). Public key cryptography—PKC 2004, proceedings of the 7th international workshop on theory and practice in public key cryptography held in Singapore, March 1-4, 2004. Lecture Notes in Computer Science 2947. Springer. ISBN 3-540-21018-0. MR 2005d:94155.

[121] Jintai Ding, Jason E. Gower, Dieter Schmidt, Christopher Wolf, Zhijun Yin. "Complexity estimates for the F_4 attack on the perturbed Matsumoto-Imai cryptosystem." MR 2007f:94036. Pages 262-277 in: Nigel P. Smart (editor). Cryptography and coding, 10th IMA international conference, Cirencester, UK, December 19-21, 2005, proceedings. Lecture Notes in Computer Science 3796. Springer. ISBN 3-540-30276-X. MR 2006m:94081.

[122] Jintai Ding, Dieter Schmidt. "Cryptanalysis of HFEv and internal perturbation of HFE." MR 2006j:94061. Pages 288-301 in: Serge Vaudenay (editor). Public key cryptography—PKC 2005: proceedings of the 8th international workshop on theory and practice in public key cryptography held in Les Diablerets, January 23-26, 2005. Lecture Notes in Computer Science 3386. Springer. ISBN 3-540-24454-9. MR 2006d:94072.

[123] Pierre-Alain Fouque, Louis Granboulan, Jacques Stern. "Differential cryptanalysis for multivariate schemes." Pages 341-353 in: Ronald Cramer (editor). Advances in cryptology—EUROCRYPT 2005. Proceedings of the 24th annual international conference on

the theory and applications of cryptographic techniques held in Aarhus, May 22-26, 2005. Lecture Notes in Computer Science 3494. Springer. ISBN 3-540-25910-4. MR 2008e:94035.

[124] Vivien Dubois, Louis Granboulan, Jacques Stern. "An efficient provable distinguisher for HFE." MR 2008b:94065. Pages 156-167 in: Michele Bugliesi, Bart Preneel, Vladimiro Sassone, Ingo Wegener (editors). Automata, languages and programming. Part II. Proceedings of the 33rd International Colloquium (ICALP 2006) held in Venice, July 10-14, 2006. Lecture Notes in Computer Science 4052. Springer. ISBN 978-3-540-35907-4.

[125] Ding J, Hodges T J. Inverting HFE Systems Is Quasi-Polynomial for All Fields[A]. In: Advances in Cryptology-CRYPTO 2011[C]. Springer, 2011: 724-742.

[126] Aline Gouget, Jacques Patarin. "Probabilistic multivariate cryptography." Pages 1-18 in: Phong Q. Nguyen (editor). Progress in cryptology—VIETCRYPT 2006, first international conference on cryptology in Vietnam, Hanoi, Vietnam, September 25-28, 2006, revised selected papers. Lecture Notes in Computer Science 4341. Springer. ISBN 3-540-68799-8.

[127] Jintai Ding, Jason E. Gower. "Inoculating multivariate schemes against differential attacks." MR 2423196. Pages 290-301 in: Moti Yung, Yevgeniy Dodis, Aggelos Kiayias, Tal Malkin (editors). Public key cryptography—PKC 2006. Proceedings of the 9th International Conference on Theory and Practice of Public-Key Cryptography held in New York, April 24-26, 2006. Lecture Notes in Computer Science 3958. Springer. MR 2009a:94034.

[128] Adama Diene, Jintai Ding, Jason E. Gower, Timothy J. Hodges, Zhijun Yin. "Dimension of the linearization equations of the Matsumoto-Imai cryptosystems." MR 2423694. Pages 242-251 in: Oyvind Ytrehus, Coding and cryptography, revised selected papers from the international workshop (WCC 2005) held in Bergen, March 14-18, 2005. Lecture Notes in Computer Science 3969. Springer. ISBN 3-540-35481-6. MR 2428233.

[129] Lih-Chung Wang, Fei-Hwang Chang. "Revision of tractable rational map cryptosystem." Version 20061227:165851. http://eprint.iacr.org/2004/046.

[130] Lih-Chung Wang, Bo-Yin Yang, Yuh-Hua Hu, Feipei Lai. "A 'medium-field' multivariate public-key encryption scheme." Pages 132-149 in: David Pointcheval (editor). Topics in cryptology—CT-RSA 2006, proceedings of the cryptographers' track at the RSA conference held in San Jose, CA, February 13-17, 2006. Lecture Notes in Computer Science 3860. Springer. ISBN 3-540-31033-9. MR 2007b:94003.

[131] Jintai Ding, Christopher Wolf, Bo-Yin Yang. "l-invertible cycles for Multivariate Quadratic (MQ) public key cryptography." MR 2404125. Pages 266-281 in: Tatsuaki Okamoto, Xiaoyun Wang (editors). Public key cryptography—PKC 2007, proceedings of the 10th international conference on practice and theory in public-key cryptography held at Tsinghua University, Beijing, April 16-20, 2007. Lecture Notes in Computer Science 4450. Springer.

ISBN 3-540-71676-9. MR 2404107.

[132] Jintai Ding, Bo-Yin Yang, Chen-Mou Cheng, Owen Chen, Vivien Dubois. "Breaking the symmetry: a way to resist the new differential attack." http://eprint.iacr.org/ 2007/366.

[133] Vivien Dubois, Louis Granboulan, Jacques Stern. "Cryptanalysis of HFE with internal perturbation." MR 2404124. Pages 249-265 in: Tatsuaki Okamoto, Xiaoyun Wang (editors). Public key cryptography—PKC 2007, proceedings of the 10th international conference on practice and theory in public-key cryptography held at Tsinghua University, Beijing, April 16-20, 2007. Lecture Notes in Computer Science 4450. Springer. ISBN 3-540-71676-9. MR 2404107.

[134] Vivien Dubois, Pierre-Alain Fouque, Jacques Stern. "Cryptanalysis of SFLASH with slightly modified parameters." Pages 264-275 in: Moni Naor (editor). Advances in Cryptology—EUROCRYPT 2007. 26th annual international conference on the theory and applications of cryptographic techniques, Barcelona, Spain, May 20-24, 2007, proceedings. Lecture Notes in Computer Science 4515. Springer. ISBN 978-3-540-72539-8.

[135] Vivien Dubois, Pierre-Alain Fouque, Adi Shamir, Jacques Stern. "Practical cryptanalysis of SFLASH." MR 2419591. Pages 1-12 in: Alfred Menezes (editor). Advances in cryptology—CRYPTO 2007, proceedings of the 27th annual international cryptology conference held in Santa Barbara, CA, August 19-23, 2007. Lecture Notes in Computer Science 4622. Springer. ISBN 978-3-540-74142-8.

[136] Cao W, Hu L, Ding J, et al. Kipnis-Shamir Attack on Unbalanced Oil-Vinegar Scheme[A]. In: Information Security Practice and Experience[C]. Springer, 2011: 168-180.

[137] Jintai Ding, Dieter Schmidt, Fabian Werner. "Algebraic attack on HFE revisited." Pages 215-227 in: Tzong-Chen Wu, Chin-Laung Lei, Vincent Rijmen, Der-Tsai Lee (editors). Information security: 11th international conference, ISC 2008, Taipei, Taiwan, September 15-18, 2008, proceedings. Lecture Notes in Computer Science 5222. Springer. ISBN 978-3-540-85884-3.

[138] Pierre-Alain Fouque, Gilles Macario-Rat, Ludovic Perret, Jacques Stern. "Total break of the l-IC signature scheme." Pages 1-17 in: Ronald Cramer (editor). Public key cryptography—PKC 2008, 11th international workshop on practice and theory in public-key cryptography, Barcelona, Spain, March 9-12, 2008, proceedings. Lecture Notes in Computer Science 4939. Springer. ISBN 978-3-540-78439-5.

[139] Shigeo Tsujii, Toshiya Itoh, Atsushi Fujioka, Kaoru Kurosawa, Tsutomu Matsumoto. "A public-key cryptosystem based on the difficulty of solving a system of nonlinear equations." Systems and Computers in Japan 19, 10-18.

[140] Adi Shamir. "Efficient signature schemes based on birational permutations." Pages 1-12

in: Douglas R. Stinson (editor). Advances in Cryptology—CRYPTO '93, 13th annual international cryptology conference, Santa Barbara, California, USA, August 22-26, 1993, proceedings. Lecture Notes in Computer Science 773. Springer. ISBN 3-540-57766-1.

[141] Don Coppersmith, Jacques Stern, Serge Vaudenay. "The security of the birational permutation signature schemes." Journal of Cryptology 10, 207-221. MR 99e:94033.

[142] Aviad Kipnis, Adi Shamir. "Cryptanalysis of the oil and vinegar signature scheme." Pages 257-266 in: Hugo Krawczyk (editor). Advances in cryptology—CRYPTO '98, 18th annual international cryptology conference, Santa Barbara, California, USA, August 23-27, 1998, proceedings. Lecture Notes in Computer Science 1462. Springer. ISBN 3-540-64892-5.

[143] Aviad Kipnis, Jacques Patarin, Louis Goubin. "Unbalanced oil and vinegar signature schemes." MR 1717470. Pages 206-222 in: Jacques Stern (editor). Advances in cryptology—EUROCRYPT '99, proceedings of the 17th international conference on the theory and application of cryptographic techniques held in Prague, May 2-6, 1999. Lecture Notes in Computer Science 1592. Springer. ISBN 3-540-65889-0. MR 2000i:94001.

[144] T. Moh. "A public key system with signature and master key functions." Communications in Algebra 27, 2207-2222.

[145] Louis Goubin, Nicolas T. Courtois. "Cryptanalysis of the TTM cryptosystem." MR 2002j:94037. Pages 44-57 in: Tatsuaki Okamoto (editor). Advances in cryptology— ASIACRYPT 2000, proceedings of the 6th annual international conference on the theory and application of cryptology and information security held in Kyoto, December 3-7, 2000. Lecture Notes in Computer Science 1976. Springer. ISBN 3-540-41404-5. MR 2002d:94046.

[146] T. Moh, Jiun-Ming Chen. "On the Goubin-Courtois attack on TTM." http://eprint.iacr.org/2001/072.

[147] Jiun-Ming Chen, Bo-Yin Yang. "A more secure and efficacious TTS signature scheme." Pages 320-338 in: Lecture Notes in Computer Science 2971. Springer. ISBN 3-540-21376-7. MR 2005d:94153.

[148] Jintai Ding, Dieter Schmidt. "The new implementation schemes of the TTM cryptosystem are not secure." MR 2005d:94100. Pages 113-127 in: Keqin Feng, Harald Niederreiter, Chaoping Xing (editors). Coding, cryptography and combinatorics. Progress in Computer Science and Applied Logic 23. Birkhauser. ISBN 3-7643-2429-5.

[149] Jintai Ding, Timothy Hodges. "Cryptanalysis of an implementation scheme of the tamed transformation method cryptosystem." Journal of Algebra and its Applications 3, 273-282. MR 2005f:94093.

[150] Masao Kasahara, Ryuichi Sakai. "A construction of public key cryptosystem for realizing ciphertext of size 100 bit and digital signature scheme." IEICE Transactions on

Fundamentals 87-A, 102-109. http://search.ieice.org/bin/summary.php?id=e87-a_1_102&category=D&year=2004&lang=E&abst=.

[151] Bo-Yin Yang, Jiun-Ming Chen, Yen-Hung Chen. "TTS: high-speed signatures on a low-cost smart card." Pages 371-385 in: Marc Joye, Jean-Jacques Quisquater (editors). Cryptographic hardware and embedded systems—CHES 2004, 6th international workshop, Cambridge, MA, USA, August 11-13, 2004, proceedings. Lecture Notes in Computer Science 3156. Springer. ISBN 3-540-22666-4.

[152] Lih-Chung Wang, Fei-Hwang Chang. "Tractable rational map cryptosystem." Version 20040221:212731. http://eprint.iacr.org/2004/046.

[153] Bo-Yin Yang, Jiun-Ming Chen. "TTS: rank attacks in tame-like multivariate PKCs." http://eprint.iacr.org/2004/061.

[154] Masao Kasahara, Ryuichi Sakai. "A construction of public-key cryptosystem based on singular simultaneous equations." IEICE Transactions on Fundamentals 88-A, 74-80. http://search.ieice.org/bin/summary.php?id=e88-a_1_74&category=D&year=2005&lang=E&abst=.

[155] Christopher Wolf, Bart Preneel. "Large superfluous keys in multivariate quadratic asymmetric systems." Pages 275-287 in: Serge Vaudenay (editor). Public key cryptography—PKC 2005: proceedings of the 8th international workshop on theory and practice in public key cryptography held in Les Diablerets, January 23-26, 2005. Lecture Notes in Computer Science 3386. Springer. ISBN 3-540-24454-9. MR 2006d:94072.

[156] An Braeken, Christopher Wolf, Bart Preneel. "A study of the security of unbalanced oil and vinegar signature schemes." MR 2006h:94169. Pages 29-43 in: Alfred Menezes (editor). Topics in cryptology—CT-RSA 2005. Proceedings of the cryptographers' track at the RSA conference held in San Francisco, CA, February 14-18, 2005. Lecture Notes in Computer Science 3376. Springer. MR 2006d:94073. ISBN 3-540-24399-2.

[157] Lih-Chung Wang, Yuh-Hua Hu, Feipei Lai, Chun-yen Chou, Bo-Yin Yang. "Tractable rational map signature." Pages 244-257 in: Serge Vaudenay (editor). Public key cryptography—PKC 2005: proceedings of the 8th international workshop on theory and practice in public key cryptography held in Les Diablerets, January 23-26, 2005. Lecture Notes in Computer Science 3386. Springer. ISBN 3-540-24454-9. MR 2006d:94072.

[158] Bo-Yin Yang, Jiun-Ming Chen. "Building secure tame-like multivariate public-key cryptosystems: the new TTS." Pages 518-531 in: Colin Boyd, Juan Manuel González Nieto (editors). Information security and privacy, 10th Australasian conference, ACISP 2005, Brisbane, Australia, July 4-6, 2005, proceedings. Lecture Notes in Computer Science 3574. Springer. ISBn 3-540-26547-3.

[159] Jintai Ding, Dieter Schmidt. "Rainbow, a new multivariable polynomial signature

scheme." Pages 164-175 in: John Ioannidis, Angelos D. Keromytis, Moti Yung (editors). Applied cryptography and network security, third international conference, ACNS 2005, New York, NY, USA, June 7-10, 2005, proceedings. Lecture Notes in Computer Science 3531. Springer. ISBN 3-540-26223-7.

[160] Olivier Billet, Henri Gilbert. "Cryptanalysis of Rainbow." Pages 336-347 in: Roberto De Prisco, Moti Yung (editors). Security and cryptography for networks, 5th international conference, SCN 2006, Maiori, Italy, September 6-8, 2006, proceedings.. Lecture Notes in Computer Science 4116. Springer. ISBN 3-540-38080-9.

[161] Xuyun Nie, Lei Hu, Jianyu Li, Crystal Updegrove, Jintai Ding. "Breaking a new instance of TTM cryptosystems." Pages 210-225 in: Jianying Zhou, Moti Yung, Feng Bao (editors). Applied cryptography and network security, 4th international conference, ACNS 2006, Singapore, June 6-9, 2006, proceedings. Lecture Notes in Computer Science 3989. Springer. ISBN 3-540-34703-8.

[162] Christopher Wolf, An Braeken, Bart Preneel. "On the security of stepwise triangular systems." Designs, Codes and Cryptography 40, 285-302. Older version: 2004. "Efficient cryptanalysis of RSE(2)PKC and RSSE(2)PKC." Pages 294-309 in: SCN 2004. Lecture Notes in Computer Science 3352. Springer.

[163] Jintai Ding, Dieter Schmidt, Zhijun Yin. "Cryptanalysis of the new TTS scheme in CHES 2004." International Journal of Information Security 5, 231-240.

[164] T. Moh. "The recent attack of Nie et al on TTM is faulty." http://eprint.iacr.org/ 2006/417

[165] Jintai Ding, Lei Hu, Xuyun Nie, Jianyu Li, John Wanger. "High order linearization equation (HOLE) attack on multivariate public key cryptosystems." MR 2404123. Pages 230-247 in: Tatsuaki Okamoto, Xiaoyun Wang (editors). Public key cryptography—PKC 2007, proceedings of the 10th international conference on practice and theory in public-key cryptography held at Tsinghua University, Beijing, April 16-20, 2007. Lecture Notes in Computer Science 4450. Springer. ISBN 3-540-71676-9. MR 2404107.

[166] Xuyun Nie, Lei Hu, Jintai Ding, Jianyu Li, John Wagner. "Cryptanalysis of the TRMC-4 public key cryptosystem." Pages 104-115 in: Jonathan Katz, Moti Yung (editors). Applied Cryptography and Network Security, 5th international conference, ACNS 2007, Zhuhai, China, June 5-8, 2007, proceedings. Lecture Notes in Computer Science 4521. Springer. ISBN 978-3-540-72737-8.

[167] T. Moh. "Two new examples of TTM." http://eprint.iacr.org/2007/144.

[168] Jintai Ding, Vivien Dubois, Bo-Yin Yang, Chia-Hsin Owen Chen, Chen-Mou Cheng. "Could SFLASH be repaired?" Pages 691-701 in: Luca Aceto, Ivan Damgård, Leslie Ann Goldberg, Magnús M. Halldórsson, Anna Ingólfsdóttir, Igor Walukiewicz (editors). Automata,

languages and programming, 35th international colloquium, ICALP 2008, Reykjavik, Iceland, July 7-11, 2008, proceedings, part II, track B: logic, semantics, and theory of programming; track C: security and cryptography foundations. Lecture Notes in Computer Science 5126. Springer. ISBN 978-3-540-70582-6.

[169] Jintai Ding, Bo-Yin Yang, Chia-Hsin Owen Chen, Ming-Shing Chen, Chen-Mou Cheng. "New differential-algebraic attacks and reparametrization of Rainbow." Pages 242-257 in: Steven M. Bellovin, Rosario Gennaro, Angelos D. Keromytis, Moti Yung (editors). Applied cryptography and network security, 6th international conference, ACNS 2008, New York, NY, USA, June 3-6, 2008, proceedings. Lecture Notes in Computer Science 5037. Springer. ISBN 978-3-540-68913-3.

[170] Jean-Charles Faugère, Françoise Levy-dit-Vehel, Ludovic Perret. "Cryptanalysis of MinRank." Pages 280-296 in: David Wagner (editor). Advances in cryptology—CRYPTO 2008, 28th annual international cryptology conference, Santa Barbara, CA, USA, August 17-21, 2008, proceedings. Lecture Notes in Computer Science 5157. Springer. ISBN 978-3-540-85173-8.

[171] Luk Bettale, Jean-Charles Faugère, Ludovic Perret: "Cryptanalysis of the TRMS signature scheme of PKC'05." Pages 143-155 in: Serge Vaudenay (editor). Progress in cryptology—AFRICACRYPT 2008, first international conference on cryptology in Africa, Casablanca, Morocco, June 11-14, 2008, proceedings. Lecture Notes in Computer Science 5023. Springer. ISBN 978-3-540-68159-5.

[172] Jintai Ding, John Wagner. "Cryptanalysis of rational multivariate public key cryptosystems." Pages 124-136 in: Johannes Buchmann, Jintai Ding (editors). Post-quantum cryptography, second international workshop, PQCrypto 2008, Cincinnati, OH, USA, October 17-19, 2008, proceedings. Lecture Notes in Computer Science 5299, Springer.

[173] John Baena, Crystal Clough, Jintai Ding. "Square-vinegar signature scheme." Pages 17-30 in: Johannes Buchmann, Jintai Ding (editors). Post-quantum cryptography, second international workshop, PQCrypto 2008, Cincinnati, OH, USA, October 17-19, 2008, proceedings. Lecture Notes in Computer Science 5299, Springer.

[174] Mehdi-Laurent Akkar, Nicolas T. Courtois, Romain Duteuil, Louis Goubin. "A fast and secure implementation of Sflash." MR 2006i:94034. Pages 267-278 in: Yvo G. Desmedt (editor). Public key cryptography—PKC 2003, proceedings of the 6th international workshop on practice and theory in public key cryptography held in Miami, FL, January 6-8, 2003. Lecture Notes in Computer Science 2567. Springer. ISBN 3-540-00324-X. MR 2006d:94071.
2003. Christopher Wolf. "Efficient public key generation for multivariate cryptosystems." http://eprint.iacr.org/2003/089.

[175] Bo-Yin Yang, Chen-Mou Cheng, Bor-Rong Chen, Jiun-Ming Chen. "Implementing minimized multivariate PKC on low-resource embedded systems." Pages 73-88 in: John A. Clark, Richard F. Paige, Fiona Polack, Phillip J. Brooke (editors). Security in pervasive computing, third international conference, SPC 2006, York, UK, April 18-21, 2006, proceedings. Lecture Notes in Computer Science 3934. Springer. ISBN 3-540-33376-2.

[176] Côme Berbain, Olivier Billet, Henri Gilbert. "Efficient implementations of multivariate quadratic systems." Pages 174-187 in: Eli Biham, Amr M. Youssef (editors). Selected areas in cryptography, 13th international workshop, SAC 2006, Montreal, Canada, August 17-18, 2006, revised selected papers. Lecture Notes in Computer Science 4356. Springer. ISBN 978-3-540-74461-0.

[177] Sundar Balasubramanian, Andrey Bogdanov, Andy Rupp, Jintai Ding, Harold W. Carter. "Fast multivariate signature generation in hardware: the case of Rainbow." ASAP 2008. IEEE.

[178] Bruno Buchberger. "Ein Algorithmus zum Auffinden der Basiselemente des Restklassenringes nach einem nulldimensionalen Polynomideal." Ph.D. thesis, University of Innsbruck.

[179] Daniel Lazard. "Gröbner-bases, Gaussian elimination and resolution of systems of algebraic equations." MR 86m:13002. Pages 146-156 in: J. A. van Hulzen (editor). Computer algebra: proceedings of the European computer algebra conference (EUROCAL) held in London, March 28-30, 1983. Lecture Notes in Computer Science 162. Springer. ISBN 3-540-12868-9. MR 86f:68004.

[180] Jean-Charles Faugère, Patrizia M. Gianni, Daniel Lazard, Teo Mora. "Efficient computation of zero-dimensional Gröbner bases by change of ordering." Journal of Symbolic Computation 16, 329-344.

[181] Jean-Charles Faugère. "A new efficient algorithm for computing Gröbner bases (F4)." Journal of Pure and Applied Algebra 139, 61-88.

[182] Nicolas T. Courtois, Alexander Klimov, Jacques Patarin, Adi Shamir. "Efficient algorithms for solving overdefined systems of multivariate polynomial equations." MR 1772028. Pages 392-407 in: Bart Preneel (editor). Advances in cryptology—EUROCRYPT 2000, proceedings of the 19th international annual conference on the theory and application of cryptographic techniques held in Bruges, May 14-18, 2000. Lecture Notes in Computer Science 1807. Springer. ISBN 3-540-67517-5. MR 2001b:94028.

[183] Jean-Charles Faugère. "A new efficient algorithm for computing Gröbner bases without reduction to zero (F5)." Pages 75-83 in: Marc Giusti (chair). Proceedings of the 2002 international symposium on symbolic and algebraic computation. ISBN 1-58113-484-3. ACM Press.

[184] Nicolas T. Courtois, Louis Goubin, Willi Meier, Jean-Daniel Tacier. "Solving underdefined systems of multivariate quadratic equations." Pages 211-227 in: David Naccache, Pascal Paillier (editors). Public key cryptography, proceedings of the 5th international workshop on practice and theory in public key cryptosystems (PKC 2002) held in Paris, February 12-14, 2002. Lecture Notes in Computer Science 2274. Springer. ISBN 3-540-43168-3. MR 2005b:94044.

[185] Nicolas T. Courtois, Jacques Patarin. "About the XL algorithm over GF(2)." MR 2080135. Pages 141-157 in: Marc Joye (editor). Topics in cryptology—CT-RSA 2003, the cryptographers' track at the RSA conference 2003, San Francisco, CA, USA, April 13-17, 2003, proceedings. Lecture Notes in Computer Science 2612. Springer. ISBN 3-540-00847-0. MR 2005b:94045.

[186] Bo-Yin Yang, Jiun-Ming Chen. "Theoretical analysis of XL over small fields." Pages 277-288 in: Huaxiong Wang, Josef Pieprzyk, Vijay Varadharajan (editors). Information security and privacy, 9th Australasian conference, ACISP 2004, Sydney, Australia, July 13-15, 2004, proceedings. Lecture Notes in Computer Science 3108. Springer. ISBN 978-3-540-22379-5.

[187] Bo-Yin Yang, Jiun-Ming Chen, Nicolas T. Courtois. "On asymptotic security estimates in XL and Gröbner bases-related algebraic cryptanalysis." Pages 401-413 in: Javier López, Sihan Qing, Eiji Okamoto (editors). Information and communications security, 6th international conference, ICICS 2004, Malaga, Spain, October 27-29, 2004, proceedings. Lecture Notes in Computer Science 3269. Springer. ISBN 978-3-540-23563-7.

[188] Claus Diem. "The XL-algorithm and a conjecture from commutative algebra." MR 2006m:12011. Pages 323-337 in: Pil Joong Lee (editor). Advances in cryptology—ASIACRYPT 2004, proceedings of the 10th international conference on the theory and application of cryptology and information security held on Jeju Island, December 5-9, 2004. Lecture Notes in Computer Science 3329. Springer. ISBN 3-540-23975-8. MR 2006b:94042.

[189] Gwénolé Ars, Jean-Charles Faugère, Hideki Imai, Mitsuru Kawazoe, Makoto Sugita. "Comparison between XL and Gröbner Basis algorithms." Pages 338-353 in: Pil Joong Lee (editor). Advances in cryptology—ASIACRYPT 2004, proceedings of the 10th international conference on the theory and application of cryptology and information security held on Jeju Island, December 5-9, 2004. Lecture Notes in Computer Science 3329. Springer. ISBN 3-540-23975-8. MR 2006b:94042.

[190] Magali Bardet, Jean-Charles Faugère, Bruno Salvy. "On the complexity of Gröbner basis computation of semi-regular overdetermined algebraic equations." http://www-calfor.lip6.fr/ICPSS/papers/43BF/43bf.htm. Pages 71-74 in: Jean-Charles Faugère, Fabrice Rouillier

(editors). Proceedings of the international conference on polynomial system solving.

[191] Bo-Yin Yang, Jiun-Ming Chen. "All in the XL family: theory and practice." MR 2006k:13060. Pages 67-86 in: Choonsik Park, Seongtaek Chee (editors). Information security and cryptology—ICISC 2004, revised selected papers from the 7th International Conference held in Seoul, December 2-3, 2004. Lecture Notes in Computer Science 3506. Springer. ISBN 3-540-26226-1. MR 2006j:94101.

[192] Magali Bardet, Jean-Charles Faugère, Bruno Salvy, Bo-Yin Yang. "Asymptotic expansion of the index of regularity of quadratic semi-regular polynomial systems." http://www-spiral.lip6.fr/~bardet/Publis/bardet_et_all_MEGA05.pdf

[193] Bo-Yin Yang, Chia-Hsin Owen Chen, Jiun-Ming Chen. "The limit of XL implemented with sparse matrices." Pages 215-225 in: http://postquantum.cr.yp.to/pqcrypto 2006record.pdf

[194] Jintai Ding, Johannes Buchmann, Mohamed Saied Emam Mohamed, Wael Said Abd Elmageed Mohamed, Ralf-Philipp Weinmann. "MutantXL." http://www.cdc.informatik.tu-darmstadt.de/reports/reports/MutantXL_Algorithm.pdf.

[195] Mohamed Saied Emam Mohamed, Wael Said Abd Elmageed Mohamed, Jintai Ding, Johannes Buchmann. "Solving polynomial equations over GF(2) using an improved mutant strategy." Pages 203-215 in: Johannes Buchmann, Jintai Ding (editors). Post-quantum cryptography, second international workshop, PQCrypto 2008, Cincinnati, OH, USA, October 17-19, 2008, proceedings. Lecture Notes in Computer Science 5299, Springer.

[196] Jacques Patarin, Louis Goubin, Nicolas T. Courtois. "Improved algorithms for isomorphisms of polynomials." Pages 184-200 in: Kaisa Nyberg (editor), Advances in cryptology—EUROCRYPT '98, international conference on the theory and application of cryptographic techniques, Espoo, Finland, May 31-June 4, 1998, proceedings. Lecture Notes in Computer Science 1403. Springer. ISBN 978-3-540-64518-4.

[197] Willi Geiselmann, Willi Meier, Rainer Steinwandt. "An attack on the Isomorphisms of Polynomials problem with one secret." http://eprint.iacr.org/2002/143.

[198] Françoise Levy-dit-Vehel, Ludovic Perret. "Polynomial equivalence problems and applications to multivariate cryptosystems." MR 2005e:94175. Pages 235-251 in: Thomas Johansson, Subhamoy Maitra (editors). Proceedings in cryptology—INDOCRYPT 2003. Proceedings of the 4th international conference on cryptology in India held in New Delhi, December 8-10, 2003. Lecture Notes in Computer Science 2904. Springer. ISBN 3-540-20609-4. MR 2005d:94154.

[199] Ludovic Perret. "A fast cryptanalysis of the isomorphism of polynomials with one secret problem." Pages 354-370 in: Ronald Cramer (editor). Advances in cryptology—EUROCRYPT 2005. Proceedings of the 24th annual international conference on the theory and applications

of cryptographic techniques held in Aarhus, May 22-26, 2005. Lecture Notes in Computer Science 3494. Springer. ISBN 3-540-25910-4. MR 2008e:94035.

[200] Jean-Charles Faugère, Ludovic Perret. "Polynomial equivalence problems: algorithmic and theoretical aspects." Pages 30-47 in: Serge Vaudenay (editor). Advances in Cryptology—EUROCRYPT 2006, 25th annual international conference on the theory and applications of cryptographic techniques, St. Petersburg, Russia, May 28-June 1, 2006, proceedings. Lecture Notes in Computer Science 4004. Springer. ISBN 3-540-34546-9.

[201] Jean-Charles Faugère, Ludovic Perret. "Cryptanalysis of 2R^- schemes." MR 2422172. Pages 357-372 in: Cynthia Dwork (editor). Advances in cryptology—CRYPTO 2006. Proceedings of the 26th annual international cryptology conference held in Santa Barbara, CA, August 20-24, 2006. Lecture Notes in Computer Science 4117. Springer. ISBN 978-3-540-37432-9. MR 2422188.

[202] Nicolas T. Courtois, Josef Pieprzyk. "Cryptanalysis of block ciphers with overdefined systems of equations." MR 2005d:94097. Pages 267-287 in: Yuliang Zheng (editor). Advances in cryptology—ASIACRYPT 2002, proceedings of the 8th international conference on the theory and application of cryptology and information security held in Queenstown, December 1-5, 2002. Lecture Notes in Computer Science 2501. Springer. ISBN 3-540-00171-9. MR 2005c:94002.

[203] Nicolas T. Courtois. "Higher order correlation attacks, XL algorithm and cryptanalysis of Toyocrypt." MR 2005d:94098. Pages 182-199 in: Pil Joong Lee, Chae Hoon Lim (editors). Information security and cryptology—ICISC 2002, papers from the 5th international conference held in Seoul, November 28-29, 2002. Lecture Notes in Computer Science 2587. Springer. ISBN 3-540-00716-4. MR 2005b:94052.

[204] Nicolas T. Courtois, Willi Meier. "Algebraic attacks on stream ciphers with linear feedback." MR 2005e:94098. Pages 345-359 in: Eli Biham (editor). Advances in cryptology—EUROCRYPT 2003, proceedings of the 22nd international conference on the theory and applications of cryptographic techniques held in Warsaw, May 4-8, 2003. Lecture Notes in Computer Science 2656. Springer. ISBN 3-540-14039-5. MR 2005c:94003.

[205] Nicolas T. Courtois. "Fast algebraic attacks on stream ciphers with linear feedback." MR 2005e:94131. Pages 176-194 in: Dan Boneh (editor). Advances in cryptology—CRYPTO 2003, proceedings of the 23rd annual international cryptology conference held in Santa Barbara, CA, August 17-21, 2003. Lecture Notes in Computer Science 2729. Springer. ISBN 3-540-40674-3. MR 2005d:94151.

[206] Frederik Armknecht, Matthias Krause. "Algebraic attacks on combiners with memory." Pages 162-176 in: Dan Boneh (editor). Advances in cryptology—CRYPTO 2003, proceedings

of the 23rd annual international cryptology conference held in Santa Barbara, CA, August 17-21, 2003. Lecture Notes in Computer Science 2729. Springer. ISBN 3-540-40674-3. MR 2005d:94151.

[207] Olivier Billet, Matthew J. B. Robshaw, Thomas Peyrin. "On building hash functions from multivariate quadratic equations." Pages 82-95 in: Josef Pieprzyk, Hossein Ghodosi, Ed Dawson (editors). Information security and privacy, 12th Australasian conference, ACISP 2007, Townsville, Australia, July 2-4, 2007, proceedings. Lecture Notes in Computer Science 4586. Springer. ISBN 978-3-540-73457-4.

[208] Makoto Sugita, Mitsuru Kawazoe, Ludovic Perret, Hideki Imai. "Algebraic cryptanalysis of 58-round SHA-1." Pages 349-365 in: Alex Biryukov (editor). Fast software encryption, 14th international workshop, FSE 2007, Luxembourg, Luxembourg, March 26-28, 2007, revised selected papers. Lecture Notes in Computer Science 4593. Springer. ISBN 978-3-540-74617-1.

[209] Jean-Philippe Aumasson, Willi Meier. "Analysis of multivariate hash functions." Pages 309-323 in: Kil-Hyun Nam and Gwangsoo Rhee (editors). Information security and cryptology—ICISC 2007, 10th international conference, Seoul, Korea, November 29-30, 2007, proceedings. Lecture Notes in Computer Science 4817. Springer. ISBN 978-3-540-76787-9.

[210] Jintai Ding, Bo-Yin Yang. "Multivariates polynomials for hashing." Pages 358-371 in: Dingyi Pei, Moti Yung, Dongdai Lin, Chuankun Wu (editors). Information security and cryptology, third SKLOIS conference, Inscrypt 2007, Xining, China, August 31-September 5, 2007, revised selected papers. Lecture Notes in Computer Science 4990. Springer. ISBN 978-3-540-79498-1.

[211] Christopher Wolf, Bart Preneel. "Taxonomy of public key schemes based on the problem of multivariate quadratic equations." http://eprint.iacr.org/2005/077.

2005. Christopher Wolf. "Multivariate quadratic polynomials in public key cryptography." http://eprint.iacr.org/2005/393.

[212] Jintai Ding, Dieter Schmidt. "Multivariable public-key cryptosystems." MR 2008a:94115. Pages 79-94 in: Dinh V. Huynh, S. K. Jain, S. R. Lopez-Permouth (editors). Algebra and its applications: papers from the international conference held at Ohio University, Athens, OH, March 22-26, 2005. Contemporary Mathematics 419. AMS. ISBN 0-8218-3842-3. MR 2007h:16002.

[213] Jintai Ding, Jason E. Gower, Dieter S. Schmidt. Multivariable public key cryptosystems. Advances in Information Security 25. Springer. ISBN 0-387-32229-9. MR 2007i:94049.

[214] Jintai Ding, Bo-Yin Yang. "Multivariate public key cryptography." Pages 193-242 in: Daniel J. Bernstein, Johannes Buchmann, Erik Dahmen (editors). Post-quantum cryptography.

Springer, Berlin. ISBN 978-3-540-88701-0.

[215] Hashimoto Y., Takagi T., Sakurai K.. General fault attacks on multivariate public key cryptosystems. Post-Quantum Cryptography, Springer, 2011, 1-18

[216] Petzoldt A, Bulygin S, Buchmann J. Linear Recurring Sequences for the UOV Key Generation[A]. In: Public Key Cryptography - PKC 2011[C]. Springer Berlin Heidelberg. 2011. Lecture Notes in Computer Science, 6571: 335-350.

[217] Clough C, Ding J. Secure Variants of the Square Encryption Scheme[A]. In: Post-Quantum Cryptography[C]. Springer Berlin Heidelberg. 2010. Lecture Notes in Computer Science, 6061: 153-164.

[218] Yasuda T, Sakurai K, Takagi T. Reducing the Key Size of Rainbow Using Non-commutative Rings[A]. In: Topics in Cryptology - CT-RSA 2012[C]. Springer Berlin Heidelberg. 2012. Lecture Notes in Computer Science, 7178: 68-83.

[219] Yasuda T, Sakurai K. A Security Analysis of Uniformly-Layered Rainbow[A]. In: Post-Quantum Cryptography[C]. Springer Berlin Heidelberg. 2011. Lecture Notes in Computer Science, 7071: 275-294.

[220] Tsujii S, Gotaishi M, Tadaki K, et al. Proposal of a Signature Scheme Based on STS Trapdoor[A]. In: Post-Quantum Cryptography[C]. Springer Berlin Heidelberg. 2010. Lecture Notes in Computer Science, 6061: 201-217.

[221] Jiao L, Li Y, Qiao S. A new scheme based on the MI scheme and its analysis[J]. Journal of Electronics (China), 2013, 30(2): 198-203.

[222] Cheng C M, Chou T, Niederhagen R, et al. Solving Quadratic Equations with XL on Parallel Architectures[A]. In: Cryptographic Hardware and Embedded Systems-CHES 2012[C]. Springer, 2012: 356-373.

[223] Bernstein D J, Duif N, Lange T, et al. High-speed high-security signatures[J]. Journal of Cryptographic Engineering, 2012, 2(2): 77-89.

[224] Huang Y J, Liu F H, Yang B Y. Public-key cryptography from new multivariate quadratic assumptions[J]. 2012: 190-205.

[225] Porras J, Baena J, Ding J. ZHFE, a New Multivariate Public Key Encryption Scheme[A]. In: PostQuantum Cryptography[C]. Springer, 2014: 229-245.

[226] Ding J, Petzoldt A, Wang L c. The Cubic Simple Matrix Encryption Scheme[A]. In: Post-Quantum Cryptography[C]. Springer, 2014: 76-87.

[227] Ding J, Ren A, Tao C. Embedded Surface Attack on Multivariate Public Key Cryptosystems from Diophantine Equations[A]. In: Information Security and Cryptology[C], 2013: 122-136.

[228] Gao S, Heindl R. Multivariate public key cryptosystems from diophantine equations[J].

Designs, codes and cryptography, 2013, 67(1): 1-18.

[229] Cao W, Nie X, Hu L, et al. Cryptanalysis of Two Quartic Encryption Schemes and One Improved MFE Scheme[A]. In: Post-Quantum Cryptography[C]. Springer, 2010: 41-60.

[230] Buchmann J, Bulygin S, Ding J, et al. Practical Algebraic Cryptanalysis for Dragon-Based Cryptosystems[A]. In: Cryptology and Network Security[C]. Springer, 2010: 140-155.

[231] Ding J, Schmidt D S. Mutant Zhuang-Zi Algorithm[A]. In: Post-Quantum Cryptography[C]. Springer, 2010: 28-40.

[232] Czypek P, Heyse S, Thomae E. Efficient Implementations of MQPKS on Constrained Devices[A]. In: Cryptographic Hardware and Embedded Systems - CHES 2012[C]. Springer Berlin Heidelberg. 2012. Lecture Notes in Computer Science, 7428: 374-389.

[233] Petzoldt A, Thomae E, Bulygin S, et al. Small Public Keys and Fast Verification for Multivariate Quadratic Public Key Systems[A]. In: Cryptographic Hardware and Embedded Systems - CHES 2011[C]. Springer Berlin Heidelberg. 2011. Lecture Notes in Computer Science, 6917: 475-490.

第 5 章 多变量公钥密码快速芯片技术

5.1 本章概述

由于信息安全应用的使用遍及生活和工作的各个方面，密码硬件的设计受到了广泛的关注。在可选择的密码算法中，多变量公钥密码算法作为少数几种能够抵御潜在量子计算机攻击的公钥密码算法，成为密码领域关注的焦点。

多变量公钥密码算法主要用于数字签名领域，设计多变量签名硬件的研究分为两个方向：

1）第一个方向是加快多变量签名硬件的运行速度，即缩短签名所需要的时间，设计时只需要考虑实现的速度而不考虑实现面积的规模；

2）第二个方向是提升多变量签名硬件的性能，即减小运行的时间——使用的面积的乘积，设计时需要同时考虑时间的消耗和实现面积的规模。

目前主要的多变量数字签名的硬件设计包括以下 6 种：

1）参考文献［1］中的 Rainbow 签名的并行硬件是目前最快的 Rainbow 签名（它的面积不是最优），它只需要 804 个时钟周期生成 Rainbow 签名；

2）参考文献［2］中的多变量数字签名的心动模型是目前最好的用心动模型设计的多变量公钥密码硬件；

3）参考文献［3］中的低资源的智能卡上的 TTS 签名是目前最好的 TTS 签名硬件；

4）参考文献［4］中的多变量公钥密码的低资源的嵌入式系统是目前最小的多变量公钥密码硬件；

5）参考文献［5］中的多变量公钥密码硬件实现了多种多变量公钥密码算法；

6）参考文献［6］中的在通用处理器上实现的多变量公钥密码硬件是目前通用处理器上最好的多变量公钥密码硬件。

本章选取的研究对象是 Rainbow 签名，它是目前最热门的多变量数字签名之一。它由多层油醋多项式结构组成，是最快的多变量数字签名之一。它可以应用于资源受限的设备，例如传感器；也可以应用于通用硬件平台，例如 FPGA。它被广泛应用于密码领域和通信领域。

我们根据第一个研究方向，设计了快速 Rainbow 签名硬件，它专注于缩短签名的时间。由于密码算法的效率依赖于有限域运算的实现方式，Rainbow 签名硬件基于有限域运算的优化而设计。通过优化组合有限域的快速求逆器、快速多元乘法器和快速求解线性方程组的硬件装置等设计以及提出新的设计思路，我们在 Altera FPGA 上实现了 Rainbow 签名的硬件。

实验结果表明这两种 Rainbow 签名硬件实现了我们预期的目标，快速 Rainbow 签名硬件将 Rainbow 签名所需的时间缩短至原来的 1/4。我们将快速 Rainbow 签名硬件与相关多变量公钥密码硬件和其他公钥密码硬件进行了对比，相关对比结果表明我们的设计更加优越。对比其他公钥密码硬件，快速 Rainbow 签名硬件在签名的速度上有绝对的优势，适合在时间敏感的环境使用。

本章余下内容包括以下 3 部分。第一，我们设计了快速的 Rainbow 签名硬件结构；第二，我们将它在 Altera FPGA 上实现，并与相关设计进行了对比；第三，我们对本章内容进行了小结。

5.2 快速多变量签名方案

在已有的多变量签名方案中，Rainbow(17,12,12) 是参考文献 [7] 建议的一种安全的方案，它的安全程度能达到 2^{80}。由于它的算法的可并行程度很高，并且适合硬件实现，所以我们选择了它作为快速多变量签名的硬件设计的方案。Rainbow(17,12,12) 中的参数指的分别是第一层的醋变量的数目（17 个）、第一层的油变量的数目（12 个）和第二层的油变量的数目（12 个），它使用的基域是有限域 $GF(256)$。

我们选择的 Rainbow 签名方案的具体实现参数如表 5-1 所示。待签名消息的散列的长度是 24 字节，消息签名后的长度是 42 字节。它由两层的非平衡油醋结构组成，第一层醋变量的数目是 17 个，油变量的数目是 12 个；第二层醋变量的数目是 30 个，油变量的数目是 12 个，其中第二层的醋变量是由第一层的醋变量（17 个）、第一层的油变量（12 个）和第二层的随机变量（1 个）组成的。私钥由两个可逆仿射变换和中心映射变换组成，公钥是其中一个可逆仿射变换、中心映射变换和另一个可逆仿射变换的依次组合。

表 5-1   快速 Rainbow 签名方案的选择

Rainbow 签名的参数	Rainbow 签名的数值
有限域	$GF(2^8)$
消息的散列长度	24 字节
签名长度	42 字节
油醋结构的层数	2
每层的醋变量的数目	17、30
每层的油变量的数目	12、12
私钥	两个可逆仿射变换和中心映射变换
公钥	一个仿射变换、中心映射变换和另一个仿射变换的依次组合

接下来我们介绍选定的 Rainbow 签名方案。我们假定消息的散列值是 $y(y_0, y_1, \ldots, y_{23})$，它的长度是 24 字节，其中 $y_0, y_1, \ldots, y_{23}$ 是 $GF(2^8)$ 的元素。而且，我们假定签名是 $x(x_0, x_1, \ldots, x_{41})$，它的长度是 42 字节，其中 $x_0, x_1, \ldots, x_{41}$ 是 $GF(2^8)$ 的元素。

为了对消息的散列值 $y(y_0, y_1, \ldots, y_{23})$ 签名，我们需要进行如下计算：

$$F \circ L_2(x_0, x_1, \ldots, x_{41}) = L_1^{-1}(y_0, y_1, \ldots, y_{23})$$

这里 F 是一个中心映射变换，L_1 和 L_2 是两个可逆的仿射变换。

为了实现上述计算，我们首先要做可逆仿射 L_1 的逆变换：

$$\bar{y} = L_1^{-1}(y_0, y_1, \ldots, y_{23})$$

这里 $\bar{y}(\bar{y}_0, \bar{y}_1, \ldots, \bar{y}_{23})$ 是可逆仿射变换的结果，$\bar{y}$ 的长度是 24 字节。

可逆仿射 L_1 的逆变换 L_1^{-1} 有如下形式：

$$\bar{y} = Ay + b$$

这里 A 是一个规模为 24×24 的矩阵，b 是一个维度为 24 的向量，A 和 b 都被当作私钥来运算。

通过求解中心映射 F 的逆变换，我们获得 $\bar{x}(\bar{x}_0, \bar{x}_1, \ldots, \bar{x}_{23})$，$\bar{x}$ 的长度是 24 字节：

$$\bar{x} = F^{-1}(\bar{y}_0, \bar{y}_1, \ldots, \bar{y}_{23})$$

中心映射变换 F 由 24 个多变量多次多项式 $(f_0, f_1, \ldots, f_{23})$ 组成，它有如下形式：

$$F(\bar{x}_0, \bar{x}_1, \ldots, \bar{x}_{41}) = (f_0, f_1, \ldots, f_{23})$$

那么，我们将 $\bar{y}(\bar{y}_0, \bar{y}_1, \ldots, \bar{y}_{23})$ 代入中心映射的变换中，它变成如下形式：

$$\bar{y}(\bar{y}_0, \bar{y}_1, \ldots, \bar{y}_{23}) = f(f_0, f_1, \ldots, f_{23})$$

中心映射 F 是一个两层的油醋结构，由 24 个多变量多次多项式组成，即 $(f_0, f_1, \ldots, f_{23})$ 被分成两层：

1）$f_i | i = 0, 1, \ldots, 11$，共 12 个多项式；
2）$f_i | i = 12, 13, \ldots, 23$，共 12 个多项式。

同时，中心映射变换 F 中的 $\bar{x}(\bar{x}_0, \bar{x}_1, \ldots, \bar{x}_{41})$ 也可以分成 4 组，分组情况如下：

1）$\bar{x}_i | i = 0, 1, \ldots, 16$，共 17 个元素组成了第一层的醋变量；
2）$\bar{x}_i | i = 17, 18, \ldots, 28$，共 12 个元素组成了第一层的油变量；
3）$\bar{x}_i | i = 0, 1, \ldots, 29$，共 30 个元素组成了第二层的醋变量（有一个随机变量 $\bar{x}_{29}$ 作为这层的醋变量）；
4）$\bar{x}_i | i = 30, 31, \ldots, 41$，共 12 个元素组成了第二层的油变量。

其中，第二层的醋变量是由第一层的醋变量、第一层的油变量和第二层的一个随机醋变量组成的。

中心映射变换 F 的 24 个多变量多次多项式 $(f_0, f_1, ..., f_{23})$ 的定义如下：

$$f(O_0, O_1, ..., O_{23}) = \sum \alpha_{ij} O_i V_j + \sum \beta_{ij} V_i V_j + \sum \gamma_i V_i + \sum \delta_i O_i + \eta$$

这里 O_i, (V_i, V_j) 分别是油变量和醋变量，α_{ij}、β_{ij}、γ_i、δ_i 和 η 是多变量多次多项式的系数，被当作私钥使用。

多变量多次多项式包括 5 项，最高次数不超过两次，若将醋变量的数值代入多变量多次多项式，它将变换成关于油变量的一次多项式。这 5 项分别为：

1）$O_i V_j$，油变量和醋变量的组合；

2）$V_i V_j$，醋变量和醋变量的组合；

3）V_i，醋变量；

4）O_i，油变量；

5）η，常数。

计算中心映射变换的步骤如下：

1）我们为第一层的 17 个醋变量 $\bar{x}_0, \bar{x}_1, ..., \bar{x}_{16}$ 选择随机的数值；

2）我们将 $\bar{x}_0, \bar{x}_1, ..., \bar{x}_{16}$ 的数值代入中心映射变换 F 的第一层 12 个多变量多次多项式 $f_0, f_1, ..., f_{11}$，并计算它们的系数；

3）$f_0, f_1, ..., f_{11}$ 被转化为关于第一层的 12 个油变量 $\bar{x}_{17}, \bar{x}_{18}, ..., \bar{x}_{28}$ 的线性方程组，它的系数矩阵规模是 12×12，然后我们求解这个线性方程组；

4）我们将第一层的 17 个醋变量、第一层的 12 个油变量和第二层的一个随机醋变量 $\bar{x}_0, \bar{x}_1, ..., \bar{x}_{29}$ 共 30 个元素作为第二层的醋变量代入中心映射变换 F 的第二层 12 个多变量多次多项式 $f_{12}, f_{13}, ..., f_{23}$ 中，并计算它们的系数；

5）$f_{12}, f_{13}, ..., f_{23}$ 被转化为关于第二层的 12 个油变量 $\bar{x}_{30}, \bar{x}_{31}, ..., \bar{x}_{41}$ 的线性方程组，它的系数矩阵规模是 12×12，然后我们求解这个线性方程组。

到此为止，$\bar{x}(\bar{x}_0, \bar{x}_1, ..., \bar{x}_{41})$ 的 42 个元素都计算完成，求得数值。

最后，我们将 $\bar{x}(\bar{x}_0, \bar{x}_1, ..., \bar{x}_{41})$ 的 42 个元素代入可逆仿射 L_2 的逆变换 L_2^{-1}，计算它的逆变换得 $x(x_0, x_1, ..., x_{41})$，即消息的散列值的 Rainbow 签名：

$$x = L_2^{-1}(\bar{x}_0, \bar{x}_1, ..., \bar{x}_{41})$$

可逆仿射的逆变换 L_2^{-1} 有如下形式：

$$x = C\bar{x} + d$$

这里，C 是一个规模为 42×42 的矩阵，d 是一个维度为 42 的向量，C 和 d 都被当作私钥来运算。

所以，$x(x_0, x_1, ..., x_{41})$ 是 $y(y_0, y_1, ..., y_{23})$ 的签名。到此，我们采用的快速 Rainbow 签名方案介绍完毕。

5.3 不可约多项式

不可约多项式在有限域运算中起的作用是保证所有运算结果均在有限域内，对于每个有限域来说，有且不止一个多项式可以被选择为不可约多项式，不同的选择会影响有限域运算的效率。

$GF(2^8)$ 的多项式可以被表示为 $x_8 + x_k + \ldots + 1 (0 < k < 8)$ 的形式，一般来说，项数越少的多项式，作为有限域的不可约多项式就越可以显著地增加密码硬件的效率。综合考虑乘法和求逆的运算，我们选取了 $(101001101)_2$ 作为 $GF(2^8)$ 的不可约多项式，即 $x^8 + x^6 + x^3 + x^2 + 1$。

5.4 加速二元和三元乘法运算

我们发现在 Rainbow 签名中，不仅需要二元乘法，还需要三元乘法。一个优化的乘法器能够显著地加快 Rainbow 签名的速度。所以，我们基于一种二元乘法设计了加速三元乘法的方法。新的设计是基于一个新的观察：在 $GF(2^8)$ 的三元乘法中，先做多项式乘法、再做求模运算的方法要比其他方法更快。需要指出的是这并不适合大的有限域。

我们假定 $a(x) = \sum_{i=0}^{7} a_i x^i$、$b(x) = \sum_{i=0}^{7} b_i x^i$ 和 $c(x) = \sum_{i=0}^{7} b_i x^i$ 是 $GF(2^8)$ 的 3 个元素，计算它们的乘积的方法如下：

$$d(x) = (a(x) \times b(x) \times c(x)) \bmod (f(x)) = \sum_{i=0}^{7} d_i x^i$$

这里，$d(x)$ 是预期的乘积，$f(x)$ 是不可约多项式。

首先，对于 $i = 0, 1, \ldots, 21$ 和 $j = 0, 1, \ldots, 7$，我们先计算 v_{ij}：

$$x^i \bmod f(x) = \sum_{j=0}^{7} v_{ij} x^j$$

其次，对于 $i = 0, 1, \ldots, 21$，我们计算 S_i：

$$S_i = \sum_{j+k+l+i} a_j b_k c_l$$

最后，对于 $i = 0, 1, \ldots, 7$，我们计算 d_i：

$$d_i = \sum_{j=0}^{21} v_{ji} S_j$$

$a(x)$、$b(x)$、$c(x)$ 的乘法结果是 $\sum_{i=0}^{7} d_i x^i$。

5.5 加速求逆运算

由于求逆算法在 Rainbow 签名中使用得并不多，为了加速求逆过程，我们设计了基于费马小定理的部分求逆算法。假定 $f(x)$ 是我们选定的 $GF(2^8)$ 的不可约多项式，β 是待求逆的有限域元素，它有如下形式：

$$\beta = \beta_7 x^7 + \beta_6 x^6 + \beta_5 x^5 + \beta_4 x^4 + \beta_3 x^3 + \beta_2 x^2 + \beta_1 x + \beta_0$$

根据费马小定理，我们可以得到如下计算：

$$\beta^{2^8} = \beta$$

$$\beta^{-1} = \beta^{2^8 - 2} = \beta^{254}$$

$$2^8 - 2 = 2 + 2^2 + 2^3 + 2^4 + 2^5 + 2^6 + 2^7$$

$$\beta^{-1} = \beta^2 \, \beta^4 \, \beta^8 \, \beta^{16} \, \beta^{32} \, \beta^{64} \, \beta^{128}$$

我们采用三元乘法加速部分求逆过程：

$$S_1 = ThreeMult(\beta^2, \beta^4, \beta^8)$$

$$S_2 = ThreeMult(\beta^{16}, \beta^{32}, \beta^{64})$$

这里，$ThreeMult(v1, v2, v3)$ 表示通过调用三元乘法器实现 $GF(2^8)$ 的 $v1, v2, v3$ 乘法，S_1、S_2 分别是乘法结果。(S_1, S_2, β^{128}) 则是 β 的部分求逆结果。

5.6 加速求解线性方程组运算

我们设计了 $GF(2^8)$ 的快速求解线性方程组的方法，它的系数矩阵的规模为 12×12。算法 5-1 是我们的设计，它是高斯约当消元法的改进算法。

算法 5-1　快速求解有限域线性方程组
Require：规模为 12×12 的矩阵 A Ensure：求解有限域线性方程组 Int i=0 A←Pivoting(A, i = 0)

5.6 加速求解线性方程组运算

```
       while i<12 do
A ← (Partial_inversion(A, i), Normalization(A, i), Elimination(A, i))
A ← Pivoting(A, i+1)
i ← i+1
end while
```

我们的算法只需要 12 次迭代，每次迭代由找主元、部分求逆、归一和消元组成。$Pivoting(A,i)$ 代表第 i 次迭代中对矩阵 A 的操作，其中 $i = 0,1,\ldots,11$。算法 5-1 中，pivoting 是找主元，$partial\_inversion$ 是部分求逆，$Normalization$ 是归一，$Elimination$ 是消元。

我们的算法与高斯约当消元法的主要不同之处有以下 4 点：

1）我们采用了三元乘法器单时钟周期实现乘法；

2）我们设计了基于费马小定理的部分求逆方法，它可以加速归一和消元的运算；

3）在找主元过程中，我们利用索引减少了消耗的时间，并且在写回数据前执行下一次的找主元；

4）我们最大限度地并行了归一和消元的两个运算，显著地加快了整个求解速度。

我们的设计可以只用 m 个时钟周期求解系数矩阵规模为 $m \times m$ 的线性方程组，每个时钟周期完成一次迭代，它的时间复杂度是 $O(m)$。

我们设计了快速求解线性方程组的硬件结构，如图 5-1 所示，其中 a_{ij} 是系数矩阵 A 的第 i 行第 j 列的元素。它由 3 类单元组成，即 I、N_l、E_{kl}，其中 $k = 1,2,\ldots,11$，$l = 1,2,\ldots,12$。I 是部分求逆单元，N_l 是编号为 l 的归一单元，E_{kl} 是编号为 kl 的消元单元。求解线性方程组的硬件结构由 1 个 I、12 个 N_l 和 132 个 E_{kl} 组成。图 5-1 中的矩阵表示每次迭代开始时矩阵的形式，我们可以看到所有迭代结束后的矩阵的最右一列是求解的结果。

为了缩短找主元的时间，我们引入索引技术。我们建立了线性方程组的系数矩阵的行的索引，即向量 $(0,1,\ldots,11)$。我们假定主元所在列是 a_i，且 $a_{ii} = 0$，则需要从主元所在列选择一个非零元素（我们假定它是 a_{ji}）作为新主元。我们在索引中交换 j 和 i 的位置，所以新的索引是 $(0,1,\ldots,i-1,j,i+1,\ldots,j-1,i,j+1,\ldots,11)$。$a_{ji}$ 作为新的主元，被传送到部分求逆器，第 j 行的所有元素被传送到归一单元，其他行的所有元素被传送到消元单元。除了使用索引技术，我们与一般高斯约当消元的另外一个不同点：在每次迭代的最后执行下一轮的找主元操作，这样在写回数据之前保证下一轮的主元不为 0。

图 5-2 所示是一个找主元的例子。在第一轮迭代的最后，我们需要为第二轮迭代找一个非零元作为主元，因为 a_{31} 不为 0，第四行被选作新的主元所在行，a_{31} 被传送给部分求逆器 I，第四行的所有元素被传送到归一单元 N_l，其他行的所有元素被传送到消元单元 E_{kl}。

第5章 多变量公钥密码快速芯片技术

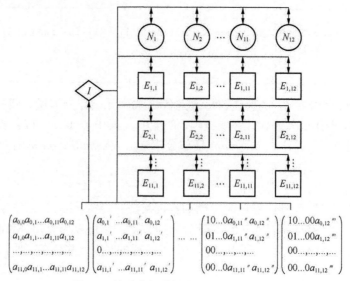

图 5-1 快速求解线性方程组的硬件结构

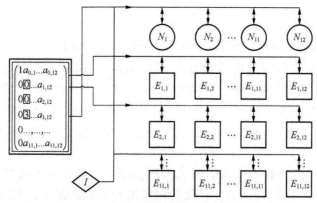

图 5-2 快速求解线性方程组中的找主元

归一操作由求逆和乘法组成,我们采用了部分求逆加快归一的速度。$GF(2^8)$ 的 β^{-1} 可以被表示成如下形式:

$$\beta^{-1} = \beta^2 \beta^4 \beta^8 \beta^{16} \beta^{32} \beta^{64} \beta^{128}$$

$$S_1 = ThreeMult(\beta^2, \beta^4, \beta^8)$$

$$S_2 = ThreeMult(\beta^{16}, \beta^{32}, \beta^{64})$$

$$S_4 = ThreeMult(\beta^2, R_j)$$

$$NOR_j = ThreeMult(S_1, S_2, S_4)$$

这里，NOR_j 是归一结果，$TwoMult(v1,v2)$ 代表 $GF(2^8)$ 的 $v1,v2$ 的二元乘法，$ThreeMult(v1,v2,v3)$ 代表 $GF(2^8)$ 的 $v1,v2,v3$ 的三元乘法。

图 5-3 所示归一计算的树状图，其中 R_j 是需要归一的元素，它是主元所在行的第 j 个元素。S_1 和 S_2 的运算在部分求逆器 I 中。S_4 和 NOR_j 的运算在归一单元 N_j 中。所以 N_j 由二元乘法器和三元乘法器等器件组成。S_1, S_2, S_4 可以并行执行。

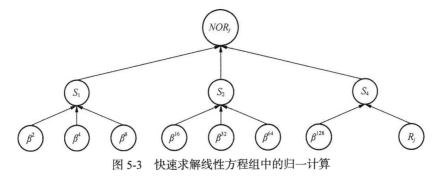

图 5-3 快速求解线性方程组中的归一计算

消元操作由乘法、求逆和加法组成，我们采用了部分求逆加快消元的速度，计算消元的过程如下：

$$S_1 = ThreeMult(\beta^2, \beta^4, \beta^8)$$
$$S_2 = ThreeMult(\beta^{16}, \beta^{32}, \beta^{64})$$
$$S_3 = ThreeMult(\beta^{128}, R_j, C_i)$$
$$ELI_{ij} = a_{ij} + ThreeMult(S_1, S_2, S_3)$$

S_1 和 S_2 的运算在部分求逆器 I 中，S_3 和 ELI_{ij} 的运算在消元单元 E_{ij} 中，$ThreeMult(v1,v2,v3)$ 代表 $GF(2^8)$ 的 $v1,v2,v3$ 的三元乘法。S_1, S_2, S_3 可以并行地执行。

图 5-4 所示为消元计算的树状图，其中 a_{ij} 是需要消元的元素，它是矩阵的第 i 行第 j 列的元素。R_j 是主元所在行的第 j 个元素，C_i 是主元所在列的第 i 个元素，ELI_{ij} 是 a_{ij} 的消元结果。

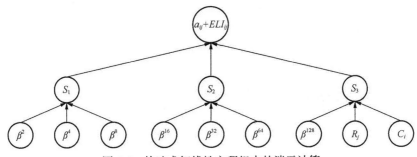

图 5-4 快速求解线性方程组中的消元计算

图 5-5 所示是原始的高斯约当消元法，图 5-6 是我们的快速求解线性方程组的方法，对比结果表明，我们把求解线性方程组的运算的关键路径从 6 个乘法和 1 个加法缩短到两个乘法和 1 个加法。我们能够在一个时钟周期内完成一次迭代，这是因为每次迭代中的运算都可以并行执行。对于求解系数矩阵的规模为 12×12 的线性方程组，我们只需要 12 个时钟周期求解。

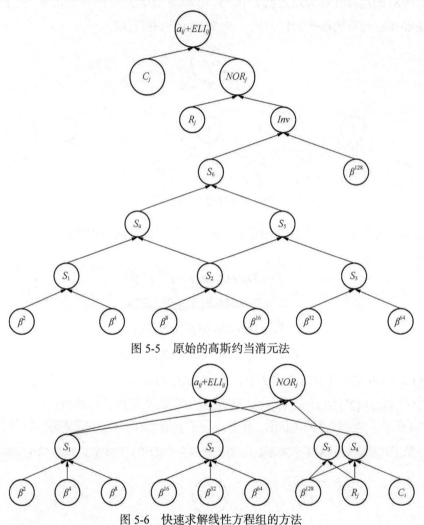

图 5-5　原始的高斯约当消元法

图 5-6　快速求解线性方程组的方法

5.7　加速可逆仿射变换运算

快速 Rainbow 签名方案有两个可逆仿射变换 L_1^{-1}：$k^{24} \to k^{24}$ 和 L_2^{-1}：$k^{42} \to k^{42}$，它

们是通过调用有限域的向量加法和矩阵-向量乘法来实现的。表 5-2 统计了可逆仿射变换中需要的乘法数目，我们通过优化乘法加速了可逆仿射变换。

表 5-2　可逆仿射变换中的乘法数目

运算	乘法数目
L_1^{-1} 可逆仿射变换	576
第一层多变量多次多项式的系数求值	6 324
第二层多变量多次多项式的系数求值	15 840
L_2^{-1} 可逆仿射变换	1 764
总共	24 504

5.8　加速多元二次多项式求值运算

两层的油醋结构由 24 个多变量多次多项式组成，每一层的多项式数目分别是 12，它的系数求值是通过调用有限域乘法实现的。表 5-3 统计了多变量多次多项式的系数求值中需要的乘法数目，我们通过优化乘法加速了多变量多次多项式的系数求值。

表 5-3　多变量多次多项式的系数求值中需要的乘法数目

	第一层	第二层
V_iO_j	2 448	4 320
V_iV_j	3 672	11 160
V_i	204	360
总共	6 324	15 840

5.9　技术实现

我们选用的编程语言是 VHDL，实现平台是 Altera Stratix II FGPA（它的型号是 EP2S130F1020I4）。表 5-4 统计了我们的设计在每一部分需要的时钟周期数，它总共只需要 198 个时钟周期实现 Rainbow 签名。实验结果表明，在时钟频率为 50 MHz 的条件下，我们的快速 Rainbow 签名硬件所需时间是 3 960 ns。所有的实验结果均是布线和布局后获得的数据。需要指出的是，我们的实现仅仅关注加快 Rainbow 签名的速度，我们的实现面积约 150 000 等效门，是参考文献［1］中的设计的 2～3 倍。

表 5-4　　　　　快速 Rainbow 签名需要的时钟周期数

步骤	运算	时钟周期数
1	L_1^{-1} 可逆仿射变换	5
2	第一层多变量多次多项式的系数求值	45
3	第一轮求解线性方程组	12
4	第二层多变量多次多项式的系数求值	111
5	第二轮求解线性方程组	12
6	L_2^{-1} 可逆仿射变换	13
总共		198

5.10　实现对比

表 5-5 和表 5-6 分别是求解线性方程组和签名方案的实现的对比，结果表明了我们将求解线性方程组的速度加快了 1 倍（12 个时钟周期对比 24 个时钟周期），将 Rainbow 签名所需的时间缩短为原来的约 1/4（198 个时钟周期对比 804 个时钟周期），而且我们的设计比 RSA 和椭圆曲线签名更快（198 个时钟周期对比 227 个时钟周期）。

表 5-5　　　　　对比求解系数矩阵规模为 12×12 的线性方程组

求解线性方程组的算法	时钟周期数
原始的高斯约当消元法	1 116
原始的高斯消元法	830
参考文献［8］中的高斯约当消元法	48
参考文献［9］中的高斯消元法	47
参考文献［10］中的高斯消元法	24
快速求解线性方程组的方法	12

表 5-6　　　　　签名方案的对比

签名方案	时钟周期数
参考文献［11］中的 ECC	30 000
参考文献［4］中的 en-TTS	16 000
参考文献［2］中的 Rainbow (42,24)	3 150
参考文献［2］中的长消息 UOV	2 260

续表

签名方案	时钟周期数
参考文献[1]中的 Rainbow	804
参考文献[2]中的短消息 UOV	630
参考文献[12]中的 RSA	227
快速 Rainbow 签名	198

5.11 本章小结

多变量公钥密码算法是少数具有抵御量子计算攻击能力的公钥密码算法之一。研究多变量公钥密码芯片不但对它有重要的理论意义和经济意义，而且对它的发展和推广起了重要作用。在多变量公钥密码中，Rainbow 签名算法是目前应用最广泛的密码算法。

在本章中，我们针对密码硬件设计的方向之一——快速实现，设计了一种快速的 Rainbow 签名硬件，它专注于缩短签名的运行时间。通过优化组合有限域的快速求逆器、快速多元乘法器和快速求解线性方程组的硬件装置等设计以及提出新的设计思路，我们在 Altera FPGA 上实现了快速 Rainbow 签名硬件。

快速 Rainbow 签名硬件显著地缩短了 Rainbow 签名所需的时间。我们采用优化多元乘法、部分求逆算法和快速求解线性方程组的算法，并且综合其他的优化，在 Altera FPGA 上实现了 Rainbow 签名算法，它只需要 198 个时钟周期。实验和对比结果表明我们将 Rainbow 签名的时间缩短为原来的约 1/4（198 时钟周期对比 804 时钟周期）。更加令人关注的是它的时间比其他的公钥签名更短（198 时钟周期对比 227 时钟周期），例如 RSA 和椭圆曲线密码。我们的设计也证明了多变量公钥密码算法有能力在时间敏感的工业应用中使用。

5.12 本章参考文献

[1] Balasubramanian S, Bogdanov A, Rupp A, et al. Fast Multivariate Signature Generation in Hardware: The Case of Rainbow[A]. In: 16th International Symposium on Field-Programmable Custom Computing Machines[C], 2008: 281-282.

[2] Bogdanov A, Eisenbarth T, Rupp A, et al. Time-Area Optimized Public-Key Engines: MQCryptosystems as Replacement for Elliptic Curves?[A]. In: Cryptographic Hardware and

Embedded Systems - CHES 2008[C]. Springer Berlin Heidelberg. 2008. Lecture Notes in Computer Science, 5154: 45-61.

[3] Yang B, Chen J, Chen Y. TTS: High-speed signatures on a low-cost smart card[J]. Cryptographic Hardware and Embedded Systems-CHES 2004, 2004: 318-348.

[4] Yang B, Cheng C, Chen B, et al. Implementing minimized multivariate PKC on low-resource embedded systems[J]. Security in Pervasive Computing, 2006: 73-88.

[5] Chen A, Chen C, Chen M, et al. Practical-Sized Instances of Multivariate PKCs: Rainbow, TTS, and ℓIC-Derivatives[J]. Post-Quantum Cryptography, 2008: 95-108.

[6] Chen A, Chen M, Chen T, et al. SSE implementation of multivariate PKCs on modern x86 CPUs[J]. Cryptographic Hardware and Embedded Systems-CHES 2009, 2009: 33-48.

[7] Ding J, Yang B Y, Chen C H, et al. New Differential-Algebraic Attacks and Reparametrization of Rainbow[A]. In: Applied Cryptography and Network Security[C].

[8] Wang C L, Lin J L. A Systolic Architecture for Computing Inverses and Divisions in Finite Fields $GF(2^m)$ [J]. IEEE Transactions on Computers, 1993, 42(9): 1141-1146.

[9] Hochet B, Quinton P, Robert Y. Systolic Gaussian elimination over $GF(p)$ with partial pivoting[J]. IEEE Transactions on Computers, 1989, 38(9): 1321-1324.

[10] Bogdanov A, Mertens M, Paar C, et al. A parallel hardware architecture for fast Gaussian elimination over $GF(2)$ [A]. In: 14th Annual IEEE Symposium on Field-Programmable Custom Computing Machines (FCCM)[C], 2006: 237-248.

[11] Alrimeih H, Rakhmatov D. Fast and flexible hardware support for ECC over multiple standard prime fields[J]. IEEE Transactions on Very Large Scale Integration (VLSI) Systems, 2014, 22(12): 2661-2674.Springer Berlin Heidelberg. 2008. Lecture Notes in Computer Science, 5037: 242-257.

[12] Großschädl J. High-speed RSA hardware based on Barret's modular reduction method[A]. In: Cryptographic Hardware and Embedded Systems—CHES 2000[C], 2000: 191-203.

第 6 章　多变量公钥密码高效芯片技术

6.1　本章概述

第 5 章提到，设计多变量签名硬件的研究分为两个方向：

1) 第一个方向是加快多变量签名硬件的运行速度，即缩短签名所需要的时间，设计时只需要考虑实现的速度，而不考虑实现面积的规模；

2) 第二个方向是提升多变量签名硬件的性能，即减小运行的时间—使用的面积的乘积，设计时需要同时考虑时间的消耗和实现面积的规模。

本章选取的研究对象是 Rainbow 签名。根据第二个研究方向，我们设计了高效 Rainbow 签名硬件，它同时对时间和面积进行了优化。

由于密码算法的效率依赖于有限域运算的实现方式，高效 Rainbow 签名硬件是基于有限域运算的优化而设计的。通过优化组合有限域的乘法器、求逆器、求解线性方程组的硬件装置等设计以及提出新的设计思路，我们在 Altera FPGA 上实现了 Rainbow 签名的硬件。实验结果表明高效 Rainbow 签名硬件实现了我们预期的目标，显著地增加了 Rainbow 签名的效率。

我们将高效 Rainbow 签名硬件与相关多变量公钥密码硬件和其他公钥密码硬件进行了对比，相关对比结果表明我们的设计更加优越。对比其他公钥密码硬件，高效 Rainbow 签名硬件的签名所需的时间-面积的乘积更小，它适合在硬件上高效实现。

本章余下内容包括以下 3 部分。首先，我们设计了高效的 Rainbow 签名硬件结构；然后，我们将它在 Altera FPGA 上实现，并与相关设计进行了对比；最后，我们对本章内容进行了小结。

6.2　高效多变量签名方案

因为 Rainbow 签名由两个可逆仿射变换和一个中心映射变换组成，其中可逆仿射变换由向量-矩阵乘法和向量加法的运算组成，中心映射变换由求解多项式的系数的值和求解线性方程组的运算组成，所以 Rainbow 的主要基本运算是有限域的乘法、求逆和求解线性方程组，它们的优化设计直接影响 Rainbow 签名的效率。我们充分考虑了 Rainbow

不同参数对这些基本运算的影响，选取了Rainbow(10,10,4,3,10)作为我们的高效Rainbow签名的硬件设计的方案。Rainbow(10,10,4,3,10)中的参数指的分别是第一层的醋变量的数目（10个）、第一层的油变量的数目（10个）、第二层的油变量的数目（4个）、第三层的油变量的数目（3个）和第四层的油变量的数目（10个）。它的基域是有限域 $GF(256)$。

我们选定的具体的 Rainbow 签名方案的参数如表 6-1 所示：待签名消息的散列的长度是 27 字节，消息签名后的长度是 37 字节；它由 4 层的非平衡油醋结构组成，第一层醋变量的数目是 10 个，油变量的数目是 10 个；第二层醋变量的数目是 20 个，油变量的数目是 4 个；第三层醋变量的数目是 24 个，油变量的数目是 3 个；第四层醋变量的数目是 27 个，油变量的数目是 10 个。其中下一层的醋变量是由上一层的醋变量和上一层的油变量组成的；私钥由两个可逆仿射变换和中心映射变换组成，公钥是其中一个可逆仿射变换、中心映射变换和另一个可逆仿射变换的依次组合。

表 6-1　　　　　　　　　　高效 Rainbow 签名方案的选择

Rainbow 签名的参数	Rainbow 签名的数值
有限域	$GF(2^8)$
消息的散列长度	27 字节
签名长度	37 字节
油醋结构的层数	4
每层的醋变量的数目	10，20，24，27
每层的油变量的数目	10，4，3，10
私钥	两个可逆仿射变换和中心映射变换
公钥	一个可逆仿射变换、中心映射变换和另一个可逆仿射变换的依次组合

接下来我们介绍选定的 Rainbow 签名方案。我们假定消息的散列值是 $y(y_0, y_1, ..., y_{26})$，它的长度是 27 字节，其中 $y(y_0, y_1, ..., y_{26})$ 是 $GF(2^8)$ 的元素。而且，签名是 $x(x_0, x_1, ..., x_{36})$，它的长度是 37 字节，其中 $x_0, x_1, ..., x_{36}$ 是 $GF(2^8)$ 的元素。

为了对消息的散列值 $y(y_0, y_1, ..., y_{26})$ 签名，我们需要进行如下计算：

$$F \circ L_2(x_0, x_1, ..., x_{36}) = L_1^{-1}(y_0, y_1, ..., y_{26})$$

这里，F 是一个中心映射变换，L_1 和 L_2 是两个可逆的仿射变换。

为了实现上述计算，我们首先要计算可逆仿射 L_1 的逆变换：

$$\overline{y} = L_1^{-1}(y_0, y_1, ..., y_{26})$$

这里，$\overline{y}(\overline{y}_0, \overline{y}_1, ..., \overline{y}_{26})$ 是可逆仿射变换的结果，$\overline{y}$ 的长度是 27 字节。

可逆仿射 L_1 的逆变换 L_1^{-1} 有如下形式：

$$\overline{y} = Ay + b$$

这里，A 是一个规模为 27×27 的矩阵，b 是一个维度为 27 的向量，A 和 b 都被当作私钥来运算。

通过求解中心映射 F 的逆变换，我们获得 $\overline{x}(\overline{x}_0, \overline{x}_1, \ldots, \overline{x}_{36})$ 的数值，$\overline{x}$ 的长度是 37 字节。它有如下形式：
$$\overline{x} = F^{-1}(\overline{y}_0, \overline{y}_1, \ldots, \overline{y}_{26})$$

中心映射变换 F 由 27 个多变量多次多项式 $(f_0, f_1, \ldots, f_{26})$ 组成，它有如下形式：
$$F\overline{x}(\overline{x}_0, \overline{x}_1, \ldots, \overline{x}_{36}) = (f_0, f_1, \ldots, f_{26})$$

那么，我们将 $\overline{y}(\overline{y}_0, \overline{y}_1, \ldots, \overline{y}_{26})$ 代入中心映射的变换中，它变成如下形式：
$$\overline{y}(\overline{y}_0, \overline{y}_1, \ldots, \overline{y}_{26}) = f(f_0, f_1, \ldots, f_{26})$$

中心映射 F 是 4 层的油醋结构，由 27 个多变量多次多项式组成，即 $(f_0, f_1, \ldots, f_{26})$，被分成 4 层：

1）$f_i | i = 0,1,\ldots,9$，共 10 个多项式；
2）$f_i | i = 10,11,12,13$，共 4 个多项式；
3）$f_i | i = 14,15,16$，共 3 个多项式；
4）$f_i | i = 17,18,\ldots,26$，共 10 个多项式。

同时，中心映射变换 F 中的 $\overline{x}(\overline{x}_0, \overline{x}_1, \ldots, \overline{x}_{36})$ 也可以分成 8 组，分组情况如下：

1）$\overline{x}_i | i = 0,1,\ldots,9$，共 10 个元素组成了第一层的醋变量；
2）$\overline{x}_i | i = 10,11,\ldots,19$，共 10 个元素组成了第一层的油变量；
3）$\overline{x}_i | i = 0,1,\ldots,19$，共 20 个元素组成了第二层的醋变量；
4）$\overline{x}_i | i = 20,21,22,23$，共 4 个元素组成了第二层的油变量；
5）$\overline{x}_i | i = 0,1,\ldots,23$，共 24 个元素组成了第三层的醋变量；
6）$\overline{x}_i | i = 24,25,26$，共 3 个元素组成了第三层的油变量；
7）$\overline{x}_i | i = 0,1,\ldots,26$，共 27 个元素组成了第四层的醋变量；
8）$\overline{x}_i | i = 27,28,\ldots,36$，共 10 个元素组成了第四层的油变量。

这里，下一层的醋变量是由上一层的醋变量和上一层的油变量组成的。

中心映射变换 F 的 27 个多变量多次多项式 $(f_0, f_1, \ldots, f_{26})$ 的定义如下：
$$f(O_0, O_1, \ldots, O_{26}) = \sum \alpha_{ij} O_i V_j + \sum \beta_{ij} V_i V_j + \sum \gamma_i V_i + \sum \delta_i O_i + \eta$$

这里，O_i、(V_i, V_j) 分别是油变量和醋变量，α_{ij}、β_{ij}、γ_i、δ_i 和 η 是多变量多次多项式的系数，被当作私钥使用。

多变量多次多项式包括 5 项，最高次数不超过两次，若将醋变量的数值代入多变量多次多项式，它将变换成关于油变量的一次多项式。这 5 项分别为：

1）$O_i V_j$，油变量和醋变量的组合；
2）$V_i V_j$，醋变量和醋变量的组合；

3）V_i，醋变量；

4）O_i，油变量；

5）η，常数。

计算中心映射变换的步骤如下。

1）我们为第一层的 10 个醋变量 $\bar{x}_0,\bar{x}_1,...,\bar{x}_9$ 选择随机的数值；

2）我们将 $\bar{x}_0,\bar{x}_1,...,\bar{x}_9$ 的数值代入中心映射变换 F 的第一层 10 个多变量多次多项式 $f_0,f_1,...,f_9$，并计算它们的系数；

3）$f_0,f_1,...,f_9$ 被转化为关于第一层的 10 个油变量 $\bar{x}_{10},\bar{x}_{11},...,\bar{x}_{19}$ 的线性方程组，它的系数矩阵规模是 10×10，然后我们求解这个线性方程组；

4）我们将第一层的 10 个醋变量和第一层的 10 个油变量 $\bar{x}_{10},\bar{x}_{11},...,\bar{x}_{19}$ 共 20 个元素作为第二层的醋变量代入中心映射变换 F 的第二层 4 个多变量多次多项式 $f_{10},f_{11},f_{12},f_{13}$ 中，并计算它们的系数；

5）$f_{10},f_{11},f_{12},f_{13}$ 被转化为关于第二层的 4 个油变量 $\bar{x}_{20},\bar{x}_{21},\bar{x}_{22},\bar{x}_{23}$ 的线性方程组，它的系数矩阵规模是 4×4，然后我们求解这个线性方程组；

6）我们将第二层的 20 个醋变量和第二层的 4 个油变量 $\bar{x}_0,\bar{x}_1,...,\bar{x}_{23}$ 共 24 个元素作为第三层的醋变量代入中心映射变换 F 的第三层 3 个多变量多次多项式 f_{14},f_{15},f_{16} 中，并计算它们的的系数；

7）f_{14},f_{15},f_{16} 被转化为关于第三层的 3 个油变量 $\bar{x}_{24},\bar{x}_{25},\bar{x}_{26}$ 的线性方程组，它的系数矩阵规模是 3×3，然后我们求解这个线性方程组；

8）我们将第三层的 24 个醋变量和第三层的 3 个油变量 $\bar{x}_0,\bar{x}_1,...,\bar{x}_{26}$ 共 27 个元素作为第四层的醋变量代入中心映射变换 F 的第四层 10 个多变量多次多项式 $f_{17},f_{18},...,f_{26}$ 中，并计算它们的系数；

9）$f_{17},f_{18},...,f_{26}$ 被转化为关于第四层的 10 个油变量 $\bar{x}_{27},\bar{x}_{28},...,\bar{x}_{36}$ 的线性方程组，它的系数矩阵规模是 10×10，然后我们求解这个线性方程组。

到此为止，$\bar{x}(\bar{x}_0,\bar{x}_1,...,\bar{x}_{36})$ 的 37 个元素都计算完成，并求得数值。

最后，我们将 $\bar{x}(\bar{x}_0,\bar{x}_1,...,\bar{x}_{36})$ 的 37 个元素代入可逆仿射 L_2 的逆变换 L_2^{-1}，计算它的逆变换得 $x(x_0,x_1,...,x_{36})$，即消息的散列值的 Rainbow 签名。

$$x = L_2^{-1}(\bar{x}_0,\bar{x}_1,...,\bar{x}_{36})$$

可逆仿射的逆变换 L_2^{-1} 有如下形式：

$$x = C\bar{x} + d$$

这里，C 是一个规模为 37×37 的矩阵，d 是一个维度为 37 的向量，C 和 d 都被当作私钥来运算。

所以 $x(x_0,x_1,...,x_{36})$ 是 $y(y_0,y_1,y_2,...,y_{26})$ 的签名。到此，我们采用的高效 Rainbow 签名方案介绍完毕。

6.3　选择特定有限域的不可约多项式

$GF(2^8)$ 的多项式可以被表示为 $x_8 + x_k + \ldots + 1(0 < k < 8)$，它的特点是第一项和最后一项分别为 1。总共有 16 个多项式可以被选择为 $GF(2^8)$ 的不可约多项式。我们用二进制来表示这些多项式的系数，并把它们统计在表 6-2 中。在这些多项式中，项数为 5 的有 12 个，项数为 7 的有 4 个。一般来说，项数越少的多项式，作为有限域的不可约多项式对密码硬件性能提升得效果越好。下面我们根据优化有限域的运算（乘法和求逆）的结果，来确定 $GF(2^8)$ 的不可约多项式。

表 6-2　　　　　　　　　　$GF(2^8)$ 的不可约多项式

项数	$GF(2^8)$ 的不可约多项式	数目
5	$(100011101)_2$　$(101110001)_2$　$(100101011)_2$　$(110101001)_2$ $(101101001)_2$　$(101100011)_2$　$(110001101)_2$　$(101001101)_2$ $(100101101)_2$　$(111000011)_2$　$(101100101)_2$　$(110000111)_2$	12
7	$(111001111)_2$　　$(111100111)_2$　$(101011111)_2$　$(111110101)_2$	4

6.4　优化特定有限域的乘法

我们选取的有限域乘法的算法是多项式基的一般乘法。我们假定 $a(x)$ 和 $b(x)$ 是 $GF(2^8)$ 的元素，那么 $c(x) = (a(x) \times b(x)) \bmod f(x)$，其中 $f(x)$ 是 $GF(2^8)$ 的不可约多项式。我们先计算多项式相乘的结果，即 $S_0, S_1, \ldots, S_{14}$。

$$S_0 = a_0 b_0$$
$$S_1 = a_1 b_0 + a_0 b_1$$
$$S_2 = a_2 b_0 + a_1 b_1 + a_0 b_2$$
$$S_3 = a_3 b_0 + a_2 b_1 + a_1 b_2 + a_0 b_3$$
$$S_4 = a_4 b_0 + a_3 b_1 + a_2 b_2 + a_1 b_3 + a_0 b_4$$
$$S_5 = a_5 b_0 + a_4 b_1 + a_3 b_2 + a_2 b_3 + a_1 b_4 + a_0 b_5$$
$$S_6 = a_6 b_0 + a_5 b_1 + a_4 b_2 + a_3 b_3 + a_2 b_4 + a_1 b_5 + a_0 b_6$$
$$S_7 = a_7 b_0 + a_6 b_1 + a_5 b_2 + a_4 b_3 + a_3 b_4 + a_2 b_5 + a_1 b_6 + a_0 b_7$$

$$S_8 = a_7b_1 + a_6b_2 + a_5b_3 + a_4b_4 + a_3b_5 + a_2b_6 + a_1b_7$$

$$S_9 = a_7b_2 + a_6b_3 + a_5b_4 + a_4b_5 + a_3b_6 + a_2b_7$$

$$S_{10} = a_7b_3 + a_6b_4 + a_5b_5 + a_4b_6 + a_3b_7$$

$$S_{11} = a_7b_4 + a_6b_5 + a_5b_6 + a_4b_7$$

$$S_{12} = a_7b_5 + a_6b_6 + a_5b_7$$

$$S_{13} = a_7b_6 + a_6b_7$$

$$S_{14} = a_7b_7$$

接下来是模运算，这里以 $GF(2^8)$ 的不可约多项式 $(101001101)_2$ 为例说明计算过程，$c(x)(c_7, c_6, \ldots, c_0)$ 是 $a(x)$ 和 $b(x)$ 的乘积。

$$c_7 = S_7 + S_9 + S_{11} + S_{12}$$

$$c_6 = S_6 + S_8 + S_{10} + S_{11}$$

$$c_5 = S_5 + S_{10} + S_{11} + S_{12} + S_{14}$$

$$c_4 = S_4 + S_9 + S_{10} + S_{11} + S_{14} + S_{14}$$

$$c_3 = S_3 + S_8 + S_9 + S_{10} + S_{12} + S_{13} + S_{14}$$

$$c_2 = S_2 + S_8 + S_{13} + S_{14}$$

$$c_1 = S_1 + S_9 + S_{11} + S_{13} + S_{14}$$

$$c_0 = S_0 + S_8 + S_{10} + S_{12} + S_{13}$$

依据相同的算法，我们分别选取了 16 种不同的多项式作为 $GF(2^8)$ 的不可约多项式，并在一款低资源的 FPGA（Altera Cyclone EP1C12Q240C8）上实现了相应的算法。

表 6-3 是乘法的实验数据，我们可以从表中得知，使用逻辑单元数最少的 3 个乘法的多项式分别是 $(101100011)_2$、$(111001111)_2$ 和 $(100101101)_2$，乘法运算时间最短的 3 个乘法的多项式分别是 $(111000011)_2$、$(100101101)_2$ 和 $(101110001)_2$。

表 6-3　不同的不可约多项式的 $GF(2^8)$ 乘法

多项式的系数	与逻辑门数	异或逻辑门数	逻辑单元数	运算时间/ns
$(100011101)_2$	64	76	55	19.542
$(101110001)_2$	64	77	53	18.426
$(100101011)_2$	64	81	51	18.852
$(110101001)_2$	64	79	55	19.610
$(100101101)_2$	64	71	49	17.647
$(101101001)_2$	64	85	52	20.287
$(101100011)_2$	64	79	47	19.473

续表

多项式的系数	与逻辑门数	异或逻辑门数	逻辑单元数	运算时间/ns
$(110001101)_2$	64	79	50	19.797
$(111001111)_2$	64	82	48	21.431
$(111100111)_2$	64	79	50	21.975
$(101001101)_2$	64	81	50	18.712
$(101100101)_2$	64	80	51	20.451
$(101011111)_2$	64	78	50	24.251
$(111110101)_2$	64	79	53	23.768
$(110000111)_2$	64	81	51	18.481
$(111000011)_2$	64	81	51	17.566

6.5 优化特定有限域的求逆

$GF(2^8)$ 的求逆在 Rainbow 签名中使用得并不是很多,主要存在于求解线性方程组中。我们采用了基于费马小定理的求逆方法,假定 β 是 $GF(2^8)$ 的元素,那么求逆过程如下:

$$\beta^{2^8} = \beta$$
$$\beta^{-1} = \beta^{2^8-2} = \beta^{254}$$
$$2^8 - 2 = 2 + 2^2 + 2^4 + 2^5 + 2^6 + 2^7$$
$$\beta^{-1} = \beta^2 \beta^4 \beta^8 \beta^{16} \beta^{32} \beta^{64} \beta^{128}$$

计算 β^2、β^4、β^8、β^{16}、β^{32}、β^{64} 和 β^{128} 是可以并行进行的,β^{-1} 是它们的乘积。

依据相同的算法,我们分别选取了 16 种不同的多项式作为 $GF(2^8)$ 的不可约多项式,并在一款低资源的 FPGA(Altera Cyclone EP1C12Q240C8)上实现了相应的算法。

表 6-4 是求逆的实验数据,使用逻辑单元数最少的 3 个求逆的多项式分别是 $(101101001)_2$、$(111100111)_2$ 和 $(101100011)_2$,求逆运算时间最短的 3 个求逆的多项式分别是 $(111000011)_2$、$(101101001)_2$ 和 $(110101001)_2$。

表 6-4　　　　不同的不可约多项式的 $GF(2^8)$ 求逆

多项式的系数	与逻辑门数	异或逻辑门数	逻辑单元数	运算时间/ns
$(100011101)_2$	384	581	446	35.104
$(101110001)_2$	384	603	474	35.074

续表

多项式的系数	与逻辑门数	异或逻辑门数	逻辑单元数	运算时间/ns
$(100101011)_2$	384	634	478	36.163
$(110101001)_2$	384	598	433	34.881
$(100101101)_2$	384	567	357	34.491
$(101101001)_2$	384	624	337	36.589
$(101100011)_2$	384	607	353	38.547
$(110001101)_2$	384	617	372	36.122
$(111001111)_2$	384	610	367	41.096
$(111100111)_2$	384	613	347	39.929
$(101001101)_2$	384	624	424	35.731
$(101100101)_2$	384	614	470	36.472
$(101011111)_2$	384	589	470	35.421
$(11111010101)_2$	384	617	469	35.292
$(110000111)_2$	384	623	466	35.531
$(111000011)_2$	384	616	463	34.385

6.6 优化特定有限域的求解线性方程组

我们选择实现的 Rainbow 签名由 4 层油醋多项式结构组成，每层都要求解一个线性方程组，所以它是提升 Rainbow 签名性能的一个瓶颈。综合优化的乘法和求逆结果，我们选取了 $(101101001)_2$ 作为 $GF(2^8)$ 的不可约多项式，即 $x^8 + x^6 + x^5 + x^3 + 1$，这是因为它在 4 项指标中有 3 项均排在前三，分别是乘法的最少逻辑单元数、乘法的最短时间和求逆的最短时间。

我们采取了有限域的高斯消元法作为求解线性方程组的方法：
1）在主元所在列选择一个非零元素作为主元；
2）交换当前行与主元所在行的所有的元素，若当前行是主元所在行则不交换；
3）对当前行所有的元素做归一操作；
4）对当前行下面所有的元素做消元操作；
5）结束本次迭代，重新选取下一列开始下一轮迭代；
6）直到所有迭代完成，线性方程组的系数矩阵成为一个等效的上三角矩阵；
7）接着使用回溯替代的方法对方程组的增广矩阵进行替代；

8）完成回溯替代后，增广矩阵的最右一列（即常数项组成的向量）是线性方程组的解。

我们的 Rainbow 签名需要求解 4 个线性方程组，这些方程组的系数矩阵的规模分别是 10×10、4×4、3×3 和 10×10。我们同时考虑了求解线性方程组所需要的时间和面积，最后采用了 11 个 $GF(2^8)$ 乘法器。由于一个乘法在一个时钟周期内完成运算，这样能保证每一行的操作能在一个时钟周期内完成。我们在一款低资源的 FPGA（Altera Cyclone EP1C12Q240C8）上实现了相应的算法，表 6-5 所示是实验数据。

表 6-5　求解高效 Rainbow 签名算法中的 $GF(2^8)$ 的线性方程组

层次	线性方程组的系数矩阵的规模	运算时间/μs
1	10×10	3.86
2	4×4	1.43
3	3×3	1.03
4	10×10	3.86
总共	—	10.18

6.7　技术实现

我们采用了硬件描述语言 VHDL 作为编程语言，Altera 公司开发的 FPGA 工具 Quartus II 64-bit version 8.0 作为编程工具。另外，实现平台我们选取的是 Altera Cyclone FPGA（芯片型号是 EP1C12Q240C8），这是一款低资源的 FPGA。

我们采用了状态机的方式在 FPGA 上实现我们的设计，一次签名所需时钟周期数是 2 570，签名的时间为 102.8μs，估计的等效门数约为 15 490。我们用密码运行的时间作为密码硬件的速度的评估指标，需要等效门数作为密码硬件的面积的评估指标，这是因为相关密码硬件使用的工艺和技术均不尽相同。

6.8　实现对比

我们将高效 Rainbow 签名硬件与相关签名算法的实现进行了对比，结果统计如表 6-6 所示，其中 en-TTS 和 UOV 都是多变量公钥密码算法。因为它们的实现方式不一，所以我们用等效门的数目来评估面积。我们用时钟周期数来评估时间，这是因为大多数的密码硬件使用的时钟频率都比较接近。我们的面积-周期的乘积被归化为 1，这项指标能够全面地体现密码硬件的性能。对比结果显示我们的实现最高效，能够在保证速度的前提下高效地进行 Rainbow 签名。

表 6-6　　　　　　　　　　　公钥密码硬件的对比

方案	面积（等效门数）	时钟周期数	面积-周期的乘积*
参考文献［1］中的 RSA	250 000	348 672	2 189.7
参考文献［2］中的椭圆曲线	191 000	88 000	422.3
参考文献［3］中的椭圆曲线	25 000	469 385	294.8
参考文献［4］中的 en-TTS	21 000	60 000	31.7
参考文献［5］中的 UOV	227 500	2 300	13.2
参考文献［6］中的 Rainbow	63 593	804	1.3
本章的 Rainbow	15 490	2 570	1

*我们的高效 Rainbow 签名硬件的面积-周期的乘积被归化为 1。

6.9　本章小结

多变量公钥密码算法是少数具有抵御量子计算攻击能力的公钥密码算法之一。研究多变量公钥密码芯片不但对它有重要的理论意义和经济意义，而且对它的发展和推广起了重要作用。在多变量公钥密码中，Rainbow 签名算法是目前应用最广泛的密码算法。

在本章中，我们针对密码硬件设计的方向之一——高效实现，设计了一种高效的 Rainbow 签名硬件，它同时对运行的时间和使用的面积进行了优化。通过优化组合有限域的乘法器、求逆器、求解线性方程组的硬件装置等设计以及提出新的设计思路，我们在 Altera FPGA 上实现了 Rainbow 签名的硬件。实验结果表明高效 Rainbow 签名硬件显著地提升了 Rainbow 签名的性能。

在评估一个密码硬件的性能时，我们一般采用密码硬件的运行时间—使用面积的乘积作为指标来衡量，即运行的时间越短、使用的面积越小，则它的性能越好。我们的设计基于时间和面积两方面的优化，既缩短了 Rainbow 签名的时间，又减小了它需要的面积。它的设计基于优化选择有限域的不可约多项式，实现高效的有限域乘法、求逆和求解线性方程组。实验结果证明了我们的设计显著地减小了 Rainbow 所需的时间-面积的乘积。与相关公钥密码硬件进行对比，结果表明我们的设计更加高效。我们的设计也证明了多变量公钥密码算法是最高效的公钥密码算法之一。

6.10 本章参考文献

[1] Großschädl J. High-speed RSA hardware based on Barret's modular reduction method[A]. In: Cryptographic Hardware and Embedded Systems—CHES 2000[C], 2000: 191-203.

[2] Schroeppel R, Beaver C, Gonzales R, et al. A low-power design for an elliptic curve digital signature chip[A]. In: Cryptographic Hardware and Embedded Systems-Ches 2002[C]. Springer, 2003: 366-380.

[3] Aigner H, Bock H, Hütter M, et al. A Low-Cost ECC Coprocessor for Smartcards[A]. In: Cryptographic Hardware and Embedded Systems-CHES 2004[C]. Springer, 2004: 107-118.

[4] Yang B, Cheng C, Chen B, et al. Implementing minimized multivariate PKC on low-resource embedded systems[J]. Security in Pervasive Computing, 2006: 73-88.

[5] Bogdanov A, Eisenbarth T, Rupp A, et al. Time-Area Optimized Public-Key Engines: MQCryptosystems as Replacement for Elliptic Curves?[A]. In: Cryptographic Hardware and Embedded Systems - CHES 2008[C]. Springer Berlin Heidelberg. 2008. Lecture Notes in Computer Science, 5154: 45-61.

[6] Balasubramanian S, Bogdanov A, Rupp A, et al. Fast Multivariate Signature Generation in Hardware: The Case of Rainbow[A]. In: 16th International Symposium on Field-Programmable Custom Computing Machines[C], 2008: 281-282.

第 7 章 多变量公钥密码处理器技术

7.1 本章概述

由于信息安全产业的发展非常迅速，密码硬件在越来越广泛的领域，例如数学、工程和通信领域发挥出重要的作用。密码算法的选择对密码硬件的设计非常关键。多变量公钥密码算法是非常重要的公钥密码算法之一，它是设计密码硬件的一个很好的选择。

在过去的 30 年中，多变量公钥密码算法是密码学术界研究的热点，其中有代表性的密码算法包括非平衡油醋签名、TTS 签名和 Rainbow 签名方案。因为多变量公钥密码算法一般运行于 $GF(2^n)$，而且它具有抵御潜在的量子计算机攻击的能力，所以非常适合设计成密码硬件。

大部分的多变量公钥密码硬件设计以优化它的运行速度作为主要目标，而很少专注于减小它的实现面积。这导致多变量公钥密码硬件的面积普遍比其他公钥密码硬件的面积大，例如 RSA 和椭圆曲线密码硬件。

在公钥密码硬件中，以面积优化为主的包括以下 4 种设计：

1）参考文献 [1] 中的多变量公钥密码硬件运行平衡油醋签名、Rainbow 签名和 TTS 签名等多变量公钥密码算法，它是一个 FPGA 上的设计；

2）参考文献 [2] 中的设计是目前 FPGA 上最小的 RSA 密码硬件；

3）参考文献 [3] 中的 MicroECC 是一个轻量级的椭圆曲线密码处理器，它是最小的椭圆曲线密码硬件之一；

4）参考文献 [4] 中的椭圆曲线密码硬件只需要 184 个 Slice，它是目前 FPGA 上最小的椭圆曲线密码硬件。

在本章中，我们设计了一种小面积的多变量公钥密码处理器。它的设计目的是缩小有限域运算的面积，减少寄存器和精简密码指令集的数目。所以，我们的设计主要包括 3 个方向：

1）我们设计了一个模运算逻辑单元，它适用于复合有限域，提供 $GF((2^n)^2)$ 的加法、乘法和求逆；

2）我们设计了一个精简的指令集和解码器；

3）我们对寄存器采取了分时复用的方式缩小实现面积。

通过其他优化，例如，设计可逆仿射变换、多变量多次多项式的系数求值和求解线性方程组的精简算法，以及整合上述设计，我们的多变量公钥密码处理器在一个低资源的 Xilinx FPGA 上实现，它只使用了非常有限的器件资源。

为了验证和测试我们的设计，我们在小面积多变量公钥密码处理器上实现了 3 种多变量公钥密码算法，即非平衡油醋签名、en-TTS 签名和 Rainbow 签名方案。选择这 3 种密码算法的原因如下：

1）到目前为止，不存在一种既安全又高效的多变量公钥加密算法，所以没有加密算法入选；

2）这 3 种密码算法是典型的多变量数字签名方案，它们能够全面地反映我们的密码处理器处理多变量公钥密码算法的能力；

3）这 3 种密码算法的实际应用非常广泛，例如在通信领域和工业领域，Rainbow 签名方案就是一个高效的选择。

这 3 种多变量数字签名方案在我们的处理器上运行的速度适中，可以满足绝大部分的应用。当然我们的密码处理器并不局限于使用这 3 种密码算法，还可以实现其他多变量公钥密码算法以及部分不同类型的密码算法。

我们将小面积多变量公钥密码处理器与其他公钥密码硬件进行了对比，结果表明我们的设计的面积比目前公认的最小的公钥密码硬件更小。

7.2 架构设计

我们设计了一个小面积的多变量公钥密码处理器，它由微控制器、总线管理、模运算逻辑单元和寄存器等器件组成，还与一个外置的 RAM 相连。我们的设计描绘在图 7-1 中，连接线上的数字表示数据宽度，实线表示数据的交换，虚线表示信号控制。

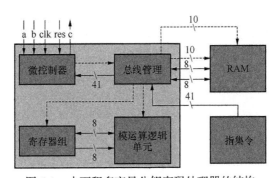

图 7-1　小面积多变量公钥密码处理器的结构

下面介绍多变量公钥密码处理器的各个组成部分：

1）处理器的指令长度是 41 比特，通过微控制器实现指令的解码译码；

2）微控制器的输入输出数据端口 A、B、C 分别用于传输待签名信息的散列值、签名和私钥 F、L、L_1、L_2 等，其中 F 是中心映射变换，L、L_1、L_2 是可逆仿射变换。它的输入端口 clk 和 res 分别是时钟信号和复位信号；

3）总线管理是 RAM 和其他部件交换数据的通道；

4）RAM 用于存储模运算逻辑单元的运算的中间结果，它地址长度是 10 比特，以 1 字节作为最小的存储单元；

5）在寄存器组中，一个寄存器存储 1 字节，主要被模运算逻辑单元使用；

6）模运算逻辑单元提供比特并行的模运算，即 $GF((2^4)^2)$ 的加法、乘法和求逆，其中我们选定的 $GF((2^4)^2)$ 的不可约多项式是 x^2+x+9。

我们的设计关注减小模运算逻辑单元、微控制器和寄存器组的实现面积，这是因为它们占据了 90% 的设计面积。

7.3 多变量数字签名方案和参数的选择

多变量数字签名算法的基本思想是利用有限域的多变量多次方程的有限集合作为密码体制的公钥，通常它是二次方程的有限集合。大部分的多变量数字签名方案都由一个中心映射以及两个用于隐藏中心映射的可逆仿射组成，其中，中心映射和两个可逆仿射是私钥，而公钥则是其中一个可逆仿射、中心映射和另一个可逆仿射的依次组合。

我们选取的用于密码处理器的密码算法，包括非平衡油醋签名、TTS 签名和 Rainbow 签名方案，它们是典型的多变量数字签名方案，能够全面地反映我们的密码处理器处理多变量公钥密码算法的能力。

接下来我们分别介绍这 3 种方案。

非平衡油醋签名算法（UOV）是由 3 位学者 Aviad Kipnis、Jacques Patarin 和 Louis Goubin 在 1999 年共同提出的，它属于油醋签名家族。在目前的非平衡油醋签名方案中，UOV(28,56)被普遍认为是一种安全的多变量数字签名方案，它的安全等级可以达到 2^{83}。UOV(28,56)的参数分别指的是油变量的数目（28 个）和醋变量的数目（56 个），它运行于有限域 $GF(256)$。

我们选择了 UOV(28,56)作为非平衡油醋签名的实现方案，具体实现参数描述在表 7-1 中。复合有限域 $GF((2^4)^2)$ 被选择为密码算法运行的基础域，它是有限域的一种特殊表现形式；待签名消息的散列的长度是 28 字节，消息签名后的长度是 84 字节；私钥包括一个可逆仿射变换和中心映射变换，公钥是中心映射变换和可逆仿射变换的组合。

7.3 多变量数字签名方案和参数的选择

表 7-1   非平衡油醋签名方案的选择

非平衡油醋签名的参数	非平衡油醋签名的数值
有限域	复合有限域 $GF((2^4)^2)$
消息的散列长度	28 字节
签名长度	84 字节
醋变量的个数	56
油变量的个数	28
私钥	可逆仿射变换和中心映射变换
公钥	可逆仿射变换和中心映射变换的组合

接下来我们介绍选定的非平衡油醋签名方案。我们假定消息的散列值是 $y(y_0, y_1, \ldots, y_{27})$，它的长度是 28 字节，其中 $y(y_0, y_1, \ldots, y_{27})$ 是 $GF((2^4)^2)$ 的元素。而且，我们假定签名是 $x(x_0, x_1, \ldots, x_{83})$，它的长度是 84 字节，其中 $x_0, x_1, \ldots, x_{83}$ 是 $GF((2^4)^2)$ 的元素。

为了对消息的散列值 $y(y_0, y_1, \ldots, y_{27})$ 签名，我们需要进行如下计算：

$$L(x_0, x_1, \ldots, x_{83}) = F^{-1}(y_0, y_1, \ldots, y_{27})$$

这里，F 是一个中心映射变换，L 是一个可逆的仿射变换。

为了实现上述计算，我们首先要做中心映射的逆变换，计算过程如下：

$$\overline{x} = F^{-1}(y_0, y_1, \ldots, y_{27})$$

这里，$\overline{x}(\overline{x}_0, \overline{x}_1, \ldots, \overline{x}_{83})$ 是变换的结果，$\overline{x}$ 的长度是 84 字节。
中心映射变换 F 由 28 个多变量多次多项式 $(f_0, f_1, \ldots, f_{27})$ 组成，它有如下的形式：

$$F(\overline{x}_0, \overline{x}_1, \ldots, \overline{x}_{83}) = (f_0, f_1, \ldots, f_{27})$$

事实上 $\overline{x}(\overline{x}_0, \overline{x}_1, \ldots, \overline{x}_{83})$ 是醋变量和油变量的有限集合：
1）$\overline{x}_0, \overline{x}_1, \ldots, \overline{x}_{55}$ 是醋变量的有限集合，共 56 个醋变量；
2）$\overline{x}_{56}, \overline{x}_{57}, \ldots, \overline{x}_{83}$ 是油变量的有限集合，共 28 个油变量。
多变量多次多项式 $f_0, f_1, \ldots, f_{27}$ 的定义如下：

$$f(O_0, O_1, \ldots, O_{27}) = \sum \alpha_{ij} O_i V_j + \sum \beta_{ij} V_i V_j + \sum \gamma_i V_i + \sum \delta_i O_i + \eta$$

这里，$O_i, (V_i, V_j)$ 分别是油变量和醋变量，α_{ij}、β_{ij}、γ_i、δ_i 和 η 是多变量多次多项式的系数，被当作私钥使用。

多变量多次多项式包括 5 项，最高次数不超过二次。若将醋变量的数值代入多变量多次多项式，它将变换成关于油变量的一次多项式。这 5 项分别为：
1）$O_i V_j$，油变量和醋变量的组合；

2)V_iV_j,醋变量和醋变量的组合;

3)V_i,醋变量;

4)O_i,油变量;

5)η,常数。

计算中心映射变换的步骤如下:

1)我们为 56 个醋变量 $\bar{x}_0,\bar{x}_1,\ldots,\bar{x}_{55}$ 选择随机的数值;

2)我们将 56 个醋变量 $\bar{x}_0,\bar{x}_1,\ldots,\bar{x}_{55}$ 的数值代入 28 个多变量多次多项式 $f_0,f_1,\ldots,f_{27}$ 中,并计算它们的系数;

3)$f_0,f_1,\ldots,f_{27}$ 被转化为只关于油变量 $\bar{x}_{56},\bar{x}_{57},\ldots,\bar{x}_{83}$ 的线性方程组,它的系数矩阵的规模是 28×28,然后我们求解这个线性方程组。

到此为止,我们计算完成并获得所有变量 $\bar{x}_0,\bar{x}_1,\ldots,\bar{x}_{83}$ 的数值。

最后,我们将 56 个醋变量和 28 个油变量 $\bar{x}(\bar{x}_0,\bar{x}_1,\ldots,\bar{x}_{83})$ 代入可逆仿射变换,计算它的逆变换得 $\bar{x}(\bar{x}_0,\bar{x}_1,\ldots,\bar{x}_{83})$,即消息的散列值的非平衡油醋签名。

$$x = L^{-1}(\bar{x}_0,\bar{x}_1,\ldots,\bar{x}_{83})$$

可逆仿射的逆变换 L^{-1} 有如下形式:

$$x = A\bar{x} + b$$

这里,A 是一个规模为 84×84 的矩阵,b 是一个维度为 84 的向量,A 和 b 都被当作私钥来运算。

所以 $x(x_0,x_1,\ldots,x_{83})$ 是 $y(y_0,y_1,\ldots,y_{27})$ 的签名。到此,我们采用的非平衡油醋签名介绍完毕。

我们下面介绍 Rainbow 签名方案,它是由两位美国学者 Jintai Ding 和 Dieter Schmidt 在 2005 年共同提出的。它是油醋签名家族的一员,可以被看成多层的非平衡油醋方案。在已有的 Rainbow 签名方案中,Rainbow(17,13,13)被普遍认为安全、可靠,它的安全程度能达到 2^{80}。Rainbow(17,13,13) 中的参数指的分别是第一层的醋变量(17 个)、第一层的油变量的数目(13 个)和第二层的油变量(13 个),它的基域是有限域 $GF(256)$。

所以我们选择了 Rainbow(17,13,13)作为 Rainbow 的实现方案,具体实现参数描述如表 7-2 所示。复合有限域 $GF((2^4)^2)$ 被选择为密码算法运行的基础域,它是有限域的一种特殊表现形式。待签名消息的散列的长度是 26 字节,消息签名后的长度是 43 字节。它由两层的非平衡油醋结构组成,第一层醋变量的数目是 17 个,油变量的数目是 13 个;第二层醋变量的数目是 30 个,油变量的数目是 13 个,其中第二层的醋变量是由第一层的醋变量和第一层的油变量组成的。私钥由两个可逆仿射变换和中心映射变换组成,公钥是其中一个可逆仿射变换、中心映射变换和另一个可逆仿射变换的依次组合。

7.3 多变量数字签名方案和参数的选择

表 7-2   Rainbow 签名方案的选择

Rainbow 签名的参数	Rainbow 签名的数值
有限域	复合有限域 $GF((2^4)^2)$
消息的散列长度	26 字节
签名长度	43 字节
油醋结构的层数	2
每层的醋油变量的数目	17, 30
每层的油变量的数目	13, 13
私钥	两个可逆仿射变换和中心映射变换
公钥	一个仿射变换、中心映射变换和另一个仿射变换的依次组合

接下来我们介绍选定的 Rainbow 签名方案。我们假定消息的散列值是 $y(y_0, y_1, \ldots, y_{25})$，它的长度是 26 字节，其中 $y_0, y_1, \ldots, y_{25}$ 是 $GF((2^4)^2)$ 的元素。而且，我们假定签名是 $x(x_0, x_1, \ldots, x_{42})$，它的长度是 43 字节，其中 $x_0, x_1, \ldots, x_{42}$ 是 $GF((2^4)^2)$ 的元素。

为了对消息的散列值 $y(y_0, y_1, \ldots, y_{25})$ 签名，我们需要进行如下计算：

$$F \circ L_2(x_0, x_1, \ldots, x_{42}) = L_1^{-1}(y_0, y_1, \ldots, y_{25})$$

这里，F 是一个中心映射变换，L_1 和 L_2 是两个可逆的仿射变换。

为了实现上述计算，我们首先要计算可逆仿射 L_1 的逆变换，计算过程如下：

$$\bar{y} = L_1^{-1}(y_0, y_1, \ldots, y_{25})$$

这里 $\bar{y}(\bar{y}_0, \bar{y}_1, \ldots, \bar{y}_{25})$ 是可逆仿射变换的结果，$\bar{y}$ 的长度是 26 字节。

可逆仿射 L_1 的逆变换 L_1^{-1} 有如下形式：

$$\bar{y} = Ay + b$$

这里，A 是一个规模为 26×26 的矩阵，b 是一个维度为 26 的向量，A 和 b 都被当作私钥来运算。

通过求解中心映射 F 的逆变换，我们获得 $\bar{x}(\bar{x}_0, \bar{x}_1, \ldots, \bar{x}_{25})$，$\bar{x}$ 的长度是 26 字节。

$$\bar{x} = F^{-1}(\bar{y}_0, \bar{y}_1, \ldots, \bar{y}_{25})$$

中心映射变换 F 由 26 个多变量多次多项式 $(f_0, f_1, \ldots, f_{25})$ 组成，它有如下形式：

$$F(\bar{x}_0, \bar{x}_1, \ldots, \bar{x}_{42}) = (f_0, f_1, \ldots, f_{25})$$

那么，我们将 $\bar{y}(\bar{y}_0, \bar{y}_1, \ldots, \bar{y}_{25})$ 代入中心映射的变换中，它变成如下形式：

$$\bar{y}(\bar{y}_0, \bar{y}_1, \ldots, \bar{y}_{25}) = f(f_0, f_1, \ldots, f_{25})$$

中心映射 F 是一个两层的油醋结构，由 26 个多变量多次多项式组成，即 $(f_0, f_1, \ldots, f_{25})$ 被分成两层：

1）$f_i | i = 0, 1, \ldots, 12$，共 13 个多项式；

2）$f_i | i = 13, 14, \ldots, 25$，共 13 个多项式。

同时，中心映射变换 F 中的 $\bar{x}(\bar{x}_0, \bar{x}_1, \ldots, \bar{x}_{42})$ 也可以分成 4 组，分组情况如下：

1）$\bar{x}_i | i = 0, 1, \ldots, 16$，共 17 个元素组成了第一层的醋变量；

2）$\bar{x}_i | i = 17, 18, \ldots, 29$，共 13 个元素组成了第一层的油变量；

3）$\bar{x}_i | i = 0, 1, \ldots, 29$，共 30 个元素组成了第二层的醋变量；

4）$\bar{x}_i | i = 30, 31, \ldots, 42$，共 13 个元素组成了第二层的油变量。

这里，第二层的醋变量是由第一层的醋变量和第一层的油变量组成的。

中心映射变换 F 的 26 个多变量多次多项式 ($f_0, f_1, \ldots, f_{25}$) 的定义如下：

$$f(O_0, O_1, \ldots, O_{25}) = \sum \alpha_{ij} O_i V_j + \sum \beta_{ij} V_i V_j + \sum \gamma_i V_i + \sum \delta_i O_i + \eta$$

这里，$O_i, (V_i, V_j)$ 分别是油变量和醋变量，α_{ij}、β_{ij}、γ_i、δ_i 和 η 是多变量多次多项式的系数，被当作私钥使用。

多变量多次多项式包括 5 项，最高次数不超过二次。若将醋变量的数值代入多变量多次多项式，它将变换成关于油变量的一次多项式。这 5 项分别为

1）$O_i V_j$，油变量和醋变量的组合；

2）$V_i V_j$，醋变量和醋变量的组合；

3）V_i，醋变量；

4）O_i，油变量；

5）η，常数。

计算中心映射变换的步骤如下：

1）我们为第一层的 17 个醋变量 $\bar{x}_0, \bar{x}_1, \ldots, \bar{x}_{16}$ 选择随机的数值；

2）我们将 $\bar{x}_0, \bar{x}_1, \ldots, \bar{x}_{16}$ 的数值代入中心映射变换 F 的第一层 13 个多变量多次多项式 $f_0, f_1, \ldots, f_{12}$，并计算它们的系数；

3）$f_0, f_1, \ldots, f_{12}$ 被转化为只关于第一层的 13 个油变量 $\bar{x}_{17}, \bar{x}_{18}, \ldots, \bar{x}_{29}$ 的线性方程组，它的系数矩阵的规模是 13×13，然后我们求解这个线性方程组；

4）我们将第一层的 17 个醋变量和第一层的 13 个油变量 $\bar{x}_0, \bar{x}_1, \ldots, \bar{x}_{29}$ 共 30 个元素作为第二层的醋变量代入中心映射变换 F 的第二层 13 个多变量多次多项式 $f_{13}, f_{14}, \ldots, f_{25}$ 中，并计算它们的系数；

5）$f_{13}, f_{14}, \ldots, f_{25}$ 被转化为只关于第二层的 13 个油变量 $\bar{x}_{30}, \bar{x}_{31}, \ldots, \bar{x}_{42}$ 的线性方程组，它的系数矩阵的规模是 13×13，然后我们求解这个线性方程组。

到此为止，$\bar{x}(\bar{x}_0, \bar{x}_1, \ldots, \bar{x}_{42})$ 的 43 个元素都计算完成，并求得数值。

最后，我们将 $\bar{x}(\bar{x}_0, \bar{x}_1, \ldots, \bar{x}_{42})$ 的 43 个元素代入可逆仿射 L_2 的逆变换 L_2^{-1}，计算它的逆变换得 $x(x_0, x_1, \ldots, x_{42})$，即消息的散列值的 Rainbow 签名。

$$x = L_2^{-1}(\bar{x}_0, \bar{x}_1, \ldots, \bar{x}_{42})$$

可逆仿射的逆变换 L_2^{-1} 有如下形式：
$$x = C\bar{x} + d$$

这里，C 是一个规模为 43×43 的矩阵，d 是一个长度为 43 的向量，C 和 d 都被当作私钥来运算。

所以 $x(x_0, x_1, \ldots, x_{42})$ 是 $y(y_0, y_1, \ldots, y_{25})$ 的签名。到此，我们采用的 Rainbow 签名介绍完毕。

我们下面介绍 en-TTS 签名方案。它是由中国台湾地区的两位学者 Bo-Yin Yang 和 Jiun-Ming Chen 在 2005 年共同提出的。它属于三角签名家族，可以被看成多变量公钥加密算法 TTM（Tame Transformation Method）的变体。在已有的 en-TTS 签名方案中，en-TTS(20,28) 被认为是目前最快的签名方案之一，它的参数指的分别是待签名消息的散列长度（20 字节）和签名的长度（28 字节），它的基域是有限域 $GF(256)$。尽管它已经被攻破，不再安全，但是我们仍然把它作为多变量公钥密码算法的例子在密码处理器上实现，主要原因是以下几点：

1) en-TTS(20,28) 作为多变量公钥密码算法的代表，是目前最快的签名方案之一；
2) en-TTS(20,28) 的相关设计和实现是多变量公钥密码领域的热点。

所以，我们选择了 en-TTS(20,28) 作为 en-TTS 的实现方案，具体实现参数描述如表 7-3 所示。复合有限域 $GF((2^4)^2)$ 被选择为密码算法运行的基础域，它是有限域的一种特殊表现形式；待签名消息的散列的长度是 20 字节，签名的长度是 28 字节；私钥由两个可逆仿射变换和中心映射变换组成，公钥是其中一个可逆仿射变换、中心映射变换和另一个可逆仿射变换的依次组合。

表 7-3   **en-TTS 签名方案的选择**

en-TTS 签名的参数	en-TTS 签名的数值
有限域	复合有限域 $GF((2^4)^2)$
消息的散列长度	20 字节
签名长度	28 字节
私钥	两个可逆仿射变换和中心映射变换
公钥	一个仿射变换、中心映射变换和另一个仿射变换的依次组合

接下来我们介绍选定的 en-TTS 签名方案。我们假定消息的散列值是 $y(y_0, y_1, \ldots, y_{19})$，它的长度是 20 字节，其中 $y(y_0, y_1, \ldots, y_{19})$ 是 $GF((2^4)^2)$ 的元素。而且，我们假定签名是 $x(x_0, x_1, \ldots, x_{27})$，它的长度是 28 字节，其中 $x_0, x_1, \ldots, x_{27}$ 是 $GF((2^4)^2)$ 的元素。

为了对消息的散列值 $y(y_0, y_1, \ldots, y_{19})$ 签名，我们需要进行如下计算：
$$F \circ L_2(x_0, x_1, \ldots, x_{27}) = L_1^{-1}(y_0, y_1, \ldots, y_{19})$$

这里，F 是一个中心映射变换，L_1 和 L_2 是两个可逆的仿射变换。

为了实现上述计算，我们首先要做可逆仿射 L_1 的逆变换，计算过程如下：

$$\overline{y} = L_1^{-1}(y_0, y_1, \ldots, y_{19})$$

这里，$\overline{y}(\overline{y}_0, \overline{y}_1, \ldots, \overline{y}_{19})$ 是变换的结果，$\overline{y}$ 的长度是 20 字节。

可逆仿射 L_1 的逆变换 L_1^{-1} 有如下形式：

$$\overline{y} = Ay + b$$

这里，A 是一个规模为 20×20 的矩阵，b 是一个维度为 20 的向量，A 和 b 都被当作私钥来运算。

通过求解中心映射 F 的逆变换，我们获得 $\overline{x}(\overline{x}_0, \overline{x}_1, \ldots, \overline{x}_{27})$，$\overline{x}$ 的长度是 28 字节。

$$\overline{x} = F^{-1}(\overline{y}_0, \overline{y}_1, \ldots, \overline{y}_{19})$$

中心映射变换 F 由 20 个多变量多次多项式 $(f_0, f_1, \ldots, f_{19})$ 组成，它有如下形式：

$$F(\overline{x}_0, \overline{x}_1, \ldots, \overline{x}_{27}) = (f_0, f_1, \ldots, f_{19})$$

那么，我们将 $\overline{y}(\overline{y}_0, \overline{y}_1, \ldots, \overline{y}_{19})$ 代入中心映射的变换中，它变成如下形式：

$$\overline{y}(\overline{y}_0, \overline{y}_1, \ldots, \overline{y}_{19}) = (f_0, f_1, \ldots, f_{19})$$

中心映射变换 F 的 20 个多变量多次多项式 $(f_0, f_1, \ldots, f_{19})$ 的定义如下：

$$f_{i-8} = \overline{x}_i + \sum_{j=1}^{7} p_{ij} \overline{x}_j \quad f_{i-8} = \overline{x} + \sum_{j=1}^{7} p_{ij} \overline{x}_j \overline{x}_{8+((i+j) \bmod 9)} \quad i = 8, 9, \ldots, 16$$

$$f_9 = \overline{x}_{17} + p_{17,1} \overline{x}_1 \overline{x}_6 + p_{17,2} \overline{x}_2 \overline{x}_5 + p_{17,3} \overline{x}_3 \overline{x}_4 + p_{17,4} \overline{x}_9 \overline{x}_{16} + p_{17,5} \overline{x}_{10} \overline{x}_{15} + p_{17,6} \overline{x}_{11} \overline{x}_{14} + p_{17,7} \overline{x}_{12} \overline{x}_{13}$$

$$f_{10} = \overline{x}_{18} + p_{18,1} \overline{x}_2 \overline{x}_7 + p_{18,2} \overline{x}_3 \overline{x}_6 + p_{18,3} \overline{x}_4 \overline{x}_5 + p_{18,4} \overline{x}_{10} \overline{x}_{17} + p_{18,5} \overline{x}_{11} \overline{x}_{16} + p_{18,6} \overline{x}_{12} \overline{x}_{15} + p_{18,7} \overline{x}_{13} \overline{x}_{14}$$

$$f_{i-8} = \overline{x}_i + p_{i,0} \overline{x}_{i-11} \overline{x}_{i-9} + \sum_{j=19}^{i} p_{i,j-18} \overline{x}_{(2i-j)} \overline{x}_j + \sum_{j=i+1}^{27} p_{i,j-18} \overline{x}_{i-j+19} x_j, i = 19, 20, \ldots, 27$$

这里，P_{ij} 是多变量多次多项式的系数，被当作私钥使用。

中心映射变换 F 的 20 个多变量多次多项式 $(f_0, f_1, \ldots, f_{19})$ 可以被分成 3 组，分组情况如下：

1）$f_i | i = 0, 1, \ldots, 8$，共 9 个多项式；
2）$f_i | i = 9, 10$，共 2 个多项式；
3）$f_i | i = 11, 12, \ldots, 19$，共 9 个多项式。

同时，中心映射变换 F 中的 $\overline{x}(\overline{x}_0, \overline{x}_1, \ldots, \overline{x}_{27})$ 也可以分成 4 组，分组情况如下：

1）$\overline{x}_i | i = 0, 1, \ldots, 7$，共 8 个元素；
2）$\overline{x}_i | i = 8, 9, \ldots, 16$，共 9 个元素；
3）$\overline{x}_i | i = 17, 18$，共 2 个元素；
4）$\overline{x}_i | i = 19, 20, \ldots, 27$，共 9 个元素。

计算中心映射变换的步骤如下：

1）我们为 $\overline{x}_i$ 的第一组的 8 个元素 $\overline{x}_0, \overline{x}_1, \ldots, \overline{x}_7$ 选择随机的数值；
2）我们将 $\overline{x}_0, \overline{x}_1, \ldots, \overline{x}_7$ 的数值代入中心映射变换 F 的第一组 9 个多变量多次多项式

$f_0, f_1, ..., f_8$，并计算它们的系数；

3）$f_0, f_1, ..., f_8$ 被转化为只关于 $\bar{x}_i$ 的第二组的 9 个元素 $\bar{x}_8, \bar{x}_9, ..., \bar{x}_{16}$ 的线性方程组，它的系数矩阵的规模是 9×9，然后我们求解这个线性方程组；

4）我们将 $\bar{x}_i$ 的第一组和第二组共 17 个元素的值代入中心映射变换 F 的第二组 2 个多变量多次多项式 f_9, f_{10} 中，就可以计算 $\bar{x}_i$ 的第三组的 2 个元素 $\bar{x}_{17}, \bar{x}_{18}$ 的值；

5）我们将 $\bar{x}_i$ 的第一组、第二组和第三组共 19 个元素的值代入中心映射变换 F 的第三组 9 个多变量多次多项式 $f_{11}, f_{12}, ..., f_{19}$ 中，并计算它们的系数；

6）$f_{11}, f_{12}, ..., f_{19}$ 被转化为只关于 $\bar{x}_i$ 的第四组的 9 个元素 $\bar{x}_{19}, \bar{x}_{20}, ..., \bar{x}_{27}$ 的线性方程组，它的系数矩阵的规模是 9×9，然后我们求解这个线性方程组。

到此为止，$\bar{x}(\bar{x}_0, \bar{x}_1, ..., \bar{x}_{27})$ 的 28 个元素都计算完成，并求得数值。

最后，我们将 $\bar{x}(\bar{x}_0, \bar{x}_1, ..., \bar{x}_{27})$ 的 28 个元素代入可逆仿射 L_2 的逆变换 L_2^{-1}，计算它的逆变换得 $\bar{x}(\bar{x}_0, \bar{x}_1, ..., \bar{x}_{27})$，即消息的散列值的 en-TTS 签名。

$$x = L_2^{-1}(\bar{x}_0, \bar{x}_1, ..., \bar{x}_{27})$$

可逆仿射的逆变换 L_2^{-1} 有如下形式：

$$x = C\bar{x} + d$$

这里，C 是一个规模为 28×28 的矩阵，d 是一个长度为 28 的向量，C 和 d 都被当作私钥来运算。

所以 $x(x_0, x_1, ..., x_{27})$ 是 $y(y_0, y_1, ..., y_{19})$ 的签名。到此，我们采用的 en-TTS 签名介绍完毕。

7.4 模运算逻辑单元

我们设计了用于计算有限域运算的模运算逻辑单元。为了运算更加高效，它使用复合有限域的表现形式，即 $GF((2^4)^2)$。复合有限域是有限域的一种特殊形式，事实上它是 $GF(2^8)$ 的等效形式。它的有限域元素可以被表示成多项式形式 $a(x) = a_h x + a_l$ 或者系数向量形式 (a_h, a_l)，其中 a_h 和 a_l 是 $GF(2^4)$ 的有限域元素。

我们组合了模运算逻辑单元的所有运算，如图 7-2 所示。它由加法器、乘法器、求逆器和寄存器组组成。它的寄存器组由 6 个寄存器组成，数据宽度均是 4 比特。而运算器又由子域加法器、子域乘法器、子域求逆器、寄存器移位器 A 和寄存器移位器 B 等组成，其中子域运算器均运行于子域 $GF(2^4)$。加法器、乘法器和求逆器的设计如下：

1）加法器由一个四位模加法器组成；

2）乘法器由 4 个子域加法器、3 个子域乘法器和寄存器移位器 A 组成；

第 7 章　多变量公钥密码处理器技术

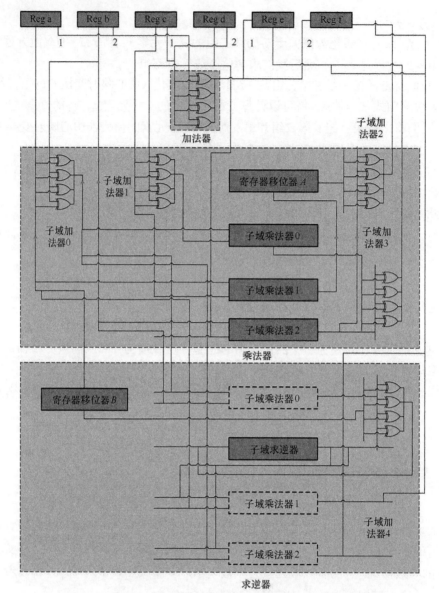

图 7-2　小面积多变量公钥密码处理器的模运算逻辑单元的结构

3）求逆器由一个子域加法器、一个子域求逆器和寄存器移位器 B 组成。

模运算逻辑单元的设计目标是减少模运算的使用面积，所以我们在 4 个方向改进了复合有限域的运算：

1）我们采取了寄存器分时复用的技术减少了寄存器的数量；

2）我们通过分时复用子域加法器和子域乘法器节约运算资源；

7.4 模运算逻辑单元

3）我们采取了分时复用的方式减小了加法器的实现面积；

4）我们设计了寄存器移位器 A 和寄存器移位器 B，用于降低计算的复杂度。

通过整合上述设计和优化，我们的模运算逻辑单元具有实现面积小、计算复杂度低和耗费资源少的特点。我们下面将逐步介绍模运算逻辑单元的设计过程。

首先我们介绍模运算逻辑单元中的运算，它每次执行一个域运算。

对于 $GF((2^4)^2)$ 的两操作数 (a,b) 和 (c,d) 的运算，模运算逻辑单元执行如下操作：

1）$(a,b)+(c,d)=(e,f)$，即复合有限域的加法；

2）$(a,b)+(c,d)=(e,f)$，即复合有限域的乘法。

(e,f) 是运算结果。

对于 $GF((2^4)^2)$ 的单操作数 (a,b) 的运算，模运算逻辑单元执行如下操作：

$$(a,b)=(e,f)$$

(e,f) 是运算结果。这个单操作数运算是 $GF((2^4)^2)$ 的求逆运算。

在介绍了模运算逻辑单元的运算之后，我们介绍子域加法器、子域乘法器、子域求逆器、寄存器移位器 A 和寄存器移位器 B 的设计方法。图 7-3 描绘了这些部件的硬件构造，其中我们选定的子域 $GF(2^4)$ 的不可约多项式是 x^4+x+1。

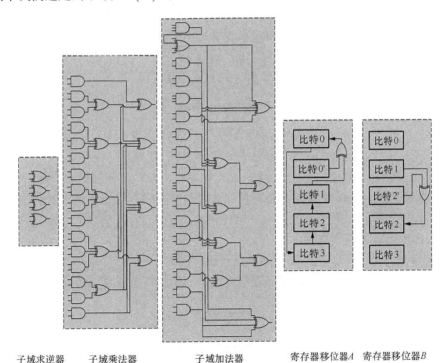

图 7-3　小面积多变量公钥密码处理器的子域加法器、子域乘法器、
子域求逆器、寄存器移位器 A 和寄存器移位器 B

子域加法器是一个 4 比特的并行加法器。

子域乘法器的设计采用一般多项式基的乘法算法，它计算乘法 $k = i \times j$ 的过程如下：

$$s_0 = i_0 j_0$$
$$s_1 = i_1 j_0 + i_0 j_1$$
$$s_2 = i_2 j_0 + i_1 j_1 + i_0 j_2$$
$$s_3 = i_3 j_0 + i_2 j_1 + i_1 j_2 + i_0 j_3$$
$$s_4 = i_3 j_1 + i_2 j_2 + i_1 j_3$$
$$s_5 = i_3 j_2 + i_2 j_3$$
$$s_6 = i_3 j_3$$
$$k_3 = s_3 + s_6$$
$$k_2 = s_2 + s_5 + s_6$$
$$k_1 = s_1 + s_4 + s_5$$
$$k_0 + s_0 + s_4$$

这里，k, i, j 均是子域 $GF(2^4)$ 的元素，$ij, i+j$ 分别是比特乘法和加法。

子域求逆器的设计采用一般多项式基的求逆算法，它计算求逆 $j = i^{-1}$ 的过程如下：

$$k = i_1 + i_2 + i_3 + i_1 i_2 i_3$$
$$j_0 = k + i_0 + i_0 i_2 + i_1 i_2 + i_0 i_1 i_2$$
$$j_1 = i_0 i_1 + i_0 i_2 + i_1 i_2 + i_3 + i_1 i_3 + i_0 i_1 i_3$$
$$j_2 = i_0 i_1 + i_0 i_2 + i_0 i_2 + i_3 + i_0 i_3 + i_0 i_2 i_3$$
$$j_3 = k + i_0 i_3 + i_1 i_3 + i_2 i_3$$

这里，i, j 均是子域 $GF(2^4)$ 的元素，k 是单比特，$ij, i+j$ 分别是比特乘法和加法。

寄存器移位器 A 是一个 4 比特寄存器，它由一个循环右移装置和一个比特加法器组成，它的作用如下：

1）4 比特寄存器循环右移；

2）第 4 比特加上第 1 比特作为新的第 4 比特的值。

寄存器移位器 B 是一个 4 比特寄存器，它由一个比特加法器组成，作用是第 3 比特加上第 2 比特作为新的第 3 比特的值。

下面介绍模运算逻辑单元的加法器、乘法器和求逆器。

模运算逻辑单元的加法器执行 $(a,b)+(c,d)=(e,f)$ 的 $GF((2^4)^2)$ 的加法运算，其中 a,b,c,d,e,f 是子域 $GF(2^4)$ 的元素。我们采用了分时复用的方法进行计算，计算过程如下：

1) $e = a + c$，即一次变量的系数相加；
2) $f = b + d$，即常数项的系数相加。

两步计算分时使用同一个 4 比特加法器，如图 7-3 所示，线上数字为 1 的运算要优先于线上数字为 0 的运算。所以相比 $GF((2^4)^2)$ 加法的算法，我们的设计节省了 50%的实现面积。

运算逻辑单元的乘法器执行 $(a,b) \times (c,d) = (e,f)$ 的 $GF((2^4)^2)$ 的乘法运算，其中 a,b,c,d,e,f 是子域 $GF(2^4)$ 的元素。乘法器采用了 4 个子域加法器、3 个子域乘法器和寄存器移位器 A。

首先，子域加法器 0 和子域加法器 1 的计算过程如下：
1) $sa0 = a + b$，在子域加法器 0 中计算；
2) $sa0 = c + d$，在子域加法器 1 中计算。

由于器件不同，这个运算可以并行。

其次，子域乘法器 0、子域乘法器 1 和子域乘法器 2 的计算过程如下：
1) $sm0 = sa0 \times sa1$，在子域乘法器 0 中计算；
2) $sm1 = a \times c$，在子域乘法器 1 中计算；
3) $sm2 = b \times d$，在子域乘法器 2 中计算。

由于器件不同，这个运算可以并行。

然后，寄存器移位器 A 的计算如下：

$$rsa(sm1_3, sm1_2, sm1_1, sm1_0) \rightarrow rsa(sm1_0, sm1_3, sm1_2, sm1_1)$$
$$rsa(sm1_0, sm1_3, sm1_2, sm1_1) \rightarrow rsa(sm1_0, sm1_3, sm1_2, sm1_1 + sm1_0)$$

即一个循环右移和一个比特加法。

最后，子域加法器 2 和子域加法器 3 的计算过程如下。
1) $e = sm0 + sm2$，在子域加法器 2 中计算；
2) $f = rsa + sm2$，在子域加法器 3 中计算。

由于器件不同，这个运算可以并行。(e,f) 是乘法 $(a,b) \times (c,d)$ 的乘积。

模运算逻辑单元的求逆器执行 $(a,b)^{-1} = (e,f)$ 的 $GF((2^4)^2)$ 的乘法运算，其中 a,b,c,d,e,f 是子域 $GF(2^4)$ 的元素。我们采用了 1 个子域加法器、1 个子域求逆器和寄存器移位器 B，并复用了 1 个子域加法器和 3 个子域乘法器。

首先，子域乘法器 0 的计算过程是 $sm0 = sa0 + 1$，其中 $sa0$ 是复用子域加法器 0 的结果。

其次，让 $c(c_3, c_2, c_1, c_0)$ 等于 $a(a_0, a_3, a_1, a_2)$。寄存器移位器 B 的计算如下：

$$rsbc(a_0, a_3, a_1, a_2) \rightarrow rsbc(a_0, a_3, a_1 + a_3, a_2)$$

即一个比特加法。

然后，子域加法器 4 和子域求逆器的计算过程如下：

1）$sa4 = sm0 + rsb$，在子域加法器 4 中计算；
2）$si = sa4^{-1}$，在子域求逆器中计算，即 $GF(2^4)$ 的求逆。
由于器件不同，这个运算可以并行。

最后，我们复用子域乘法器 1 和子域乘法器 2，它们的计算过程如下：
1）$e = a \times si$，在子域乘法器 1 中计算；
2）$f = sa0 \times si$，在子域乘法器 2 中计算，其中 $sa0$ 是复用子域加法器 0 的结果。
由于器件不同，这个运算可以并行。(e,f) 是求逆 $(a,b)^{-1}$ 的结果。

到此为止，我们介绍完模运算逻辑单元。我们采用了多种策略优化模运算逻辑单元，减少了使用的逻辑门和寄存器的数目。

接下来我们分析模运算逻辑单元的资源使用情况。

首先，我们分析模运算逻辑单元的基础部件。它由 5 个子域加法器、3 个子域乘法器、1 个子域求逆器、寄存器移位器 A 和寄存器移位器 B 等器件组成。这些部件的资源使用情况统计如表 7-4 所示，其中，子域乘法器和子域求逆器的总逻辑门数分别是 31 和 40，这说明乘法和求逆占用模运算逻辑单元的大部分的资源。

表 7-4　小面积多变量公钥密码处理器的基本器件的资源分配

	子域加法器	子域乘法器	子域求逆器	寄存器移位器 A	寄存器移位器 B
寄存器（比特）	12	12	6	5	5
异或逻辑门	4	15	20	1	1
与逻辑门	0	16	20	0	0
总逻辑门数	4	31	40	1	1

我们接下来分析模运算逻辑单元的运算，即加法、乘法和求逆。表 7-5 和图 7-4 所示是它们的资源使用情况，其中乘法运算使用了 69% 的逻辑门数，求逆运算使用了 28% 的逻辑门数。

表 7-5　小面积多变量公钥密码处理器的模运算逻辑单元的资源分配

	加法	加法占比	乘法	乘法占比	求逆	求逆占比
异或逻辑门数	4	4%	62	68%	25	28%
与逻辑门数	0	0%	48	71%	20	29%
总逻辑门数	4	3%	110	69%	45	28%

就模运算逻辑单元的整体设计而言，它只使用了 24 比特的寄存器、91 个异或逻辑门和 68 个与逻辑门。

最后，我们将模运算逻辑单元与原始有限域运算的算法进行了对比，结果统计如表 7-6 所示。其中，模运算逻辑单元分别减少了 62.5%、36.8% 和 41.4% 的寄存器数目、异或逻辑门数目和与逻辑门数目。

7.5 RAM 和寄存器

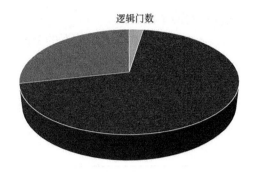

图 7-4 小面积多变量公钥密码处理器的模运算逻辑单元的资源占用

表 7-6 小面积多变量公钥密码处理器的模运算逻辑单元与原始有限域运算的对比

	原始有限域运算				模运算逻辑单元	
	加法	乘法	求逆	所有运算	本章的	减小的比率
寄存器（比特）	24	24	16	64	24	62.5%
异或逻辑门数	8	62	74	144	91	36.8%
与逻辑门数	0	48	68	116	68	41.4%
总逻辑门数	8	110	142	260	159	38.8%

7.5 RAM 和寄存器

通过使用 FPGA 内置的 IP 核，我们设计了一个双端口 RAM。它的输入输出端口的分布和设计如下：

1）$ra0$ 和 $ra1$ 是地址端口，地址的长度是 10 比特；

2）$rd0$ 和 $rd1$ 是数据输入端口，数据宽度是 8 比特，因为我们不需要同时写两个数据，所以 $rd1$ 不使用；

3）$ro0$ 和 $ro1$ 是数据输出端口，数据宽度是 8 比特；

4）$re0$ 和 $re1$ 是使能信号端口，用单比特表示，数值为 0 表示读数据，数值为 1 表示写数据。

我们可以同时从 RAM 读取一个或者两个数据，这是因为 RAM 有两个数据端口。我们每次写一个数据到 RAM，这是因为我们不需要同时写两个数据到 RAM。双端口 RAM 的使用情况如下：

1）从 RAM 读取一个数据：$re0 = 0$，$ra0$ 是 RAM 的地址，$ro0$ 是读取的数据；

2）同时从 RAM 读取两个数据：$re0 = 0$，$re1 = 0$，$ra0$ 和 $ra1$ 分别是要读取数据在

RAM 的地址，*ro*0 和 *ro*1 是读取的两个数据；

3）写一个数据到 RAM：*re*0 = 1，*ra*0 是要写数据的 RAM 的地址，*rd*0 是要写的数据。

寄存器一般被模运算逻辑单元（MALU）、微控制器和 RAM 使用。我们在表 7-7 中统计了寄存器的使用情况。

表 7-7　　　　　　　　小面积多变量公钥密码处理器的寄存器组

寄存器	描述	宽度（比特）
MALU Reg.(a,b)	模运算逻辑单元的第一个操作数	8
MALU Reg.(c,d)	模运算逻辑单元的第二个操作数	8
MALU Reg.(e,f)	模运算逻辑单元的运算结果	8
Reg. Inst.	指令	41
RAM Reg. ra0	RAM 的第一个地址	10
RAM Reg. ra1	RAM 的第二个地址	10
RAM Reg. rd0	RAM 的第一个数据输入	8
RAM Reg. rd1	RAM 的第二个数据输入	8
RAM Reg. ro0	RAM 的第一个数据输出	8
RAM Reg. ro1	RAM 的第二个数据输出	8
RAM Reg. re0	RAM 的第一个使能信号	1
RAM Reg. re1	RAM 的第二个使能信号	1

7.6　微控制器和指令集

我们的多变量公钥密码处理器的指令格式由操作代码、目标地址、第一个源操作数、第二个源操作数和位移量组成，具体的设计如表 7-8 所示。

表 7-8　　　　　　　　小面积多变量公钥密码处理器的指令格式

0	1	2	3	4
操作代码	目标地址	第一个源操作数	第二个源操作数	位移量

为了降低解码译码的复杂度，在指令集中，我们只设计了 5 条基础指令，如表 7-9 所示，它们均在一个时钟周期内执行完毕。

表 7-9　　　　　　　　　小面积多变量公钥密码处理器的基础指令

	寄存器 (a,b)	寄存器 (c,d)	地址 ra0	地址 ra1	数据 rd0	使能 re0	输出端口 C
空操作	-	-	-	-	-	-	-
读 RAM	-	-	第一个源地址	第二个源地址	-	0	-
写 RAM（赋值）	-	-	第一个源地址	-	0	1	-
写 RAM（模运算）	-	-	第一个源地址	-	寄存器 (e,f)	1	-
模运算（第一种）	输入端口 A	输入端口 B	-	-	-	-	-
模运算（第二种）	输入端口 A	读数据 ro1	-	-	-	-	-
模运算（第三种）	读数据 ro0	输入端口 B	-	-	-	-	-
模运算（第四种）	读数据 ro0	读数据 ro1	-	-	-	-	-
输出数据	-	-	-	-	-	-	读数据 ro0

1）空操作，用于等待或者空置一个时钟周期的操作；
2）读 RAM，读取 RAM 中的一个数据或者两个数据；
3）写 RAM，将一个数据写入 RAM 中；
4）模运算逻辑单元，执行一个复合有限域的运算；
5）输出数据，将数据输出到处理器的输出端口 C。

其中写 RAM 指令分成两种情况：
1）赋值，即写寄存器，本条语句是将一个数值赋值给一个寄存器，无需使用 RAM；
2）模运算，即将模运算逻辑单元的计算结果写入 RAM 中。

其中模运算逻辑单元指令分成 4 种情况：
1）两个源操作数分别来自输入端口 A 和输入端口 B；
2）两个源操作数分别来自输入端口 A 和 RAM；
3）两个源操作数分别来自 RAM 和输入端口 B；
4）两个源操作数均来自 RAM。

我们介绍了多变量公钥密码处理器中的 5 种基础指令，然后在它的基础上定义了处理器的 4 种基本运算，如表 7-10 所示。

第 7 章　多变量公钥密码处理器技术

表 7-10　　　　　　　　小面积多变量公钥密码处理器的基本运算

	第一条指令	第二条指令	第三条指令	指令总数
空运算	空操作	-	-	1
赋值运算	写 RAM（赋值）	-	-	1
输出数据运算	读 RAM	输出数据	-	2
模运算逻辑单元运算（第一种）	模运算逻辑单元（第一种）	写 RAM（模运算）	-	2
模运算逻辑单元运算（第二种）	读 RAM	模运算逻辑单元（第二种）	写 RAM（模运算）	3
模运算逻辑单元运算（第三种）	读 RAM	模运算逻辑单元（第三种）	写 RAM（模运算）	3
模运算逻辑单元运算（第四种）	读 RAM	模运算逻辑单元（第四种）	写 RAM（模运算）	3

1）空运算，即执行空操作指令，总共一条指令；

2）赋值运算，即执行写 RAM（赋值）指令，总共一条指令。作用是给寄存器赋值；

3）输出数据运算，即首先执行读 RAM 指令，然后执行输出数据指令，将数据输出到输出端口 C，总共两条指令；

4）模运算逻辑单元运算，它分为两种情况：第一种是源操作数来自于输入端口，不来自于 RAM，那么先执行模运算逻辑单元指令，然后执行写 RAM 指令（模运算），总共两条指令；第二种是源操作数部分来自 RAM 或全来自 RAM，那么先执行读 RAM 指令，然后执行模运算逻辑单元指令，最后执行写 RAM 指令（模运算），总共 3 条指令。

7.7　多变量公钥密码的基本密码运算

多变量公钥密码算法使用一个中心映射和一个或两个用于隐藏中心映射的可逆仿射。其中，中心映射是多变量多次多项式，一般次数不超过两次，可逆仿射一般是以矩阵和偏移的形式出现。所以，多变量公钥密码的基本密码运算由以下 3 种运算组成：

1）可逆仿射变换，即矩阵-向量乘法和向量之间的加法；

2）多变量多次多项式的系数求解，即计算多项式的系数矩阵；

3）求解线性方程组，即求解中心映射中所有变量的值。

多变量公钥密码的后两种基本密码运算组成了中心映射的变换。我们下面介绍如何利用小面积多变量公钥密码处理器的指令来实现这 3 种基本密码运算。

7.7 多变量公钥密码的基本密码运算

可逆仿射变换是通过调用矩阵-向量乘法和向量之间的加法实现的，所有的运算都在复合有限域 $GF((2^4)^2)$。

$$a \times b = c$$
$$c + d = e$$

其中我们假定矩阵 a 的规模是 $n \times n$，向量 b, d 的维度是 n。可逆仿射变换的执行过程如下。

1）矩阵—向量乘法可以被分解成 n^2 对模运算逻辑单元中的乘法和加法，因为模运算逻辑单元中的运算一般都使用 3 条指令，所以每一对运算需要 6 条指令。总地来说，矩阵—向量乘法需要 $6n^2$ 条指令。

2）向量之间的加法由 n 个模运算逻辑单元中的加法组成，因为模运算逻辑单元中的运算一般都使用 3 条指令，所以它总共需要 $3n$ 条指令。

所以，在我们假定它的矩阵规模是 $n \times n$ 的情况下，可逆仿射变换执行的总指令数是 $6n^2 + 3n$，详情统计如表 7-11 所示。

表 7-11　　　矩阵规模为 $n \times n$ 的可逆仿射变换的指令数

	模运算逻辑单元的乘法	模运算逻辑单元的加法	指令数
矩阵-向量的乘法	n^2	n^2	$6n^2$
向量之间的加法	0	n	$3n$
总计	n^2	$n^2 + n$	$6n^2 + 3n$

接下来我们介绍多变量多次多项式的系数求解，即计算多项式的系数矩阵。

我们假定多变量公钥密码算法中的一般多变量多次多项式的未知变量的数目是 n，已知变量的数目是 m，则它可以被定义为以下形式：

$$f(x_0, x_1, \ldots, x_{n-1}) = \sum \alpha_{ij} y_i x_j + \sum \delta_i x_i + \sum \beta_{ij} y_i y_j + \sum \gamma_i y_i + \eta$$

其中 $x(x_0, x_1, \ldots, x_{n-1})$ 是待求解的未知变量，$y(y_0, y_1, \ldots, y_{m-1})$ 是需要代入方程组的已知元素，α_{ij}、β_{ij}、γ_i、δ_i 和 η 是多项式的系数。

对于多项式的第一项和第二项 $\sum \alpha_{ij} y_i x_j$ 和 $\sum \delta_i x_i$，它们需要如下的计算：

$$\sum_{i=0}^{i=n-1} \left(\sum_{j=0}^{j=m-1} \alpha_{ij} y_j + \delta_i \right)$$

模运算逻辑单元的乘法和加法的数目分别是 mn 和 $mn + n$，因为模运算逻辑单元中的运算一般都使用 3 条指令，多项式的第一项和第二项总共需要 $3n(m+1)$ 条指令。

对于多项式剩余的 3 项：$\sum \beta_{ij} y_i y_j, \sum \gamma_i y_i, \eta$，它们需要如下的计算：

$$\sum_{i=0}^{i=n-1} \left(\sum_{j=0}^{j=m-1} y_i y_j + \sum_{j=0}^{j=m-1} \gamma_i y_j \right) + \eta$$

模运算逻辑单元的乘法和加法的数目分别是 $2mn$ 和 $2mn+n+1$,因为模运算逻辑单元中的运算一般都使用 3 条指令,多项式剩余的 3 项总共需要 $12mn+3n+3$ 条指令。

总地来说,我们假定它的未知变量的数目是 n,已知变量的数目是 m,一个多变量多次多项式的系数求解需要 $15mn+3n+3$ 条指令,详细情况统计如表 7-12 所示。

表 7-12　未知变量的数目是 n 和已知变量的数目是 m 的多变量多次多项式的指令数

	模运算逻辑单元的乘法	模运算逻辑单元的加法	指令数
前两项的计算	mn	$mn+n$	$3mn+3n$
后三项的计算	$2mn$	$2mn+n+1$	$12mn+3n+3$
总计	$3mn$	$3mn+2n+1$	$15mn+6n+3$

在计算完多变量多次多项式的系数矩阵之后,它变成等效的线性方程组,我们下面介绍求解有限域的线性方程组。

我们假定有限域的线性方程组的系数矩阵规模是 $n \times n$,则求解线性方程组的计算复杂度是 $O(n^3)$。它要计算 n 次迭代,每次迭代由找主元、归一和消元运算组成。我们下面介绍这 3 种运算:

1)找主元是选择一个非零元素作为主元,然后计算它的逆元。例如计算 $b=a^{-1}$ 的逆元,我们需要调用模运算逻辑单元的求逆运算,其中 a 和 b 均是域元素;

2)归一是对迭代当前行所有的元素进行归一计算,例如 $c=a \times b$,我们需要调用模运算逻辑单元的乘法运算,其中 a,b,c 均是域元素;

3)消元是对迭代当前行以外的所有的元素进行消元计算,例如 $c=a \times b$ 和 $e=c+d$,总共两个运算,我们需要调用模运算逻辑单元的乘法和加法运算,其中 a,b,c,d,e 均是域元素。

在求解线性方程组中,每次迭代进行的找主元、归一和消元的数目统计如表 7-13 所示,分别是 n、$n(n+1)/2$ 和 $(n^2-1)n/2$。因为每次迭代的操作数目并不尽相同,所以需要的指令数也不同,我们在表 7-14 中统计了需要的指令数。

表 7-13　求解系数矩阵规模为 $n \times n$ 的线性方程组的过程

	找主元的数目	归一的数目	消元的数目
第一轮迭代	1	n	$(n-1)n$
第二轮迭代	1	$n-1$	$(n-1)^2$
…	…	…	…
最后一轮迭代	1	1	$n-1$
总共	n	$n(n+1)/2$	$(n^2-1)n/2$

表 7-14　　求解系数矩阵规模为 $n \times n$ 的线性方程组的指令数

	模运算逻辑单元的乘法	模运算逻辑单元的加法	模运算逻辑单元的求逆	指令数
第一轮迭代	n^2	$(n-1)n$	1	$6n^2 - 3n + 3$
第二轮迭代	$(n-1)n$	$(n-1)^2$	1	$6n^2 - 9n + 6$
...	...	...	...	...
最后一轮迭代	n	$n-1$	1	$6n$
总共	$n^2(n+1)/2$	$n(n+1)/2$	n	$3n^3/2 + 3n^2 + 9n/2$

总地来说，我们假定线性方程组的系数矩阵规模是 $n \times n$，则它需要的模运算逻辑单元的乘法、加法和求逆的数目分别是 $n^2(n+1)/2$、$n(n+1)/2$ 和 n，总共需要的指令数是 $3n^3/2 + 3n^2 + 9n/2$。

到此为止，我们介绍完毕多变量公钥密码的基本密码运算。

7.8　技术实现

为了验证和测试我们的设计，我们在硬件上实现了小面积多变量公钥密码处理器。我们选择的编程语言是 Verilog HDL，实现的硬件是 Xilinx FPGA XC3S50。由于我们的设计所需的资源很少，所以它也适用于其他 FPGA。

根据密码算法的应用的不同，我们设计了 3 种类型的处理器，它们使用的 RAM 的大小分别是 1 000 字节、200 字节和 100 字节。

1）第一类处理器适合处理计算复杂和中间计算结果多的多变量公钥密码算法；
2）第二类处理器适合处理计算和存储适中的多变量公钥密码算法；
3）第三类处理器适合处理计算较少和中间计算结果少的多变量公钥密码算法。

我们实现了这三类多变量公钥密码处理器，实验结果统计如表 7-15 所示。所有的实验数据均是取自布线和布局之后的结果。实验结果表明我们的多变量公钥密码处理器只使用了很少的 FPGA 资源，即实现面积非常小，例如第三类的处理器仅仅使用了 47 个触发器、155 个 LUT 和 92 个 Slice。我们的多变量公钥密码处理器非常适合用于资源受限的环境。

表 7-15　　小面积多变量公钥密码处理器的的实现结果

类型	RAM/B	指令/bit	触发器	LUT	Slice	频率/MHz	功耗/mW
1	1 000	41	50	170	99	60.39	27.34
2	200	35	48	167	96	60.39	27.34
3	100	32	47	155	92	60.39	27.34

为了评估和验证模运算逻辑单元的设计，我们在 FPGA 上单独实现了这个部件。实验结果表明模运算逻辑单元使用了 95 个 LUT 和 51 个 Slice，它运算一次乘法和求逆的时间分别是 15.002 ns 和 16.559 ns。如果用 Slice 的数目作为面积评估的一个指标，那么模运算逻辑单元在我们的设计中占约 55%的面积，它的直观的表示体现如图 7-5 所示。

我们在小面积多变量公钥密码处理器上实现了 3 个有代表性的多变量数字签名算法，即非平衡油醋签名、en-TTS 签名和 Rainbow 签名算法。

首先，我们实现了非平衡油醋签名算法，它的设计实现流程如图 7-6 所示。

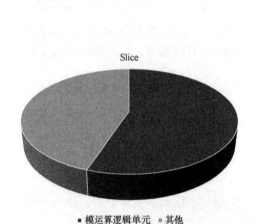

图 7-5 小面积多变量公钥密码处理器的模运算逻辑单元的资源占用

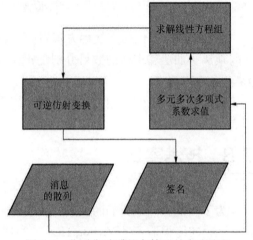

图 7-6 非平衡油醋签名算法的实现流程

1）消息通过散列函数进行变换；
2）计算多变量多次多项式的系数矩阵；
3）求解有限域的线性方程组；
4）线性可逆仿射变换。

非平衡油醋签名的实验结果统计如表 7-16 所示，UOV 运行于第一类的密码处理器上，即采用了 1 000 字节的 RAM，这是因为它需要求解一个系数矩阵规模为 28×28 的线性方程组，使用 RAM 的面积较大。消息的散列值的长度是 28 字节，签名的长度是 84 字节，可逆仿射变换中的矩阵的规模是 84×84，需要计算 28 个多变量多次多项式的系数，指令的总数为 1 038 366。在 60.39 MHz 的时钟频率下运行，需要 17.19 ms 实现一次非平衡油醋签名。

表 7-16　　非平衡油醋签名、en-TTS 签名和 Rainbow 签名的实验结果

签名方案	处理器类型	消息/B	签名/B	仿射变换	求解系数	求解方程组	指令数	时间/ms
UOV	1	28	84	84×84	28	28×28	10^6	17.19

续表

签名方案	处理器类型	消息/B	签名/B	仿射变换	求解系数	求解方程组	指令数	时间/ms
Rainbow	2	26	43	26×26 43×43	26	13×13 13×13	10^5	2.39
en-TTS	3	20	28	20×20 28×28	20	9×9 9×9	10^4	0.17

在非平衡油醋签名的所有运算中，计算多变量多次多项式的系数的运算时间最长，其次是求解有限域的线性方程组，直观的描述如图 7-7 所示。

然后，我们实现了 Rainbow 签名，它的实现流程如图 7-8 所示。

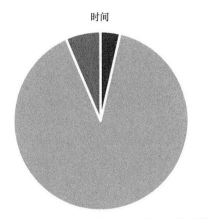
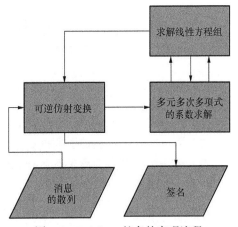

图 7-7 非平衡油醋签名的实现时间　　图 7-8 Rainbow 签名的实现流程

1）消息通过散列函数进行变换；
2）第一次线性可逆仿射变换；
3）第一次计算多变量多次多项式的系数矩阵；
4）第一次求解有限域的线性方程组；
5）第二次计算多变量多次多项式的系数矩阵；
6）第二次求解有限域的线性方程组；
7）第二次线性可逆仿射变换。

Rainbow 签名的实验结果统计如表 7-16 所示，Rainbow 运行于第二类的密码处理器上，即采用了 200 字节的 RAM，这是因为它需要求解两个系数矩阵规模同为 13×13 的线性方程组，使用 RAM 的面积适中。消息的散列值的长度是 26 字节，签名的长度是 43 字节，两个可逆仿射变换中的矩阵的规模分别是 26×26 和 43×43，需要计算 26 个多变量多次多项式的系数。指令的总数为 144 330，在 60.39 MHz 的时钟频率下运行，需要 2.39 ms 实现一次 Rainbow 签名。

在 Rainbow 签名的所有运算中，计算多变量多次多项式的系数的运算时间最长，其次是求逆仿射变换，直观的描述如图 7-9 所示。

最后，我们实现了 en-TTS 签名，它的实现流程如图 7-10 所示。

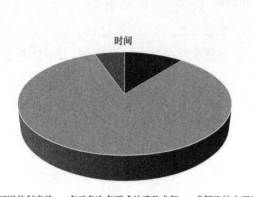

图 7-9　Rainbow 签名的实现时间

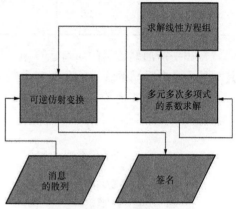

图 7-10　en-TTS 签名的实现流程

1) 消息通过散列函数进行变换；
2) 第一次线性可逆仿射变换；
3) 第一次计算多变量多次多项式的系数矩阵；
4) 第一次求解有限域的线性方程组；
5) 第二次计算多变量多次多项式的系数矩阵；
6) 第三次计算多变量多次多项式的系数矩阵；
7) 第二次求解有限域的线性方程组；
8) 第二次线性可逆仿射变换。

en-TTS 签名的实验结果统计如表 7-16 所示，en-TTS 运行于第三类的密码处理器上，即采用了 100 字节的 RAM，这是因为它需要求解两个系数矩阵规模同为 9×9 的线性方程组，使用 RAM 的面积较少。消息的散列值的长度是 20 字节，签名的长度是 28 字节，两个可逆仿射变换中的矩阵的规模分别是 20×20 和 28×28，需要计算 20 个多变量多次多项式的系数。指令的总数为 10 265，在 60.39MHz 的时钟频率下运行，需要 0.17 ms 实现一次 en-TTS 签名。

在 en-TTS 签名的所有运算中，计算可逆仿射变换的运算时间最长，其次是求解有限域的线性方程组，直观的描述如图 7-11 所示。

综上所述，多变量公钥密码算法的代表方案非平衡油醋签名、Rainbow 签名和 en-TTS 签名均在我们设计的处理器上高效地运行，它们的实现所需要的资源较少，非常适合在资源受限的情况下使用,并且它们的运行时间适中。需要指出的是,除了这 3 种多变量数字签名方案以外,我们的处理器还可以应用其他的密码算法。

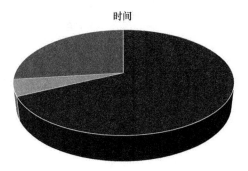

图 7-11　en-TTS 签名的实现时间

7.9　实现对比

目前最流行的公钥密码是 RSA 和椭圆曲线密码。参考文献 [2,3,4,9,10,11,12] 中的设计被认为分别是目前 FPGA 上最好的 RSA 和椭圆曲线密码硬件之一。我们将小面积多变量公钥密码处理器与这些密码硬件和相关的多变量公钥密码硬件进行对比,结果统计如表 7-17 所示。

表 7-17　FPGA 的公钥密码硬件对比

签名方案	触发器	LUT	Slice	FPGA 型号
参考文献 [1] 中的 UOV(60,20)	5 665	16 694	9 821	Virtex 3
参考文献 [1] 中的 Rainbow(42,24)	2 332	7 173	4 123	Virtex 3
参考文献 [1] 中的 am-TTS(24,34)	1 697	5 434	3 139	Virtex 3
参考文献 [1] 中的 en-TTS(20,24)	1 649	5 304	3 060	Virtex 3
参考文献 [10] 中的 RSA 密码硬件	-	-	27 597	Virtex 3
参考文献 [11] 中的 RSA 密码硬件	14 338	25 342	-	Virtex 2
参考文献 [2] 中的 RSA 密码硬件	-	5 696	4 144	Virtex 5
参考文献 [2] 中的 RSA 密码硬件	-	-	2 469	Virtex 2
参考文献 [12] 中的椭圆曲线密码硬件	-	22 936	6 150	Virtex 5
参考文献 [9] 中的椭圆曲线密码硬件	-	3 815	2 431	Virtex 5
参考文献 [3] 中的椭圆曲线密码硬件	-	-	773	Virtex 2

续表

签名方案	触发器	LUT	Slice	FPGA 型号
参考文献［4］中的椭圆曲线密码硬件	83	253	184	Virtex 3
本章的（第一类处理器）	50	170	99	Virtex 3
本章的（第二类处理器）	48	167	96	Virtex 3
本章的（第三类处理器）	47	155	92	Virtex 3

表明我们的设计是目前最小的处理多变量公钥密码算法的硬件之一，并且比 RSA 密码硬件小 96%（92 个 Slice 对比 2469 个 Slice），比椭圆曲线密码硬件小 50%（92 个 Slice 对比 184 个 Slice）。

7.10 本章小结

密码处理器一直是密码硬件设计领域研究的重点，它比一般专用密码硬件的灵活性和通用性更高，而且它的适用范围更加广泛。

在本章中，我们希望研究一种通用的多变量公钥密码处理器，它能够处理绝大部分的多变量公钥密码算法，甚至能够处理其他公钥密码算法。所以，我们设计了一种小面积的多变量公钥密码处理器。之所以把面积优化作为第一目标，是因为我们希望将它应用到资源受限的领域，例如传感器、智能芯片和射频识别标签等。

我们的设计基于降低模运算、指令解码译码的复杂度和减少使用寄存器的数目。所以，我们的设计主要在 3 个方向优化：

1）我们设计了一个模运算逻辑单元，它适用于复合有限域，提供 $GF((2^4)^2)$ 的加法、乘法和求逆；

2）我们设计了一个精简的指令集和解码器；

3）我们对寄存器采取了分时复用的方式减少实现的面积。

通过其他优化，例如，可逆仿射变换、多变量多次多项式的系数求值和求解线性方程组的精简算法，以及整合上述设计，我们的多变量公钥密码处理器在低资源的 Xilinx xc3s50 FPGA 上实现，只使用了非常有限的器件资源，即 47 个触发器、155 个 LUT 和 92 个 Slice。

我们在小面积多变量公钥密码处理器上实现了 3 种多变量数字签名方案，即非平衡油醋签名、Rainbow 签名和 en-TTS 签名方案。这 3 种多变量签名方案在我们的处理器上运行的速度适中，它们可以满足绝大部分的应用，例如非平衡油醋签名在 60.39 MHz 的时钟频率的条件下，需要 17.19 ms 的运算时间。当然我们的密码处理器并不局限于使用这 3 种密码算法，还可以实现其他多变量公钥密码算法和其他密码算法。

我们将小面积多变量公钥密码处理器与其他公钥密码硬件进行了对比，结果表明我们的设计是目前最小的处理多变量公钥密码算法的硬件之一，并且比 RSA 密码硬件小 96%（92 个 Slice 对比 2 469 个 Slice），比椭圆曲线密码硬件小 50%（92 个 Slice 对比 184 个 Slice）。这也证明了多变量公钥密码算法有能力成为低端设备上一个好的选择。

7.11 本章参考文献

[1] Bogdanov A, Eisenbarth T, Rupp A, et al. Time-Area Optimized Public-Key Engines: MQCryptosystems as Replacement for Elliptic Curves?[A]. In: Cryptographic Hardware and Embedded Systems - CHES 2008[C]. Springer Berlin Heidelberg. 2008. Lecture Notes in Computer Science, 5154: 45-61.

[2] Sutter G D, Deschamps J, Imaña J L. Modular multiplication and exponentiation architectures for fast RSA cryptosystem based on digit serial computation[J]. IEEE Trans. Ind. Electron., 2011, 58(7): 3101-3109.

[3] Varchola M, Guneysu T, Mischke O. MicroECC: a lightweight reconfigurable elliptic curve cryptoprocessor[A]. In: 2011 International Conference on Reconfigurable Computing and FPGAs (ReConFig)[C], 2011: 204-210.

[4] Hassan M N, Benaissa M. Small Footprint Implementations of Scalable ECC Point Multiplication on FPGA[A]. In: 2010 IEEE International Conference on Communications (ICC)[C], 2010: 1-4.

[5] Thomae E, Wolf C. Solving Underdetermined Systems of Multivariate Quadratic Equations Revisited[A]. In: Public Key Cryptography - PKC 2012[C]. Springer Berlin Heidelberg. 2012. Lecture Notes in Computer Science, 7293: 156-171.

[6] Moh T. A public key system with signature and master key functions[J]. Communications in Algebra, 1999, 27: 2207-2222.

[7] Petzoldt A, Bulygin S, Buchmann J. Selecting parameters for the Rainbow signature scheme[A]. In: Post-Quantum Cryptography[C]. Springer, 2010: 218-240.

[8] Thomae E, Wolf C. Cryptanalysis of Enhanced TTS, STS and All Its Variants, or: Why Cross-Terms Are Important[A]. In: Progress in Cryptology - AFRICACRYPT 2012[C]. Springer Berlin Heidelberg.2012. Lecture Notes in Computer Science, 7374: 188-202.

[9] Loi K C, Ko S B. High performance scalable elliptic curve cryptosystem processor for Koblitz curves[J]. Microprocessors and Microsystems, 2013, 37(4): 394-406.

[10] Sakiyama K, Mentens N, Batina L, et al. Reconfigurable modular arithmetic logic unit supporting high-performance RSA and ECC over GF(p)[J]. International Journal of Electronics, 2007, 94(5): 501-514.

[11] Cilardo A, Mazzeo A, Romano L, et al. Exploring the design-space for FPGA-based implementation of RSA[J]. Microprocessors and Microsystems, 2004, 28(4): 183-191.

[12] Sutter G, Deschamps J, Imana J. Efficient Elliptic Curve Point Multiplication Using Digit-Serial Binary Field Operations[J]. IEEE Trans. Ind. Electron., 2013, 60(1): 217-225.